# Reactivity and Structure Concepts in Organic Chemistry

Volume 9

J. R. Blackborow    D. Young

# Metal Vapour Synthesis in Organometallic Chemistry

With 36 Figures and 32 Tables

Springer-Verlag
Berlin Heidelberg New York 1979

John R. Blackborow
Université Louis Pasteur
Institut de Chimie
4, rue Blaise Pascal
F-67008 Strasbourg Cedex

Dennis Young
N. T. D.
BP Research Centre
Chertsey Road
Sunbury-on-Thames
Middlesex TW16 7LN
Great Britain

ISBN 3-540-09330-3 Springer-Verlag Berlin Heidelberg New York
ISBN 0-387-09330-3 Springer-Verlag New York Heidelberg Berlin

Library of Congress Cataloging in Publication Data. Blackborow, J. 1942–. Metal vapour synthesis in organometallic chemistry; (Reactivity and structure; v. 9). Bibliography: p. Includes indexes. 1. Organometallic compounds. 2. Chemistry, Organic – Synthesis. I. Young, Dennis, 1949– joint author. II. Title. III. Series. QD411.B55 547'.05. 79-9844


Printed in Germany

Typesetting: Elsner & Behrens, Oftersheim
Printing and binding: Konrad Triltsch, Würzburg
2152/3140-5 4 3 2 1 0

*For Hilary and La Tendre Poulet*

# Preface

Metal Vapour Synthesis (MVS) can be defined as; "The use in synthesis of high temperature gaseous species such as metal atoms by their reactions with themselves or other materials in a condensed phase." This short book, covering the literature up to the middle of 1979, describes MVS in organic chemistry; i.e. the reactions of metal atoms with various, predominantly organic, substrates in the synthesis and reactivity studies of organic and organometallic compounds. In order to effectively describe all the underlying principles and to present a coesive picture of pertinent metal atom processes in condensed organic phases, some inorganic substrates such as rare gases, dinitrogen, dioxygen, dihalogens, and inorganic halides have been included. For similar reasons, we have used, where relevent, information provided by the closely related technique of Matrix Isolation Spectroscopy (MIS).

After an introductory chapter which gives the basis principles and includes a brief critique of the technique, the book is divided into three further chapters dealing respectively with

(a) experimental techniques,
(b) behaviour of metal atoms in matrices, and
(c) results of preparative experiments.

While not being encyclopaedic the book describes or refers to all noteworthy areas if not in the deliberately short text in the many tables and figures. Therefore we hope that chemists, fresh or experienced, will find this book useful as a comprehensive introduction to the technique, a starting point for the setting up of MVS equipment, a guide to the metal atom processes, and a synopsis of metal atom reactions to date.

On the present evidence we can see a considerable future for MVS both in the research laboratory and in the commercial production of catalysis and related metal coated systems.

J. R. Blackborow
D. Young

# Acknowledgements

We extend our thanks to all who have been instrumental in the preparation of this book. We are especially grateful to those who gave us our start and guided us in our formative years in MVS, the late Dr. E. A. Koerner von Gustorf (JRB) and Dr. M. L. H. Green (DY). One of us (DY) should like to thank his wife for her support and understanding during the writing of this book. Professor J. A. Osborn is thanked for giving time and materials for the presentation of the manuscript (JRB).

# Contents

# Abbreviations

| | |
|---|---|
| Ac | acetyl |
| acac | acetylacetonate |
| acacH | acetylacetone |
| An | acrylonitrile |
| Bd | buta-1,3-diene |
| bipy | dipyridyl |
| bp | boiling point |
| Bu | Butyl |
| $Bu^n$ | n-butyl |
| $Bu^i$ | iso-butyl |
| $Bu^s$ | sec-butyl |
| $Bu^t$ | tert-butyl |
| c | cis |
| Cdt | cyclododeca-1,5,9-triene |
| Cht | cycloheptatrienyl |
| ChtH | cycloheptatriene |
| $ChtH_2$ | cycloheptadienyl |
| $ChtH_3$ | cycloheptadiene |
| Cod | cyclooctadiene |
| Cot | cyclooctatetraene |
| Cp | cyclopentadienyl |
| CpH | cyclopentadiene |
| Cy | cyclohexyl |
| CyH | cyclohexane |
| Dba | dibenzylideneacetone |
| Ddt | dodecateraene |
| dec | decomposition |
| diphos | 1,2-bis(diphenylphosphino)ethane |
| e | electron |
| esr | electron spin resonance |
| Et | ethyl |
| glc | gas liquid chromatography |
| Hal | halogen |
| Hex | hexyl |
| I.P. | ionisation potential |
| ir | infrared (spectroscopy) |
| L | 2-electron ligand |
| Me | methyl |
| Mht | methylheptatriene |
| MIS | matrix isolation spectroscopy |
| M.O. | molecular orbital |
| mp | melting point |
| MVS | metal vapour synthesis |
| Nbd | norbornadiene |
| Nbe | norbornene |
| nmr | nuclear magnetic resonance |
| Ot | octatriene |
| Pbd | polybutadiene |
| pes | photoelectron spectroscopy |
| Ph | phenyl |
| PhH | benzene |
| PhMe | toluene |
| Pr | propyl |
| $Pr^n$ | n-propyl |
| $Pr^i$ | iso-propyl |
| py | pyridine |
| t | trans |
| TCNQ | tetracyanoquinodimethane |
| thf | tetrahydrofuran |
| uv | ultraviolet (spectroscopy) |
| uv-vis | ultraviolet-visible (spectroscopy) |
| X | 1-electron ligand |
| xs | excess |
| $\Delta$ | heat, warm |

# List of Editors

Professor Dr. Klaus Hafner
Institut für Organische Chemie der TH
Petersenstr. 15
D-6100 Darmstadt

Professor Dr. Jean-Marie Lehn
Institut de Chimie
Université de Strasbourg
1, rue Blaise Pascal
B. P. 296/R8
F-67008 Strasbourg-Cedex

Professor Dr. Charles W. Rees, F. R. S.
Hofmann Professor of Organic Chemistry
Department of Chemistry
Imperial College of Science and Technology
South Kensington
London, SW 2 AY
England

Professor Dr. Paul v. R. Schleyer
Lehrstuhl für Organische Chemie
der Universität Erlangen-Nürnberg
Henkestr. 42
D-8520 Erlangen

Professor Barry M. Trost
Department of Chemistry
The University of Wisconsin
1101 University Avenue
Madison, Wisconsin 53706
USA

Prof. Dr. Rudolf Zahradník
Tschechoslowakische Akademie der Wissenschaften
J.-Heyrovský-Institut für Physikal. Chemie und Elektrochemie
Máchova 7
121 38 Praha 2
C. S. S. R.

# 1 The Basic Pinciples

## 1.1 Introduction

It is now possible to use vapours such as atoms of transition metals[1, 2], carbon atoms[3], and molecules which may only exist in the gaseous state, e.g. $SiF_2$[4], Cr-Fe[5], or TiO[6] as routine reagents in synthesis[1, 2], catalysis[6], or other reactivity studies[7]. Recent research has already shown that interesting and novel compounds such as $(Cod)_2Fe$[8], $(Bd)_3Mo$[9], $(PhH)_2Ti$[10], and $C_6F_5PdBr$[11], which have not been prepared by conventional methods may be readily made by vapour synthesis.

The technique owes its origins to experiments conducted by Pimentel[12] and others in the late 1950s who showed that highly reactive species such as radicals could be trapped in rare gas matrices at liquid $H_2$ (20 K) and liquid He (4 K) temperatures, and their spectroscopic properties determined at leisure. Following this, Linevsky[13] and others showed that species generated only at high temperatures could be similarly trapped, while Skell[14] and collaborators found that species such as carbon atoms when trapped in reactive matrices at liquid $N_2$ temperatures (77 K) formed compounds in gram quantities. Timms logically extended the scope to syntheses using transition metal atoms[15, 16].

The highly reactive atoms or molecules are generated at high temperatures in a vacuum and then brought together with the chosen coreactants on a cold surface. The use of species such as atoms can provide synthetic advantage from both kinetic and thermodynamic considerations. Indeed atoms have virtually no kinetic barrier towards ligand addition whereas, even for a spontaneous reaction, the massive metal may be inert due to a relatively low surface area, inhibition by thin oxide films, or the strong cohesive forces of the metal lattice. Figure 1.1 shows the synthetic advantages of metal atoms compared with the solid metal. For example gaseous Cr or Ti atoms with benzene form bis($\eta$-benzene) chromium in 60% yield[16] or bis($\eta$-benzene)titanium in 40% yield[10]. These compounds cannot be made directly from the solid metal: $(\eta\text{-PhH})_2Cr$ must be synthesised by a circuitous route[17], while $(\eta\text{-PhH})_2Ti$ cannot even be similarly prepared[18].

It is not firmly known at what stage the isolated products are formed, but the generally held belief is that there is some initial reaction between atom and coreactant in the nascent cocondensate, otherwise solid metal will be formed. Reaction is presumed to continue through a series of non-isolable intermediates, either at the cocondensation temperature or on warming, till sufficiently inert or stable products are formed which are then extracted. These points are considered in further detail in Chap. 4.

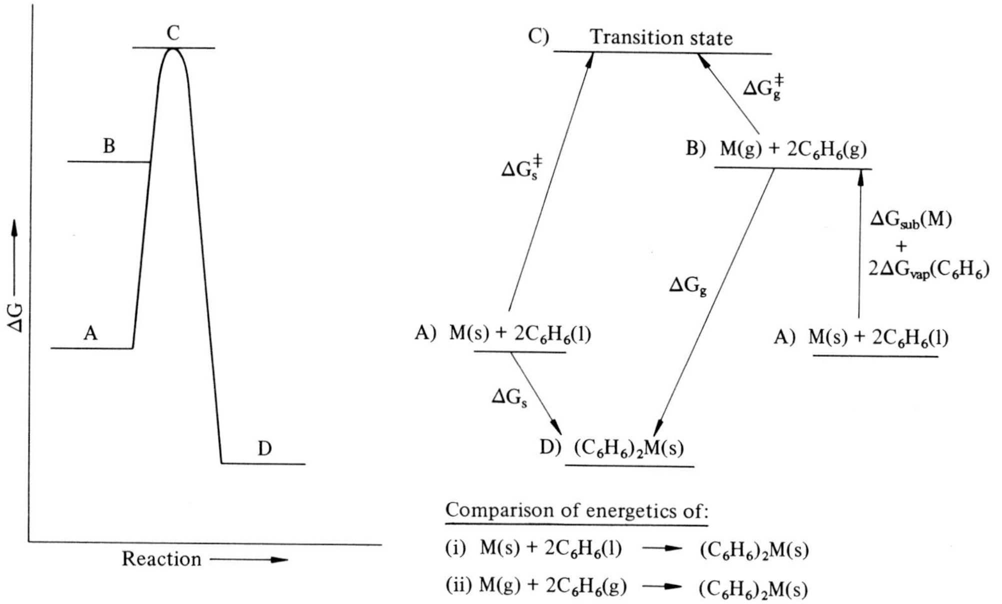


**Fig. 1.1.** Synthetic advantage of metal atoms
$\Delta G_s^\ddagger$ is the energy required to activate reaction (i).
$\Delta G_g^\ddagger$ is the energy required to activate reaction (ii).
$\Delta G_g^\ddagger < \Delta G_s^\ddagger$ therefore reaction (ii) is more favourable kinetically.
$\Delta G_s$ is the energy gained by forming $(C_6H_6)_2M(s)$ in reaction (i).
$\Delta G_g$ is the energy gained by forming $(C_6H_6)_2M(s)$ in reaction (ii).
$\Delta G_g > \Delta G_s$ therefore reaction (ii) is more favourable thermodynamically.
$-(\Delta G_g^\ddagger - \Delta G_s^\ddagger) \triangleq \Delta G_g - \Delta G_s \triangleq \Delta G_{sub}$, the sublimation energy of the metal (to a reasonable approximation) and is a reflection of the synthetic advantage of the metal atom over the massive metal

## 1.2 The Method

The principle of the technique is simple. The metal atoms (e.g., Cr atoms) are brought into contact with the other reactants (e.g., $PF_3$ molecules) at a reaction site where the product (i.e. $Cr(PF_3)_6$)[16] is formed and extracted. This is shown in Fig. 1.2.

### *1.2.1 Metal Atom Source*

The most convenient and attractive source of metal atoms is a furnace inside a vacuum system. The furnace is heated to the required vaporisation temperature and in the low pressure environment (pressure lower than $10^{-3}$ to $10^{-4}$ torr) the metal atoms will proceed by essentially collision-free paths from the furnace to the reaction site. Furnaces which have been successfully used in vapour synthesis include electron beam guns[6, 10, 20],

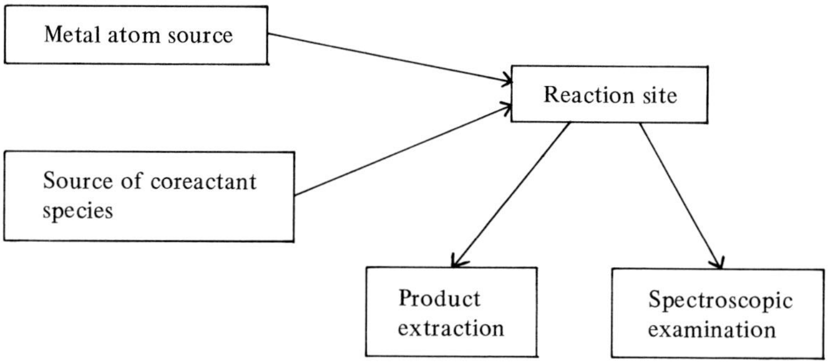


**Fig. 1.2.** The principal requirements of the method

resistance heated furnaces[1, 21, 22], laser beam heating[23, 24], hot filaments (resistance heating)[19, 22], electric arcs[3, 25], and sputtering[26]. Each source may have special advantage for a particular situation but the first two methods are likely to prove to be the most versatile and convenient. A fuller discussion can be found in Chap. 2.

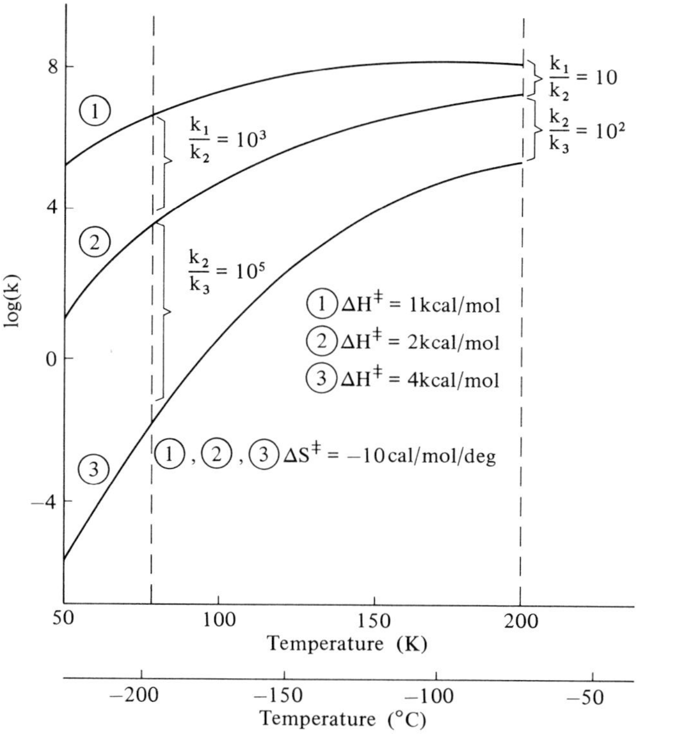


**Fig. 1.3.** Variation of reaction velocities with temperature

$$k \text{ (reaction velocity)} = \text{const. } e^{-\Delta G^{\ddagger}/RT} = \text{const. } e^{\Delta S^{\ddagger}/R} \cdot e^{-\Delta H^{\ddagger}/RT}$$
$$= \text{const. } e^{-\Delta H^{\ddagger}/RT} \quad \text{(at constant } \Delta S^{\ddagger}\text{)}$$

$k_1/k_2 = \text{const. } \exp[(\Delta H_2^{\ddagger} - \Delta H_1^{\ddagger})/RT]$

as T increases, $k_1/k_2$ decreases

### *1.2.2 The Reaction Site*

Unless the reaction site (conveniently the vacuum vessel wall) is at the same temperature as the furnace, metal atoms in contact with wall will polymerise back to a metal lattice. Fortunately, a large excess of coreactant (10- to 100-fold) intimately mixed with the metal atoms at the surface can greatly reduce atom-atom contact, while a very low temperature (less than $-200\ ^\circ C$) which freezes the atoms in a solid coreactant cage minimises atom diffusion. Unfortunately such temperatures can prevent atom-coreactant interactions[1a)]. By raising the temperature, the difference between the atom-atom and atom-coreactant reaction rates can be reduced, as explained in Fig. 1.3. However, the optimum reaction site temperature is usually not determined by kinetic but by other factors.

In order to maintain the low pressure requirements of metal atom generation, the vapour pressure of coreactants must be ideally less than $10^{-4}$ torr. For most coreactants this is achieved within the temperature range $-78\ ^\circ C$ (solid $CO_2$) to $-196\ ^\circ C$ (liquid $N_2$), but in some favourable cases, room temperature is adequate. For example Ni atoms condensed into a solution of $PPh_3$ in a silicone oil at $0\ ^\circ C$ form $Ni(PPh_3)_3$[8)]. However, low temperatures can also be required because the product is thermally unstable; e.g. $Ni(PH_3)_2(PF_3)_2$[19)].

### *1.2.3 Source of Coreactants*

The principal requirement of a coreactant source is to intimately mix a large excess of coreactants with the metal atoms in order to effect favourable reaction conditions.

The most frequently used technique to date has been to continuously introduce coreactants as a uniform vapour beam through a nozzle onto the reaction site[19, 27)]. Rotation of the vacuum vessels allows a larger surface area to be used as reaction site; and as the two vapour beams need not then be coincident – concondensation being effected by rotation – gas-phase reactions are minimised[10,20)].

For the introduction of involatile coreactants, metal atom beams can be condensed down onto a cooled, stirred solution[28)] or up onto a thin film of solution carried around by a rotating vessel from a bulk solution which is at the bottom of the vessel and which constantly renews the film[8, 10)]. A more versatile method is to introduce the involatile material as a beam of microdroplets through a rotary atomiser device which can be used in the same convenient way as a vapour inlet, or the two may be coupled to give a very versatile multiple inlet system. Thus $PPh_3$, $Et_2AlCl$, PhMe and Bd have been simultaneously introduced through such a system[6, 20)].

## 1.3 The Scope of the Technique

Although the technique is a simple one, a certain amount of money, time and effort is required for initial installation and maintenance. Therefore the technique must justify itself by having definite, substantial advantages over other methods, and this it does in three important areas: synthesis of new and interesting compounds; improved synthesis of known compounds; and providing a quick, convenient research tool.

### *1.3.1 Synthesis of New Compounds*

The conditions of the technique make it especially suitable for the synthesis of new compounds which are high reactive or at least unusual. Recently, several completely new types of organometallic compounds have been prepared from metal atoms; tris($\eta$-butadiene)molybdenum *1.1*[9], bis($\eta$-cycloocta-1,5-diene)iron *1.2*[8], bis($\eta$-benzene)titanium *1.3*[10], pentafluorophenyl-palladiumbromide *1.4*[11], and tris(carbon disulphide)nickel *1.5*[29] (Eqs. (1.1) to (1.5)).

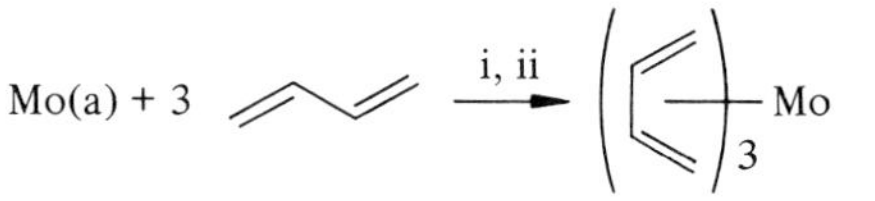


*1.1* (1.1)

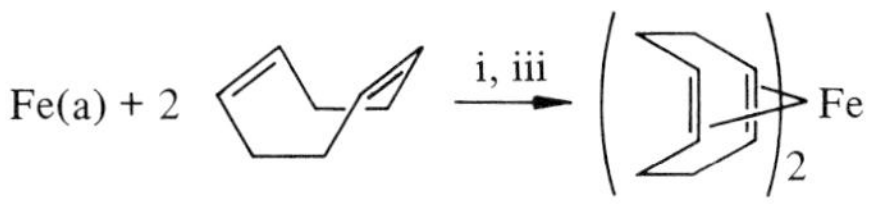


*1.2* (1.2)

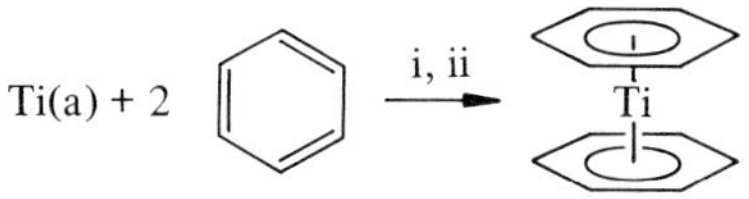


*1.3* (1.3)

$$Pd(a) + C_6F_5Br \xrightarrow{i,ii} C_6F_5PdBr \; \mathit{1.4} \qquad (1.4)$$

$$Ni(a) + CS_2 \xrightarrow{iv} Ni(CS_2)_3 \; \mathit{1.5} \qquad (1.5)$$

(i) cocondense, 77 K; (ii) warm to room temperature; (iii) warm to – 20 °C; (iv) cocondense, 15 K

The use of low temperatures can be exploited to synthesise thermally unstable new compounds such as $Ni(PH_3)_2(PF_3)_2$[19], unsolvated perfluoroalkylzincfluorides[30] and $Ti(CO)_6$[31].

The high lability of some reaction products shows great promise for the introduction of sensitive ligands into complexes. For example the Fe-arene cocondensate *1.6*[32, 33], the Cr-butadiene cocondensate *1.7*[18] and the compound of approximate stoichiometry Ni-$C_6H_6$ *1.8*[34] will react with ligand precurosors with facility under mild conditions (e.g. *ca.* 77 K) (Eqs. (1.6) to (1.9)).

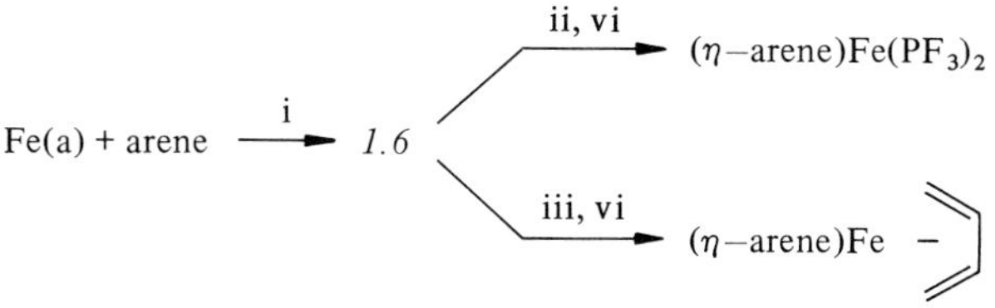


(1.6)

(1.7)

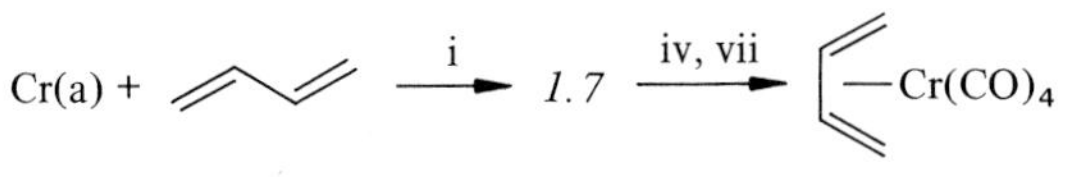

(1.8)

$$Ni(a) + C_6F_6 \xrightarrow{i} \mathit{1.8} \xrightarrow{v, vi} NiL_4 \qquad (1.9)$$

(i) cocondense, 77 K; (ii) $PF_3$, 77 K; (iii) Bd, 77 K; (iv) CO, 77 K; (v) L, 77 K (L = CO, $PEt_3$); (vi) warm to 25 °C; (vii) warm to −20 °C

High temperature species can also be used to form new compounds. Thus condensation of BF or BCl with acetylenes is the only known route to 1,4-diboracyclo hexadienes *1.9*[35] (Eq. (1.10)): other methods give *clovo*-carborane structures. Similarly $SiF_2$ is a very versatile reagent for introducing Si into organic and inorganic compounds[4]. Its reaction with acetylene is shown in Eq. (1.11)[36].

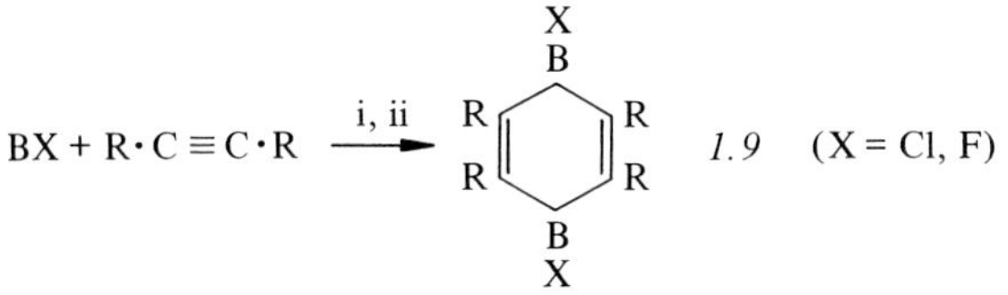

(1.10)

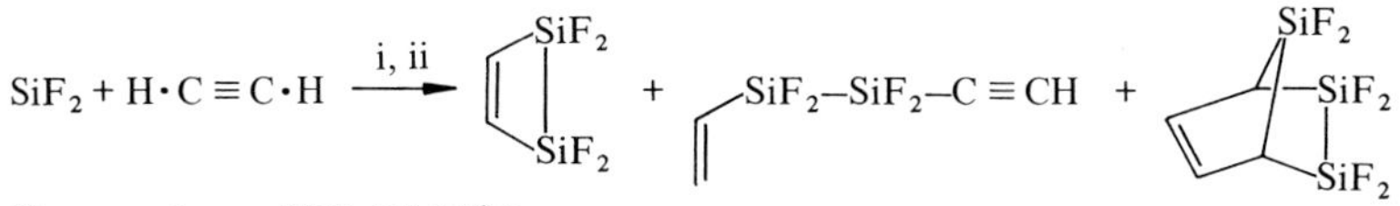

(1.11)

(i) cocondense, 77K; (ii) 25°C

### *1.3.2 Improved Syntheses of Known Materials*

The reaction of Cu atoms with $BCl_3$ has been scaled up to produce 10 g/h of $B_2Cl_4$[37] (Eq. (1.12)) which is ten times faster than the conventional discharge method[38]. Condensation of $SiF_2$ with $I_2$ is the best synthesis of $SiF_2I_2$[39], while pentaborane-9 reacts with carbon vapour to give carbahexaborane-7 in a yield twenty times greater than the previous best route[40].

$$2\,Cu(a) + 2\,BCl_3 \xrightarrow{i, ii} B_2Cl_4 + 2\,CuCL \qquad (1.12)$$

(i) cocondense, 77 K; (ii) 25 °C

Many vapour syntheses offer attractive alternative small scale routes to known compounds. Synthesis of square-planar Ni or Pd complexes of type *1.10* can be conveniently made by the route shown in Eq. (1.13)[11, 41].

$$M(a) + RHal \xrightarrow{i} [R\text{-}M\text{-}Hal] \xrightarrow{ii, iii} R\text{-}\underset{\displaystyle PR'_3}{\overset{\displaystyle PR'_3}{M}}\text{-}Hal \;\; \mathit{1.10} \qquad (1.13)$$

(i) cocondense, 77 K; (ii) $PR'_3$, 77 K; (iii) 25 °C
e.g. R = $C_6F_5$, Hal = Cl, R′ = Et, M = Ni, yield = 55%

Compounds of the type $M(PF_3)_n$ (n = 6, M = Cr[19], Mo[42]; n = 5, M = Fe; n = 8, M = $Co_2$; n = 4, M = Ni, Pd[19]) have been made from metal atoms in good yield, and more conveniently than by the alternative high pressure technique[43].

Cocondensations of arenes and atoms of Ti, V, Nb, Cr, Mo and W give bis($\eta$-arene) metal complexes (see Table 4.13, Chap. 4). The complexes containing Ti[10] or Nb[44] are new and are yet to be made by conventional methods, while $(\eta\text{-arene})_2W$ complexes[45] are made in much higher yields than the 2% yields of other routes[46]. Even in the cases of the well established syntheses of $(\eta\text{-arene})_2M$ (M = V, Cr, Mo)[47], advantage can be gained from the fact that conventional syntheses take a matter of days compared with the hours of the MVS technique. In addition, as conventional methods often use reducing Friedel-Crafts conditions (e.g. Eq. (1.14)), Lewis acid catalysed scrambling of alkyl groups on substituted benzenes can occur. However, isomerically pure $(\eta\text{-arene})_2Cr$ compounds have been synthesised from Cr atoms and a large number of alkylbenzenes[32, 48, 49]. Furthermore, arenes with electron-withdrawing groups such as Cl are inert under conventional conditions but, for example, $(\eta\text{-PhCl})_2Cr$ (47%) has been synthesised from Cr atoms[49, 50]. These points are expanded in Chap. 4.

$$3\ MCl_3 + 2\ Al + AlCl_3 + 6\ \text{arene} \longrightarrow 3[(\eta^6\text{-arene})_2M]^+[AlCl_4]^- \qquad (1.14)$$

### *1.3.3 A Quick Convenient Research Tool*

The technique is suitable for the study of unstable species formed and frozen into very cold matrices[51] where they can be studied spectroscopically (matrix isolation spectroscopy) as shown in Eqs. (1.15) to (1.20).

$$Ta(a) + CO \xrightarrow{i} Ta(CO)_6 \qquad (1.15)^{52)}$$

$$V(a) + N_2 \xrightarrow{ii} V(N_2)_6 \qquad (1.16)^{53)}$$

$$Ni(a) + O_2 \xrightarrow{ii} Ni(O_2)_2 \qquad (1.17)^{54)}$$

$$Ni(a) + Ar \xrightarrow{iii,\ iv} Ni_3 \qquad (1.18)^{55)}$$

$$M(a) + NO + Ar \xrightarrow{v} M^+(NO)^- (M = Li, Na, K, Rb, Cs) \qquad (1.19)^{56)}$$

$$Li(a) + CCl_4 + Ar \xrightarrow{v} Cl^-Li^+ \cdots\cdots CCl_2\ (\textit{inter alia}) \qquad (1.20)^{57)}$$

(i) cocondense, 4 K; (ii) cocondense, 12 K; (iii) cocondense, 10 K, Ni : Ar = 1 : 20 to 100; (iv) 35 K; (v) cocondense 15 K, large excess of Ar

In one series of experiments[58] ingenious use has been made of rotation to follow the course of unstable species such as radicals in their reactions with other substrates at low temperatures (for example Eq. (1.21)). The concept is shown in Fig. 1.4.

$$Et^{\cdot} + CH_2{=}CH_2 \longrightarrow Et\ CH_2CH_2^{\cdot} \qquad (1.21)$$

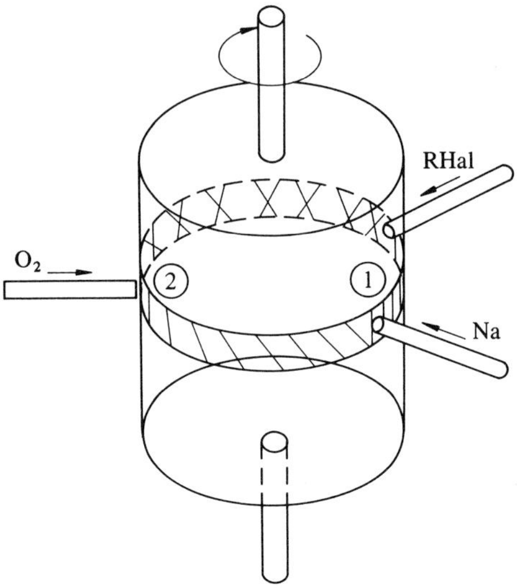


Step ① RHal + Na$^\bullet$ ⟶ R$^\bullet$ + NaHal

Step ② R$^\bullet$ + $O_2$ ⟶ ROO$^\bullet$

**Fig. 1.4.** Rotating cryostat for Consecutive reactions

The oligomerisation or polymerisation of unsaturated hydrocarbon monomers such as ethylene or butadiene can be catalysed by transition metal complexes *via* some incipient or induced unsaturation in the complexes. Metal atoms are the ultimately unsaturated organometallic species and as such can be expected in some cases to act as catalyst precursors when condensed with appropriate monomers. Indeed butadiene (Bd) telomerisation is observed when atoms of the first row transition metals are condensed with Bd and an arene such as toluene[6)]. The very basic nature of the starting materials, metal atoms and monomers allows a fundamental approach to catalysis study. For example it allows such questions as what are the essential constituents of a particular catalyst to be answered. Thus by the permutation of various additives to a basic Ti/Bd cocondensate it can be shown that Ti, Al and Cl are sufficient to catalyse formation of cyclododecatriene (Cdt) from Bd, and that if one component is absent, no activity is observed[6)]. In addition, by judicious addition of various cocatalysts to the basic cocondensate it is possible to "build up" catalysts of specific, selective activity. For example condenstion of Ti atoms, Bd, toluene, diethylaluminium chloride, and triphenylphosphine gives a highly active catalyst for the cyclotrimerisation of Bd to *trans,trans,trans*-Cdt.

It is also possible to use unstable complexes formed by reaction of metal atoms with various monomers and trapped in very low temperature matrices as models for catalysts and catalytic processes[59)]. Thus, for example, the great similarity of the ir spectra of ($\eta$-$C_2H_4$)Pd[60)] and $C_2H_4$ absorbed on silica-supported Pd[61)] suggests a Π-bonding model for the adsorbed $C_2H_4$ and that ($\eta$-$C_2H_4$)Pd may act as a model for the system.

$$\mathrm{Pd(a)} + C_2H_4 + \mathrm{Xe} \xrightarrow{i} (\eta\text{-}C_2H_4)\mathrm{Pd} \qquad (1.22)$$

(i) cocondense, 15 K; Pd : $C_2H_4$ : Xe = $1 : 10^3 : 10^5$

Furthermore, in certain cases, reactions can be followed (Eq. (1.23))[62] which may act as models for catalytic cycles (in this case the oxidation of CO on Au surfaces).

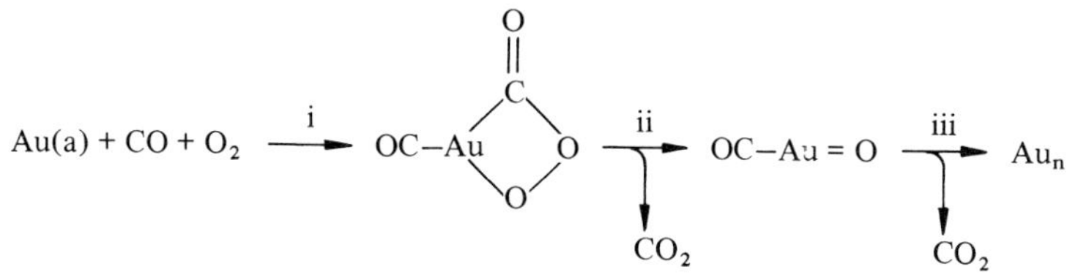


(1.23)

(i) cocondense 10K; (ii) CO : $O_2$ = 1 : 1; 30K; (iii) 40K

One of the vexing side-reactions of the cocondensation experiment is the formation of metal aggregates. Not only is the metal usually lost to the desired reaction, it can, in some cases, cause decomposition of the required product[1, 10, 63]. However, such aggregates can have their own interesting desirable chemistry. For example the slurry formed on warming a Mg/thf cocondensate will react vigorously with organic halides to form Grignard reagents[64].

$$\text{Mg(a)} + \text{thf} \xrightarrow{\text{i,ii}} \text{Mg/thf slurry} \xrightarrow{\text{iii}} \text{R MgHal} \qquad (1.24)$$

(i) cocondense, 77 K; (ii) warm; (iii) RHal

The high dispersion of such slurries can impart catalytic activity[65]. Furthermore the variation of cocondensant solvent can lead to different aggregate sizes and crystallinities leading to differing catalytic properties. Thus the Ni/n-hexane slurry contains rough spheres which act as a better hydrogenation catalyst than Raney nickel while the Ni/thf slurry contains smaller spheres which act as a poor hydrogenation catalyst but as a better catalyst for the disproportionation of cyclohexene to benzene and cyclohexane. Preliminary experiments show that metal crystallites can be deposited onto typical catalysts supports such as alumina when unstable metal atom/solvent cocondensates are allowed to warm in the presence of such solids[65, 66]. Recently, small Pt crystallites have been formed on an alumina support by direct vapour deposition of Pt atoms without use of solvent[67]. A study of the catalytic properties of these materials is underway.

## References


1. (a) Timms, P. L.: Adv. Inorg. Chem. Radiochem. *14*, 121 (1972); (b) Timms, P. L., Turney, T. W.: Adv. Organometal. Chem. *15*, 53 (1977)
2. The Symposium "Metal Atoms in Chemical Synthesis," held under the auspices of Merck'sche Gesellschaft für Kunst und Wissenschaft e. V. at Darmstadt (May 12–15, 1974), published in Angew. Chem., Int. Ed. *14* (1975)
3. Skell, P. S., Havel, J. J., McGlinchey, M. J.: Acc. Chem. Res. *6*, 97 (1973)
4. Margrave, J. L., Wilson, P. W.: Acc. Chem. Res. *4*, 145 (1971); Timms, P. L.: Acc. Chem. Res. *6*, 118 (1973); Perry, D. L., Margrave, J. L.: J. Chem. Ed. *53*, 696 (1976)
5. Middleton, R.: Ph. D. Thesis, Univ. Bristol 1974

6. Akhmedov, V. M., Anthony, M. T., Green, M. L. H., Young, D.: J. C. S. Dalton *1975*, 1412
7. E.g. Bennett, J. E., Mile, B., Thomas A., Ward, B.: Adv. Phys. Org. Chem. *8*, 1 (1970)
8. McKenzie, R., Timms, P. L.: J. C. S. Chem. Comm. *1974*, 650
9. Skell, P. S., Van Dam, E. M., Silvon, M. P.: J. Amer. Chem. Soc. *96*, 626 (1974)
10. Anthony, M. T., Green, M. L. H., Young, D.: J. C. S. Dalton, *1975*, 1419
11. Klabunde, K. J., Low, J. Y. F.: J. Organometal. Chem. *51*, C35 (1973); J. Amer. Chem. Soc. *96*, 1984 (1974)
12. Pimentel, G. C.: Spectrochim. Acta *12*, 94 (1958); Pure Appl. Chem. *4*, 61 (1962); Pimentel, G. C., Charles, S. W.: Pure Appl. Chem. *7*, 111 (1963)
13. Linevsky, M. J.: J. Chem. Phys. *34*, 587 (1961)
14. Skell, P. S.: J. Amer. Chem. Soc. *85*, 1023 (1963); Skell, P. S., Wescott, L. D. jr., Goldstein, J.-P., Engel, R. R.: J. Amer. Chem. Soc. *87*, 2829 (1965); Skell, P. S., Engel, R. R.: J. Amer. Chem. Soc. *88*, 3749 and 4883 (1966)
15. Timms, P. L.: Endeavour *27*, 133 (1968)
16. Timms, P. L.: Chem. Comm. *1969*, 1033
17. Fischer, E. O.: Inorg. Synth. *6*, 132 (1960)
18. E.g. see Vohwinkel, V.: Trans. N. Y. Acad. Sci. *24*, 446 (1964) and references therein
19. Timms, P. L.: J. Chem. Soc. (A) *1970*, 2526
20. Green, M. L. H., Young, D.: Phys. Bull., *1975*, 350
21. Klabunde, K. J.: Acc. Chem. Res. *8*, 393 (1975)
22. Klabunde, K. J., Murdock, T. O.: Chemtech. *5*, 624 (1975)
23. Koerner von Gustorf, E. A., Jaenicke, O., Polansky, O. E.: Angew. Chem., Int. Ed. *11*, 532 (1972)
24. Koerner von Gustorf, E. A., Jaenicke, O., Wolfbeis, O., Eady, C. R.: Angew. Chem. Int. Ed. *14*, 278 (1975)
25. Skell, P. S., McGlinchey, M. J.: Angew. Chem., Int. Ed. *14*, 195 (1975)
26. Boyd, J. W., Lavoie, J. M., Gruen, D. M.: J. Chem. Phys. *60*, 4088 (1974); Carstens, D. H. W., Kozlowski, J. F., Gruen, D. M.: High. Temp. Sci. *4*, 301 (1972)
27. Skell, P. S., Wescott, L. D. jr., Goldstein, J.-P., Engel, R. R.: J. Amer. Chem. Soc. *87*, 2829 (1965)
28. Skell, P. S., Cholod, M. S.: J. Amer. Chem. Soc. *91*, 6035 (1969)
29. Huber, H., Ozin, G. A., Power, W. J.: Inorg. Chem. *16*, 2234 (1977)
30. Klabunde, K. J., Scott, M., Low, J. Y. F.: J. Amer. Chem. Soc. *94*, 999 (1972)
31. Lever, A. B. P., Ozin, G. A.: Inorg. Chem. *16*, 2012 (1977)
32. Middleton, R., Hull, J. R., Simpson, S. R., Tomlinson, C. M., Timms, P. L.: J. C. S. Dalton *1973*, 120
33. Williams-Smith, D. L., Wolf, L. R., Skell, P. S.: J. Amer. Chem. Soc. *94*, 4042 (1972)
34. Klabunde, K. J., Efner, H. F.: J. Fluorine Chem. *4*, 114 (1974)
35. Timms, P. L.: J. Amer. Chem. Soc. *90*, 4585 (1968)
36. Margrave, J. L., Perry, D. L.: Inorg. Chem. *16*, 1820 (1977) and references therein
37. Timms, P. L.: Chem. Comm. *1968*, 1525; J. C. S. Dalton, *1972* 830
38. Wartik, T., Rosenberg, R., Fox, W. B.: Inorg. Synth. *10*, 118 (1967); Urg. G., Wartik, T., Moore, R. E., Schlesinger, H. I.: J. Amer. Chem. Soc. *76*, 5293 (1954)
39. Margrave, J. L., Sharp, K. G., Wilson, P. W.: J. Inorg. Nucl. Chem. *32*, 1817 (1970).
40. Price, S. R., Schaefer, R.: Chem. Comm. *1968*, 41
41. Klabunde, K. J., Low, J. Y. F.: J. Amer. Chem. Soc. *96*, 7675 (1974); Roberts, J. S., Klabunde, K. J.: J. Organometal. Chem. *85*, C13 (1975); J. Amer. Chem. Soc. *99*, 2509 (1977)
42. Young, D.: D. Phil. Thesis, Univ. Oxford 1974; P. L. Timms: unpublished, see Ref. 1(b)
43. Kruck, T.: Angew. Chem., Int. Ed. *6*, 53 (1967)
44. Gruen, D. M.: personal communication to Timms, P. L., Turney, T. W.: Ref. 1(b)
45. Silvon, M. P., Van Dam, E. M., Skell, P. S.: J. Amer. Chem. Soc. *96*, 1945 (1974); Cloke, F. G. N., Green, M. L. H.: J. C. S. Chem. Comm. *1978*
46. Fischer, E. O., Scherer, F., Stahl, H. O.: Chem. Ber. *93*, 2065 (1960)
47. Silverthorn, W.: Adv. Organometal. Chem. *13*, 48 (1975)
48. See Evans, S., Green, J. C., Jackson, S. E., Higginson, B.: J. C. S. Dalton *1974*, 301
49. Graves, V., Lagowski, J. J.: Inorg. Chem. *15*, 597 (1976)
50. Skell, P. S., Williams-Smith, D. L., McGlinchey, M. J.: J. Amer. Chem. Soc. *95*, 3337 (1973)

51. Meyer, B.: Low Temperature Spectroscopy. New York: Elsevier 1971; Hallam, H.: Vibrational Spectroscopy of Trapped Species. New York: Wiley 1972; Moskovits, M., Ozin, G. A. (eds.): Cryochemistry. New York: Wiley 1976
52. Dekock, R. L.: Inorg. Chem. *10,* 1205 (1971)
53. Huber, H., Ford, T. A., Klotzbücher, W. E., Ozin G. A.: J. Amer. Chem. Soc. *98,* 3176 (1976)
54. Huber, H., Ozin, G. A.: Can. J. Chem. *50,* 3746 (1972); Ozin, G. A., Klotzbücher, W. E.: J. Amer. Chem. Soc. *97,* 3965 (1975)
55. Moskovits, M., Hulse, J. E.: J. Chem. Phys. *66,* 3988 (1977)
56. Andrews, W. L. S., Pimentel, G. C.: J. Chem. Phys. *44,* 2361 (1966); Tevault, D. E., Andrews, L.: J. Phys. Chem. *77,* 1640 and 1646 (1973)
57. Hutzenbühler, D. A., Andrews, L., Carey, F. A.: J. Amer. Chem. Soc. *97,* 187 (1975)
58. Mile, B.: Angew. Chem., Int. Ed. *7,* 507 (1968); Bennett, J. E., Mile, B., Thomas, A., Ward, B.: Adv. Phys. Org. Chem. *8,* 1 (1970)
59. Ozin, G. A.: Acc. Chem. Res. *10,* 21 (1977)
60. Ozin, G. A., Power, W. J.: Inorg. Chem. *16,* 212 (1977); Huber, H., Ozin, G. A., Power, W. J.: *Ibid.,* p. 979
61. Prentice J. D., Lesiunas, A., Sheppard, N.: J. C. S. Chem. Comm. *1976,* 76.
62. Huber, H., McIntosh, D., Ozin, G. A.: Inorg. Chem. *16,* 975 (1977)
63. Blackborow, J. R., Grubbs, R. H., Miyashita, A., Scrivanti, A., Koerner v. Gustorf, E. A.: J. Organometal. Chem. *122,* C6 (1976)
64. Klabunde, K. J., Efner, H. F., Satek, L., Donley, W.: J. Organometal. Chem. *71,* 309 (1976)
65. Klabunde, K. J., Efner, H. F., Murdock, T. O., Ropple R.: J. Amer. Chem. Soc. *98,* 1021 (1976)
66. Blackborow, J. R.: private communication
67. Imperial Chemical Industries, Belgian Pat. 841924 (1977)

# 2 Practical Aspects of Metal Vapour Synthesis

## 2.1 General Considerations

The success of an MVS experiment depends principly on maintaining high vacuum ($10^{-4} \sim 10^{-5}$ torr) within a vessel where a high temperature evaporation furnace usually above,1,600 °C is not far from a cold surface onto which considerable quantities of volatile material are being condensed for periods of up to two or three hours. Loss of vacuum halts the evaporation process and usually terminates the experiment. The best possible pumping system is required under such conditions, as well as effective vacuum seals. The pumping system invariably involves three stages, a rough rotary oil pump followed by an oil diffusion pump and finally a cryostatic pump (cold trap) (Fig. 2.1). The size of the pumps is determined mainly by size and working conditions of the vacuum chamber. (The pumping speed of most oil diffusion pumps is a function of their size only.)

The vapour pressure above a matrix depends, for a given substrate, on the temperature of the coolant surrounding the reaction vessel and the extent of thermal contact between the coolant and the matrix. The rates of substrate condensation and of absorption of radiative energy from the evaporation source determine the temperature gradient between the condensation surface and the coolant; obviously this should be minimised, and the thermal contact between the matrix and coolant maximised, within the bounds of the experiment. Thin walled glass or steel vacuum systems are therefore necessary with good radiative shielding of the evaporation source. The coolant is usually liquid nitrogen.

Normally the vaporisation vessel is designed to be as large as the pumping efficiency, and hence as *m*ean *f*ree *p*ath considerations will allow (MFP $\geqslant$ distance between heating source and vessel wall). This maximises the metal atom dispersion per unit condensation area for a given metal flux, and allows the maximum condensation efficiency of the coolant. System pressures tend initially to increase (due to out-gassing of the metal) and subsequently decrease (due to gettering by the condensing metal vapour) at the commencement of an experiment. "Gettering" depends on the metal and on the residual gases in the vacuum chamber. Pressures may drop by as much as a factor or two due to gettering before substrate cocondensation is initiated.

Pressure is monitored continuously during an MVS experiment with either a high voltage Penning gauge or a hot filament gauge; the pressure reading is very dependent on the position of the gauge within the vacuum system. The optimal position is usually well removed from the cold surface and the metal vapour source but not too near the cold trap.

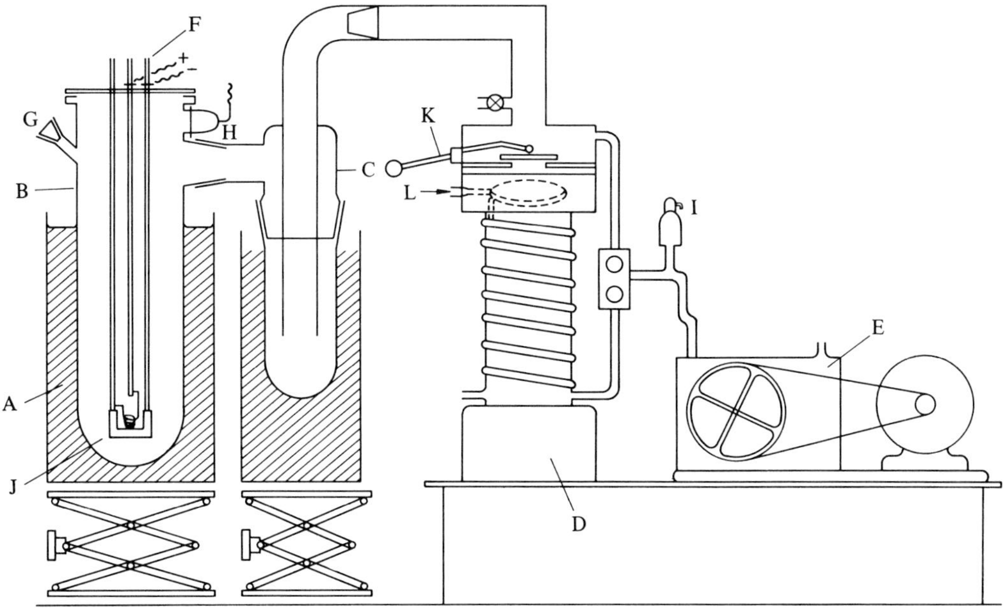


**Fig. 2.1.** Simple metal evaporation apparatus. *A* Coolant; *B* evaporation vessel (5 to 10 litres); *C* cryopump; *D* diffusion pump; *E* rotary pump; *F* heating source and substrate inlets; *G* argon inlet and product extraction port; *H* penning head; *I* pirani head; *J* crucible and water cooled radiation shield; *K* butterfly value; *L* water cooling

A registered pressure may not correspond to the vapour pressure over the matrix but the readings may be used empirically after some experiences to judge the probable effectiveness and the rate of the evaporation process. Pressure near the metal vapour source is difficult to determine (and not very meaningful). Often however when using high voltage heating sources, electric glow discharges from the condensing substrate gases are observed usually indicating a pressure of $10^{-3}$ torr or higher.

The maintenance of a good vacuum, provided the matrix remains intact, is often determined by the quality of the vacuum seals for the various inlet ports required. There appears to be little problem in this respect provided there is not a large temperature gradient across the seal. Seals of glass, metal, rubber or plastic with or without lubricants have all been used successfully and manual or electric vacuum valves or stopcocks are readily available.

Product recovery under oxygen free conditions is an important aspect of MVS experiments. Usually products are extracted just after the matrix has melted after removal of the coolant ($-160 \sim 0$ °C). The material is often filtered immediately to remove a large proportion of metal aggregates which may be active enough to decompose the products at higher temperatures. A preparative vacuum line attached directly to the MVS apparatus is often useful for the isolation and characterisation of volatile products.

Effective cleaning of the whole vacuum system avoids long periods of out-gassing at the commencement of new experiments. The proximity of hydrocarbon matrices to molten metals may represent a hazard if vacuum is lost and oxygen enters the system. This hazard is augmented by hydrocarbon coolant baths which may sometimes be used. Usually liquid nitrogen is the coolant and implosion of a vacuum system is accompanied by the inrush of very cold inert gas. There is far more risk of fire from pyrophoric metal powders left inadvertantly on discarded filter pads. Iron, chromium and cobalt are particularly bad in this respect.

*Warning*: Argon condenses in contact with liquid nitrogen. It is important when filling vacuum vessels with argon to remove the liquid nitrogen coolant first. Many flasks and fittings have been destroyed by boiling argon under pressure.

## 2.2 Pumping Systems

Most commercial pumps have very similar characteristics[1, 2)]. The pumping speeds of rotary oil or oil diffusion pumps within their optimum working range do not usually change with change of pressure and are solely a function of pump size. The size combination diffusion pump-rotary oil pump is determined by the criterion that the critical backing pressure of the diffusion pump is never exceeded by the rotary pump pressure under working conditions.

$$\text{Pumping speed of rotary} > \frac{\text{maximum throughput of gas*}}{\text{critical backing pressure}}$$

Double stage rotary oil pumps without gas ballasting and with cryostatic pumping will give vacuums of $10^{-3}$ torr or better in small vessels and it is possible to usefully evaporate alkali, alkali earth metals and Cr, Mn, Cu, Zn, Ag and Cd without a diffusion pump, as they have high rates of evaporation at relatively low temperature (Table 2–3). In principle, any metal may be evaporated in system without a diffusion pump but such high temperatures are often required that heat dissipation problems become insurmountable.

The speed of pumping of a commercial oil diffusion pump is determined by the diameter of the pump inlet, for example, pumps of diameter 4, 6.5, 10 and 15 cm will pump at 30, 170, 400 and 1,000 l/s.

The vacuum vessel size and reaction conditions determine the diffusion pump size. Usually vessels of less than five litre capacity will be evacuated quite adequately by a 4 cm diameter diffusion pump. The authors have used a 6.5 cm diameter pump to evacuate a rotating seal cryostat of 20 l capacity to $4 \times 10^{-5}$ torr and a 2.5 cm pump to evacuate a five litre static vessel to the same pressure.

The cross-sectional area of the outlet from the vacuum vessel through the cryopump should match that of the diffusion pump or pumping efficiency is lost.

---

* The lower limit (vacuum) of an oil diffusion pump is taken as $10^{-2}$ torr therefore the maximum throughput of gas is: diffusion pump speed $\times 10^{-2}$ l torr/sec. The critical backing pressure is determined by the diffusion pump oil pressure. Using a diffusion pump of speed 80 l/s and silicone 704 oil (Edwards, Dow Corning):

$$\text{Pumping speed of rotary } \frac{80 \times 10^{-2}}{0.8} = 1 \text{ l/s.}$$

Allowing a 100% error, a rotary pump of speed 2 l/s might be used.

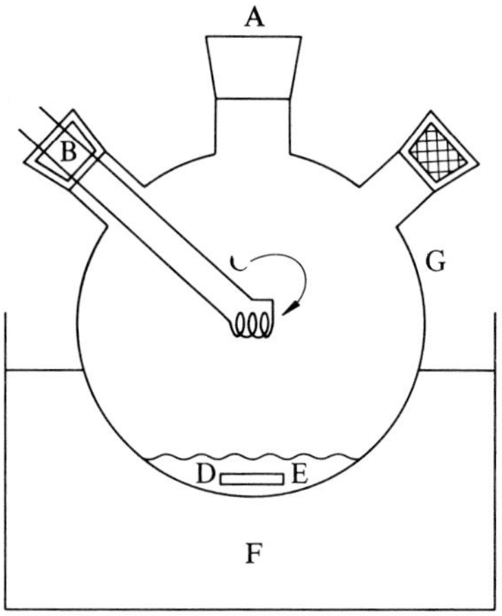


**Fig. 2.2.** Vessel for preparing resin or gel supported chromium or for evaporating chromium into non-volatile liquids. *A* Pump out to normal laboratory rotory oil pump; *B* copper electrodes glued into a rubber bung; *C* tungsten basket containing chromium; *D* silica gel and or liquid reactant; *E* stirrer; *F* coolant bath; *G* 500 ml glass flask

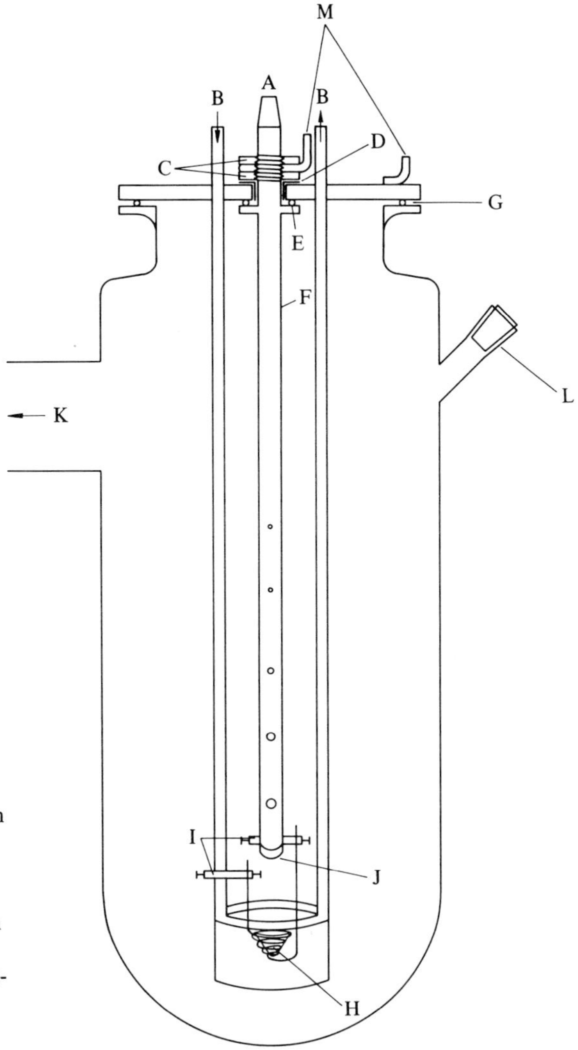


**Fig. 2.3.** Simple resistance heated vaporisor in glass vacuum vessel. *A* Cocondensate inlet; *B* coolant water; *C* fixing screw; *D* teflon flange; *E* O Ring seal; *F* spray-out tube for cocondensate; *G* seal surface for glass vessel (O ring, adhesive, or grease); *H* tungsten basket with crucible; *I* retainers for tungsten basket; *J* heat shield for substrate inlet; *K* pump and pressure gauge; *L* inlet for product extraction; *M* electrical leads

## 2.3 Vacuum Vessels

Vacuum vessels are usually made of glass or steel; the latter material, although more robust, is not as easily manipulated and suffers from outgassing problems, it is also much more costly. The advantage of glass vessels is their transparency (Steel vessels have to be fitted with glass windows). The design of reaction vessels has depended mainly on the type of experiment undertaken and owes much to Skell[3], Timms[4], Green[5] and Koerner von Gustorf[6].

### *2.3.1 Static Vessels*

Static vessels are the simplest and cheapest to design and use and are therefore popular. They are most frequently used to cocondense vapour beams of metal atoms and substrate molecules as a matrix on the cooled walls of the vessel. Any form of heat source may be

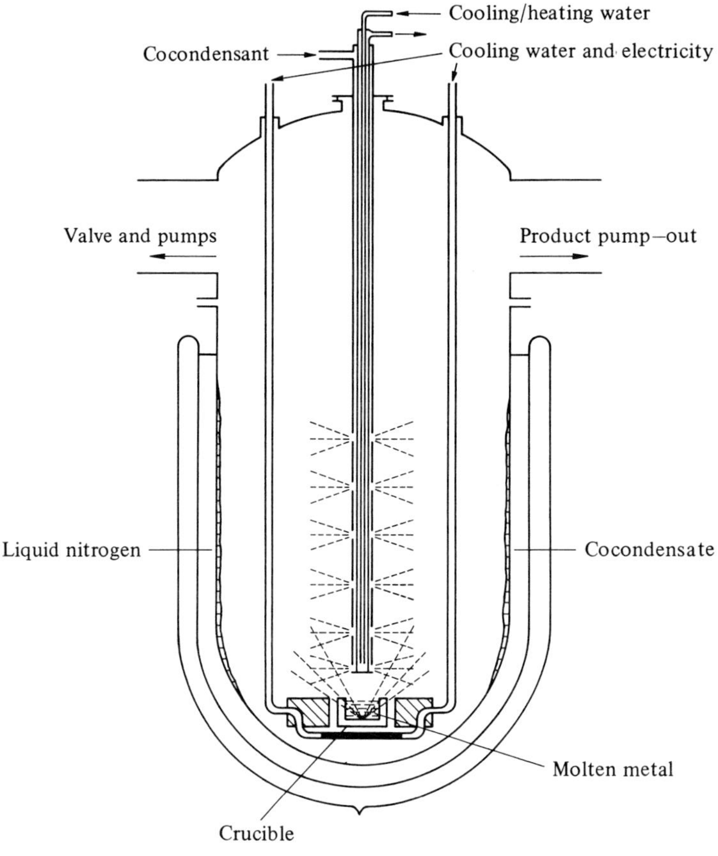


**Fig. 2.4.** Apparatus encorporating heated or cooled inlet for substrates and pump out to vacuum line assembly (Timms, P.L.: Adv. Inorg. Chem. Radiochem. *14*, 132 (1972); with permission of Academic Press)

used, and solids, liquid or gaseous substrates may be introduced provided they can be vaporised. In certain cases, metal atoms can be condensed down into a liquid whose surface is being constantly renewed by agitation (Fig. 2.2). Many versions of the static glass vacuum vessel have appeared in the literature and differences in design are usually to accommodate different heating sources.

We illustrate here a selection of the more important designs, starting with the simplest and progressing to more sophisticated versions (Figs. 2.2 to 2.6).

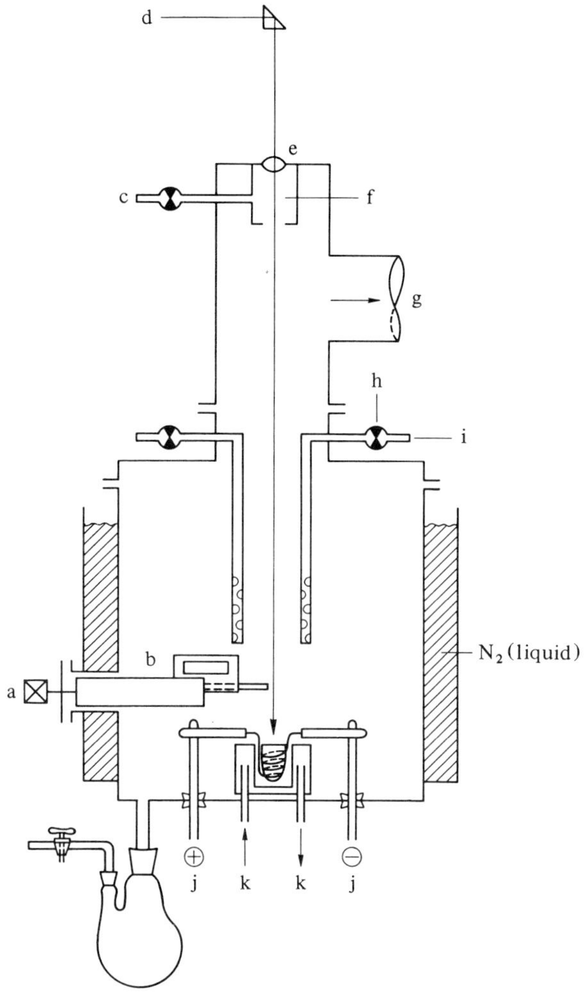


**Fig. 2.5.** Diagram of an apparatus for the evaporation (by laser and resistive heating) and co-condensation of metals with organic substrates. *a* Motor and gearbox; *b* wire feeding mechanism; *c* gas flow, control valve; *d* laser beam; *e* lens; *f* gas window; *g* to vacuum pump; *h* magnetic valve; *i* substrate inlet; *j* current feedthroughs for crucible heating; *k* cooling water. Scale 1:9 (Koerner von Gustorf, E. A., et al.: Angew. Chem., Int. Ed. *14,* 278 (1975))[7]

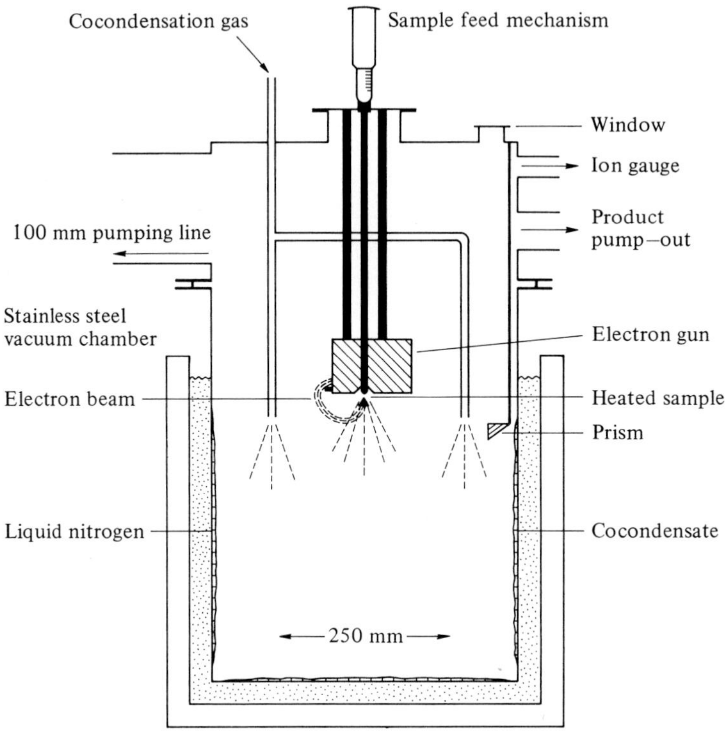


**Fig. 2.6.** Steel apparatus for electron beam evaporation of sublimable materials[8]. Molten materials would be evaporated upwards with an electron gun placed at the bottom of the vessel (see Fig. 2.9) (Timms, P. L.: Adv. Inorg. Radio Chem. *14*, 136 (1972); with permission of Academic Press)

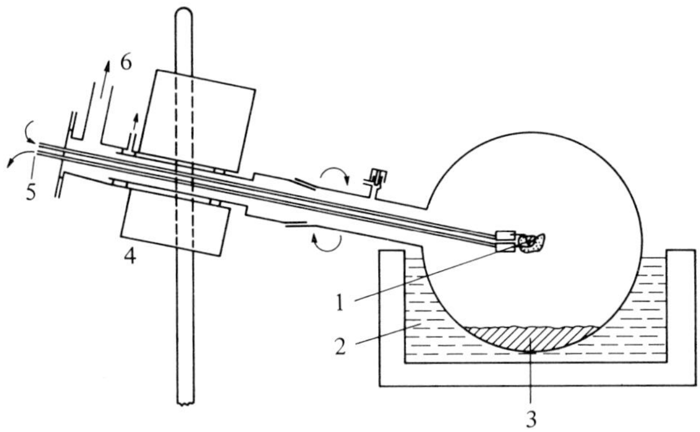


**Fig. 2.7.** Rotary apparatus for reacting metal atoms with cooled solutions. *1* Evaporating metal; *2* cooling bath; *3* substrate solution; *4* rotary drive mechanism; *5* water and electrical power; *6* high vacuum (Timms, P. L.; Angew. Chem., Int. Ed. *14*, 195 (1975))

## *2.3.2 Rotating Vessels*

Rotating MVS vessels were originally developed in order to separate the metal and substrate vapour beams as this reduced the substrate vapour pressure in the region of the furnace and consequently reduced contamination by substrate pyrolysis on the furnace[9]. The low pressure requirements of an electron beam furnace ($10^{-5 \sim -4}$ torr) were more easily obtained and lower substrate pressures in the metal vapour flux ensured a longer mean free path (Fig. 2.9). The development of vessels with rotating vacuum seals proved surprisingly easy, even for pressures of $10^{-5}$ torr or less. Previous experiments had shown that condensation of metal atoms onto solid reactant surfaces, constantly renewed by rotation lead to reaction product yields comparable to those of the coincident cocondensation of static vessels[51]. This concept has since been developed to allow reaction of metals with involatile reactants.

The vessels vary in size from two to twenty litres. The larger apparatus are sufficiently heavy to require incorporation of supportive ball-races and reinforced vessels. The rotating seals can vary from home-made designs based on laboratory rotary evaporators to proprietary seals such as Crane seals[10]. Rotation is controllable up to at least 200 revolutions per minute.

Three designs are given (Figs. 2.7 to 2.11).

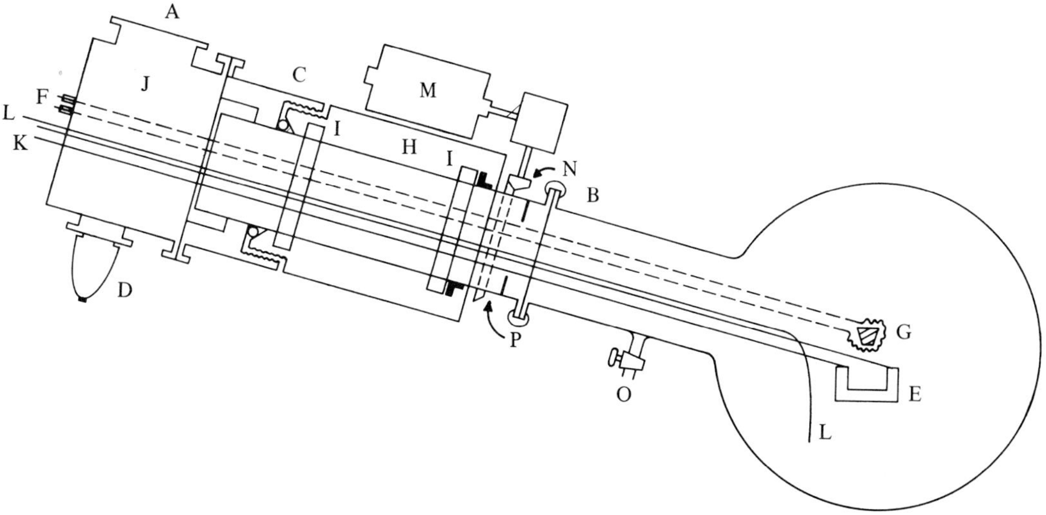


**Fig. 2.8.** Five to twenty litre "heavy duty" metal evaporating rotating cryostat (with key design features) for use with resistance or electron beam heating. *A* Pump out; *B* glass-metal conical or straight flange seal; *C* lubricated O ring seal (The O ring rests in a triangular grove which becomes a vacuum seal when the two parts of the body are screwed together. A contra-rotating locking nut can be interposed between the two parts ensuring stability for rotating in either direction); *D* pressure gauge; *E* water cooled radiation shield; *F* electrical inlets; *G* crucible, or electron gun; *H* brass body and rotating sleeve; *I* ball bearings; *J* demountable head; *K* water cooling; *L* substrate inlet; *M* drive motor (1/2 H. P.); *N* cog drive; *O* argon inlet; *P* removable teflon support for feedthroughs

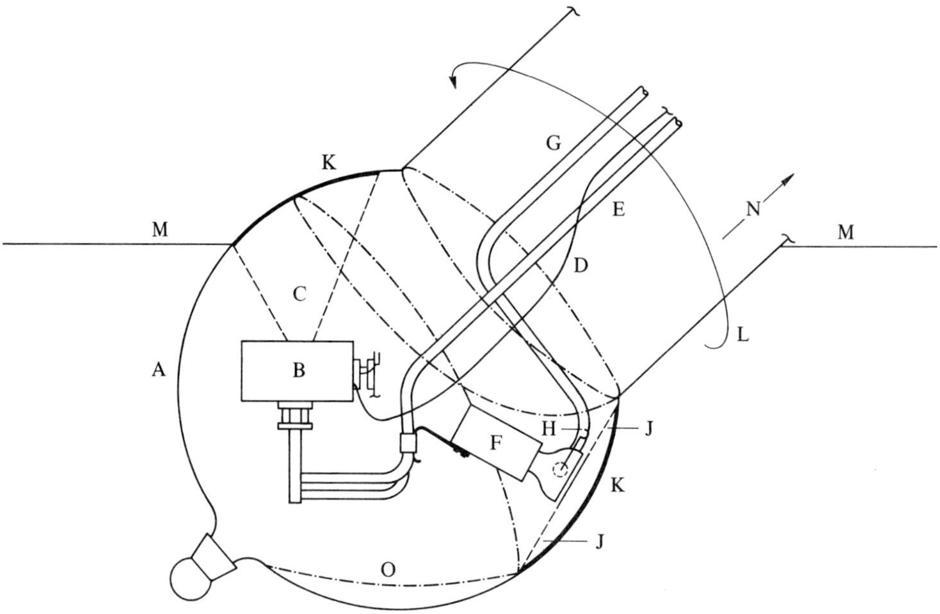


**Fig. 2.9.** Electron gun evaporator encorporating solution sprayer[9,10]. *A* Reaction vessel; *B* furnace; *C* metal vapour beam; *D* one of two furnace electrical connectors; *E* one of two cooling water pipes; *F* combined coreactant solution and vapour dispersion device; *G* vapour inlet pipe (containing the solution inlet pipe); *H* solution inlet pipe, emerging from G; *J* combined vapour and solution microdroplet beam; *K* cocondensate; *L* rotation; *M* liquid nitrogen coolant level; *N* to rotating seal and high vacuum; *O* PVC tubing, shaped to raise, on rotation, the coreactant, as an involatile solution (from the lowest point of the reaction vessel where it drains), as a film into which metal atoms are condensed

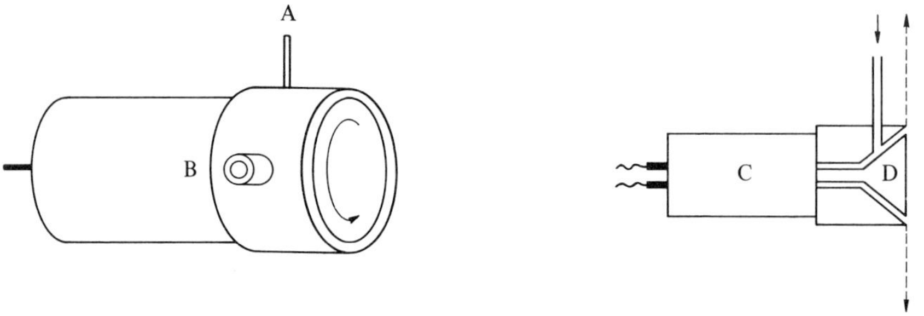


**Fig. 2.10.** Solution sprayer used in rotating cryostats (Fig. 2.7 to 2.9). *A* Feed for solution; *B* feed for other cocondensates; *C* small electric motor; *D* rotating brass conic acting as disperser

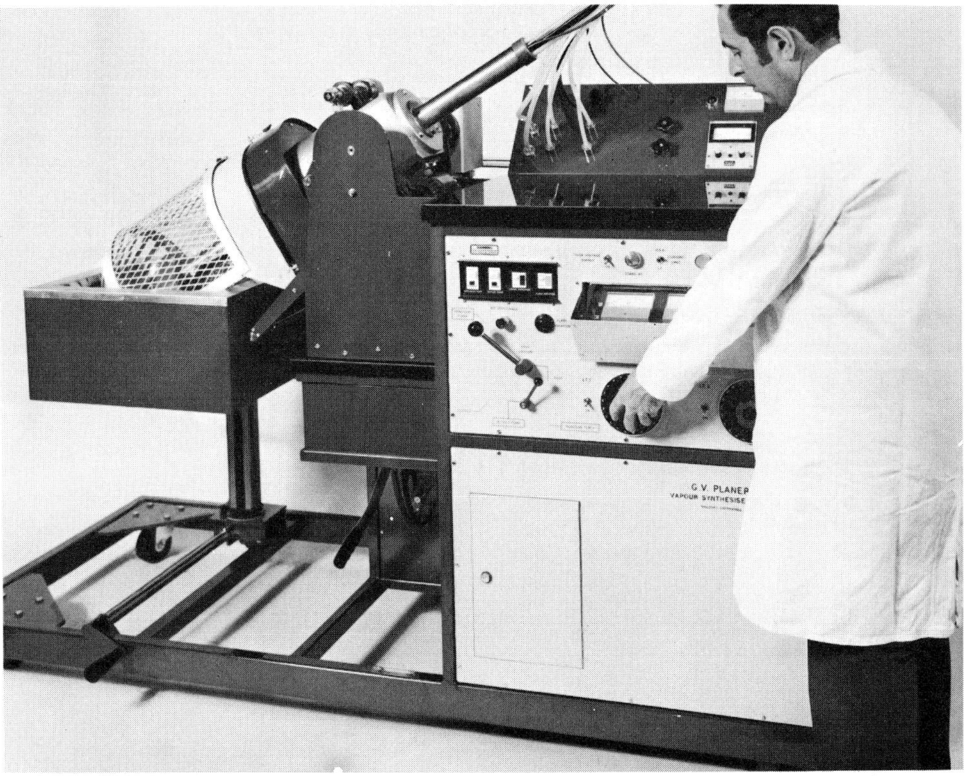


**Fig. 2.11.** Commercial rotating vacuum system utilising electron beam heating[9,10]
(Photograph reproduced with permission of G. V. Planer Ltd.)

## 2.4 Reactant Inlet Systems

Inlet systems for reactants are usually simple in design; with the exception of the motor driven solution sprayer for non-volatiles shown in Fig. 2.10, all reactants enter the reaction vessel as gases, and are dispersed to the walls of the cooled vessel by a culinder type nossel (Figs. 2.4, 2.5). The inlets may be part of the vessel or detachable vacuum sealable components. The rate of introduction of reactants is controlled by either needle valves, stopcocks or electrically timed on-off valves. Reactants may be evaporated or sublimed through such valves without much difficulty.

## 2.5 Vacuum Sealed Inlets for Electrical Supply

### *2.5.1 Low Tension*

The majority of MVS devices utilise low tension leads for resistance heating. Typical powers of 2 kwatts may be obtained at 20 volts and 100 amps. a.c. The leads into the vacuum vessel have to be insulated from each other and their metal halter (teflon, bakalite, viton; see Figs. 2.2 to 2.6). This insulation usually forms the vacuum seal. A 100 amp supply uses water cooling for the leads and the radiation shield; at 20 volts no insulation for the cooling water is necessary. Inlets for low power auxiliary electrical circuits (sublimers or electric motors within the vacuum system) typically use 0.5 to 5 amps at 10 volts and can be copper wires sealed in glass or araldite.

### *2.5.2 High Tension*

Electron beam heating or sputtering can require high voltages (up to 7,000 volts) thus extremely well insulated electrical leads (mica or glass coated) are essential. Normal circuitry involves an automatic cut out when shorting occurs for safety reasons[10)]. The vacuum seal for the cathode is glass or ceramic.

## 2.6 Pressure Measurement

Pressure measurement below $10^{-1}$ torr is usually accomplished by gas thermal conductivity or ionisation gauges; these properties vary from gas to gas and the gauges should be calibrated for each gas[2)]. A Pirani thermal conductivity gauge is used to measure pressure between the backing pump and the diffusion pump and usually a cold cathode Penning ionisation gauge is used to measure pressures within the vacuum vessel. Occasionally a hot cathode three electrode ionisation head may be used. The Penning gauge is a two electrode ionisation gauge using a magnetic field to increase the probability of ionisation. The electrical discharge current is usually proportional to the pressure in the range $10^{-2}$ to $10^{-7}$ torr. Contamination usually results in a reduction in ionisation current. Typical calibration characteristics are given below. The evolution of hydrogen during cocondensation may affect pressure readings quite considerably, as the calibration factor differs from that of air by a factor of two.

Penning Head

$$\text{True pressure} = \frac{\text{meter reading}}{\text{gas calibration factor}}$$

Gas calibration factor: – Air 1.0; $N_2$ 1.0; $CO_2$ 1.6; CO 1.1; $H_2$ 0.4; Ar 1.4

The hot cathode ionisation gauge also has linear characteristics over the range $10^{-7} \sim 10^{-2}$ torr and has the added advantage of in situ rapid degassing. It is however less robust than the Penning gauge.

## 2.7 Vaporisation Sources

The amount of energy required to vaporise metals is a surprisingly small fraction of the energy supplied to the source. For example, an evaporation rate for tungsten ($\Delta H$ vap = 185 kcal/mol) of 10 g/h/cm$^2$ requires:

$$\frac{185 \times 10 \times 4176}{60 \times 60 \times 183} = 12 \text{ watts.}$$

W mol. wt. = 183; 1 kcal/s = 4.176 kwatts

However, to maintain a temperature of 3,670 °C (at which point tungsten may be evaporated at a rate of 10 g/h/cm$^2$ at a pressure of 0.3 torr) requires considerably more energy.

The net rate of radiative heat loss per unit area at 3,670 °C is given by Stefan's equation:

$$R = \sigma A\,(T^4 - T_0^4)$$

$\sigma$ = Stefan's constant ($5.6697 \times 10^{-12}$ watts/cm$^2$/K$^4$); A = surface area (cm$^2$);
T = temperature (3,943 K); $T_0$ = background temperature (radiation shield ~ 283 K);
then $R = 5.6697 \times 10^{-12} \times 1 \times (3{,}943^4 - 283^4) = 1{,}370$ watts/cm$^2$

The energy used to vaporise tungsten represents, therefore, less than 1% of the total energy required.

The rate of heat loss per unit surface area from a black body is proportional to $T^4$; the rate of evaporation $R^e$, per unit surface area is related to T as

$$\log R^e = A - B \log T - C/T \text{ (A, B and C are constants).}$$

It is therefore more efficient to evaporate from a small sample at high temperature than a larger sample at a lower temperature. For example, Zr evaporates at $10^{-2}$ mol/cm$^2$/h at *ca.* 2,820 K and $10^{-1}$ mol/cm$^2$/h at *ca.* 3,100 K. Equal amounts of Zr will therefore be evaporated from 10 cm$^2$ at 2,820 K or 1 cm$^2$ at 3,100 K. Assuming that radiation is the dominant heat loss (using Stefan's equation), 10 cm$^2$ at 2,820 K radiate *ca.* 3.59 kW and 1 cm$^2$ at 3,100 K radiates *ca.* 0.52 kW. Thus *ca.* 7 times as much energy is radiated from the cooler but larger surface (Table 2.3)[48].

### *2.7.1 Resistance Heating*

Resistance heating is the simplest and most widely used for producing metal vapours in laboratory experiments. A boat or spiral of a refractory metal (W, Mo, Ta) or a metal wound ceramic crucible or Knudsen Cell are the most common forms of source[11–16]. The sample usually in the form of a metal powder or chips[17] is placed in the source and current supplied until melting (or sublimation) is observed. Evaporation rates are then controlled by current variation. It is necessary to shield the source with a water-cooled jacket to avoid radiation impinging on the cold reaction surface.

Typical geometries are illustrated (Figs. 2.12 to 2.13).

Ceramic crucibles can be made or bought, and are constructed from alumina, beryllium oxide or boron nitride. Premade crucibles must fit exactly into their wire baskets for efficient heating. Wire baskets can be painted with alumina or beryllium oxide cement and baked out in vacuum. The crucibles withstand evaporation of Mg, Ca, Cr, Mn, Fe, Co, Ni, Cu, Ag, Zn, Cd and Al without much difficulty. The life time of the crucible depends on its construction and power input used; four or five evaporations are common with a good crucible[4,12].

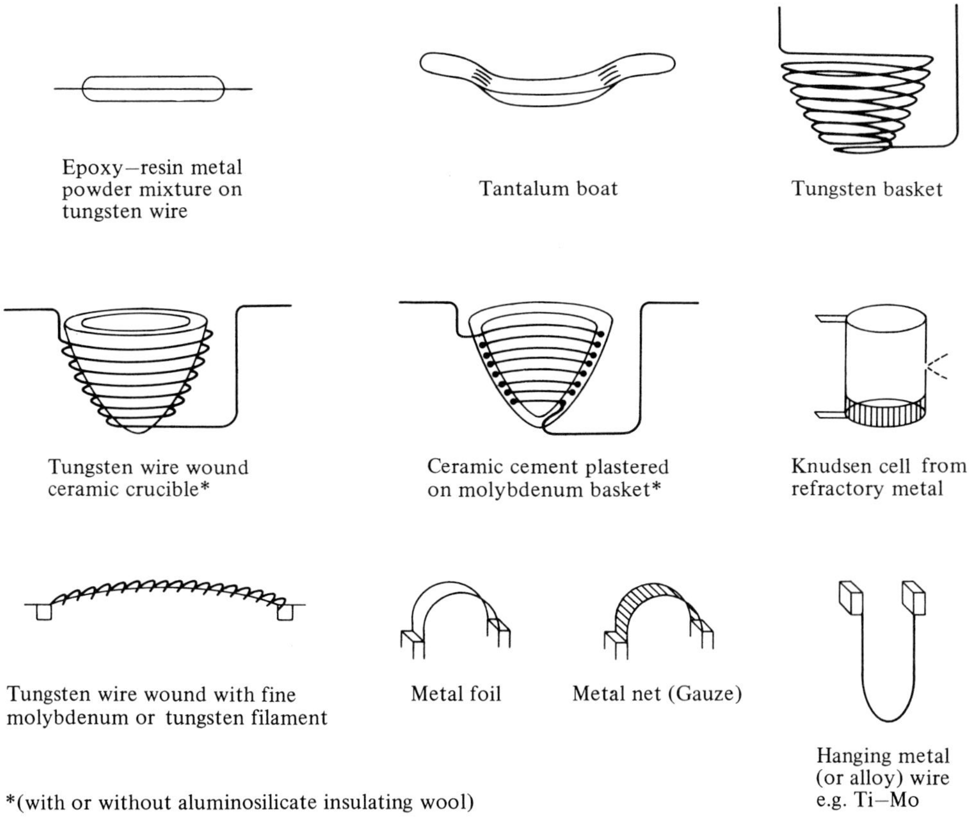


**Fig. 2.12.** Resistance heated sources

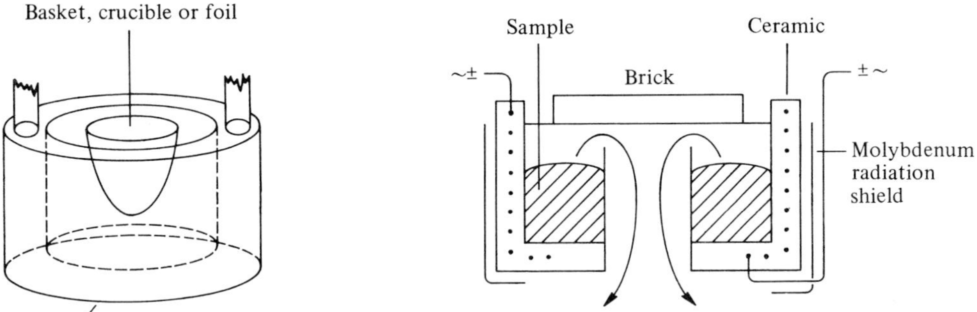


Radiation shield from brass with water cooling for "upward" evaporation of metal

Ceramic baked on molybdenum wire for "downwards" evaporation of metal vapour (after Middleton [45]). Radiation shield can be water cooled if required.

**Fig. 2.13.** Geometry of radiation shields and sources

A common reason for crucible failure is thermal shock from temperature cycling. Chromium is the most electropositive metal that can be evaporated from an alumina crucible: titanium for example will react with alumina at elevated temperatures. Most of the metals of the second and third transition series are not amenable to vaporisation by crucible heating because the crucibles are thermally unstable at the temperatures required to give useful evaporation rates. Heated crucibles can, however, be useful in assisting laser heating (see 2.7.3). The crucible is used to heat metal to approximately 1,400 °C and the laser focused onto the metal surface to produce a melt a few millimeters in diameter.

Direct evaporation from a resistively heated filament has been used but may be unreliable as filaments may burn out due to non uniformity in construction. A more reliable method is to wind filaments on a refractory metal wire and to heat the wire. This has been used successfully to evaporate a molybdenum filament from a tungsten wire[18]. The degree of contamination from tungsten was negligible. Electrically heated metal foils or refractory metal rods or foils electrochemically coated with another metal (the evaporating metal) may be used[19].

### *2.7.2 Electron Beam Heating*

An electron beam furnace in its simplest form consists of a source of electrons which are accelerated and focused onto an evaporant. Electron beam bombardment heats the evaporant surface by transfer of kinetic energy where the electrons impinge, thus the evaporant's base is not directly heated and can in fact be held in a water-cooled hearth without great loss of surface heating efficiency. This nicely avoids the often insoluble problem of finding a crucible which remains both inert to the evaporant and mechanically sound at the temperatures required for reasonable evaporation rates. This, combined with the fact that high power densities (e.g. 10 $kW/cm^2$) can be applied to the beam, means that much higher temperatures than those achieved by resistance heating can be obtained. $(\eta\text{-PhMe})_2W$ has been made in gram quantities from W vapour formed by heating tungsten metal to *ca.* 3,500 °C in an electron beam furnace)[20]. Furthermore as the water-cooled portions of the evaporant, in effect, form a crucible around any directly heated molten portion, metals with a long liquid range, such as niobium, can be evaporated. Niobium is too involatile to be evaporated in conventional resistively heated crucibles or filaments as it has a vapour pressure at its melting point well below that required for preparative evaporation rates and has to be heated above the temperature at which most crucibles decompose[21].

The are two methods for electron beam production, hot-filament (which is more applicable to vapour synthesis) and hollow-cathode.

## Hot-filament Gun

A commercially available hot-filament gun[22] which has been used in a number of vapour syntheses[23] is shown in Fig. 2.14. Thermionic electrons are produced by the hot-filament cathode then electrostatically accelerated and focused onto the evaporant as either a small spot or as a diffuse heating zone of several $mm^2$. For example an electron beam of 100 mA accelerated by a 3 kV electrostatic field repeatably evaporates Ti at *ca.* 10 mmol/h from a hot zone of *ca.* 100 $mm^2$. In larger furnaces, the spot can be scanned across the evaporant, especially for high evaporation rates (e.g. tens of kg/h)[24]. Although the evaporant atoms pass through the electron beam, the electron energies are too high for efficient interaction to ionise the atoms which therefore maintain the usual energies of thermally evaporated species. Ionisation of a gas by electrons is most efficient at electron energies of *ca.* 100 eV while, for example, in the example already cited the electron energies used to evaporate Ti are 3 keV.

Similar furnaces to that cited have been used by others for synthetic purposes[8,25] and a number of not insurmountable disadvantages have been recognised. Hot filament guns require quite a hard vacuum (typically $10^{-4}$ to $10^{-5}$ torr), which means careful planning

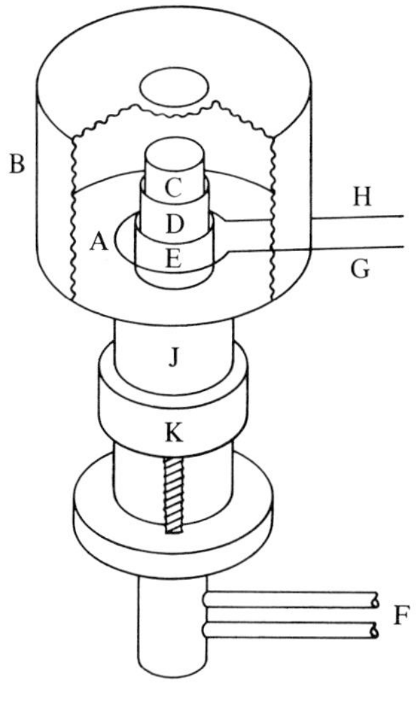


**Fig. 2.14.** Hot filament electron gun. *A* Molybdenum hot-filament cathode; *B* cylinder to which is applied a large negative potential to accelerate and focus the thermionic electron beam onto C; *C* evaporant sample; *D* water cooled copper hearth; *E* tantalum heat shield; *F* cooling water; *G* electrical connector carrying emitter current and accelerating potential; *H* electrical connector carring emitter current; *J* ceramic insulator bush; *K* support and focussing clamp

of equipment. Furthermore, ensuing reaction of the evaporated atoms may be adversely affected by electrical discharges from the high accelerating potentials, radical formation caused by the electron bombardment of the evaporant by secondary (reflected) electrons or photoreaction with ultraviolet light. (X-Rays are unlikely at less than 10 keV electron energy.) Such problems to a large extent can be avoided by maintaining a hard vacuum close to the furnace[20,21] and by having the evaporant surrounded by a highly charged cage as in the Planer evaporation source (Fig. 2.11) or having the evaporant at a positive potential thus trapping the charged particles. With such precautions ($\eta$-arene)$_2$W complexes[20] and ($\eta$-arene)$_2$Nb complexes[21] can be made very cleanly.

Hollow-Cathode Gun

This mode has not yet been used in vapour synthesis and will be mentioned only briefly[26]. Ionised argon at pressures of $10^{-3}$ torr or higher is the source of electrons which can then be employed as either finely focused or broadly diffuse beams which can be biased to any potential relative to the evaporant, as the necessary electron acceleration is done in the source rather than between source and evaporant, as in hot filament guns.

Hollow-cathode guns can be used hot or cold. Cold hollow-cathode guns work at high discharge voltages and low Ar pressures (*ca.* $10^{-3}$ torr) and emit well-collimated electron beams. Guns in the hot mode have a high current, low voltage discharge producing a diffuse electron beam which in turn generates an evaporated beam containing a large number of ionised species.

### *2.7.3 Laser Heating*

Laser heating sources have been used in both the pulsed and continuous wave modes[7,27]. The efficiency of laser heating depends to a large extent on the optical reflectivity of the heated surface. Metals possess a markedly wavelength dependent reflectivity in the range of linear optics when considered as black body absorbers. Grey body characteristics[28] are observed near their melting point which lead to lower reflectivity. Flux densities of greater than $10^8$ W/cm$^2$ involve non linear photon metal interaction and increasing absorption. Non linear interactions are probable only for pulses of a few joules in the nanosecond range[29] in well focused beams. Metal evaporation in the pulsed or C.W. (continuous wave) mode has been attempted with $CO_2$ (10.6 $\mu$), $Nd^{3+}$/YAG (1.06 $\mu$) and Ruby (0.7 $\mu$) lasers operating in the range of linear optics. 7 joule pulses from a ruby laser lead to loss of weight of a steel target of 1 mg/pulse. A $CO_2$ laser in the C.W. mode (550 W) evaporated chromium at 5 g/h and iron at 0.4 g/h. A $Nd^{3+}$/YAG laser operating in the C.W. mode (200 W) evaporated iron at 0.54 g/h[30]. Figures 2.15 and 2.16 illustrates reflectivity as a function of wavelength and flux density. Clearly pulse modes at high flux densities seem more efficient.

The experimental arrangement for C.W. laser operation is illustrated (Fig. 2.6). The laser may be operated alone or in tandem with resistive heating; high rates of evaporation can be obtained by this method, also some of the second row transition metals with melting points

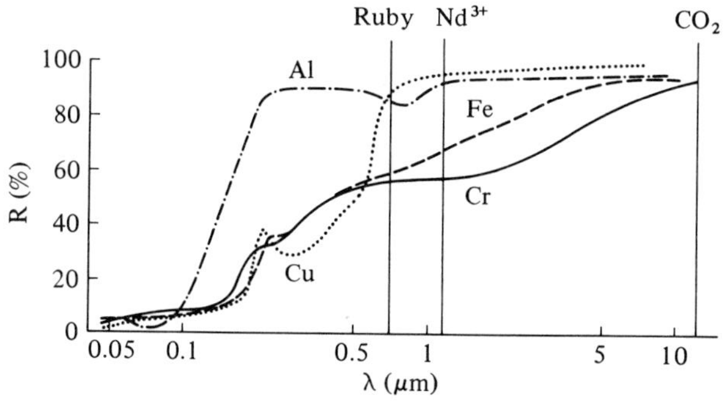


**Fig. 2.15.** Reflectivity R of metals as a function of wavelength, λ (Angew. Chem., Int. Ed. *14*, 278 (1975))

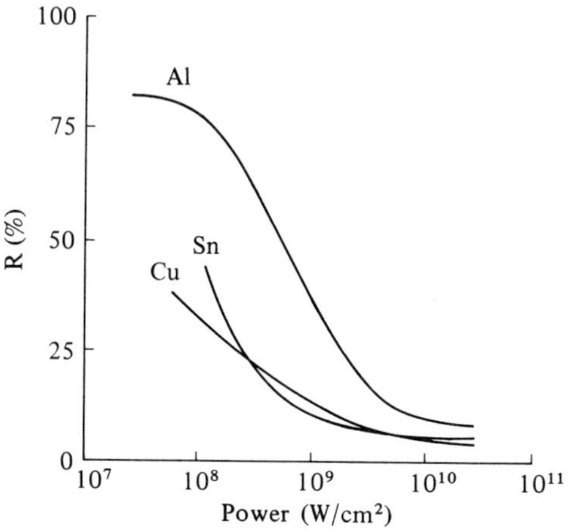


**Fig. 2.16.** Reflectivity R at 1.06 μm of aluminum, tin, and copper as a function of energy flux density (Angew. Chem., Int. Ed. *14*, 278 (1975))

higher than 1,800 °C can be evaporated without crucible damage. In some cases evaporation of metal wires passing slowly into the focused laser beam has been successful.

A preheated metal in a crucible subject to focused laser heating will tend to develop a crater from which the metal evaporates; when this becomes too deep the focus point can be moved by adjusting a reflectance prism external to the vacuum system. The focusing lens must be kept clear of metal, this is done by leaking argon at $10^{-2}$ torr across the lens (Fig. 2.5).

### *2.7.4 Inductive Heating*

There has been relatively little use of inductive heating sources for MVS on the laboratory scale. Designs for the production of boron and silicon subhalides, and for evaporation of actinides and lanthanides in MIS are documented[4, 31].

### 2.7.5 *Cathode Sputtering*

Cathode sputtering is by far the most popular and oldest method of coating surfaces with atoms or molecules[32]. Its potential in MVS has not been realised but with respect to the production of metal coated particles, perhaps for catalytic uses, it is worth briefly discussing this technique.

Two electrodes are mounted in a vacuum chamber ($10^{-2} - 10^{-3}$ torr) containing an inert gas (argon). Application of an electric field of between 1 and 5 kV produces a self-sustained gas discharge. Positive argon ions after acceleration impinge on the cathode. The loss of kinetic energy of the ions causes the ejection of the surface layer of the cathode into the vacuum chamber. Although sputtering does not involve the heat energies of thermal vaporisation processes, targets are usually water cooled. The incorporation of self biasing and an R.F. glow discharge[33] permits the coating of dielectric, insulating or conducting materials. Films of quartz, aluminosilicate, glass, alumina, titanium dioxide, silicon carbide, nichrome, gold and tantalum can be produced by this method. Sputtering rates are generally low and simple sputtering devices may be impractical for producing large quantities of metal (Table 2.1).

High rates of metal evaporation can be obtained using a Ring Gap Plasmatron[34] as a sputtering source. Typical performance parameters for a high rate planar plasmatron sputtering source are: a discharge voltage of 500 V, a discharge current of 10 A operating at a pressure of $3 \times 10^{-3}$ torr. Sputtering rates are Cu: 42 g/h, Aluminium Molybdenum, Tantalum 6 g/h. A major problem is that ferromagnetic targets create a magnetic shunt across the field gap unless the target thickness is limited to the order of a millimeter (thus ensuring magnetic saturation of the material prior to sputtering). This method seems a promising way of producing metal aggregates[49].

The main advantage of sputtering over other methods is that it is essentially a surface bombardment technique where vaporisation takes place under thermodynamic non-equilibrium conditions via direct transfer of kinetic energy from the accelerated plasma to the surface molecules. Hence concurrent sputtering of two metals from an alloy surface may occur even though the metals have vastly differing equilibrium (thermal) vaporisation rates (Table 2.3).

### 2.7.6 *Reactive and Activated Reactive Vaporisation*

Reactive vaporisation is a term used to describe the process whereby a vapour is mixed with a reactive gas prior to condensation on a surface[32]. Activated reactive vaporisation is a similar process but involving a gas *plasma.* Reactive sputtering necessarily involves a plasma and is widely used commercially. The sputtered species and the sputtering plasma can co-condense onto the substrate at the same time. The plasma may absorb reversably or irreversably; for example a plasma of argon and acetylene sputtering silicon yields a $\beta$-SiC[35] condensate with deabsorption of argon. Oxygen in the plasma gas can act either as a reactive co-condensate or as an oxidant for the surface of the sputtering species. Controlled deposition of titanium oxide by this method is well documented[36].

### *2.7.7 Ionised Cluster Beam Deposition*

This technique is used at the present time for producing metal and semiconductor films. The clusters containing 5 x $10^{-2} \sim 10^3$ atoms are formed by the adiabatic expansion of vaporised metal through the nozzle of an electron beam heated furnace[37]. They are ionised by electron bombardment and are accelerated towards the substrate in a region of high vacuum. The kinetic energy imparted to the substrate surface by the ions causes secondary surfaces processes at the substrate. Migration, substrate sputtering, re-evaporation, self sputtering and absorption all occur.

### *2.7.8 Electric Arc Heating*

Arc heating is potentially a large scale method of producing metal atoms but has been relatively little investigated. The vaporisation of carbon with low voltage (16 ~ 24 V A.C.)[3] has been described. Self sputtering, high temperatures, and low energy electron bombardment contribute to the vaporisation process, in which atoms, oligomers and ions are produced. Control of the arcing process may sometimes prove difficult with less conducting materials. Arcs however, have been used for decades in atomic spectroscopy studies and considerable experience has been obtained[47].

## 2.8 Vaporisation Sources for Matrix Isolation Spectroscopy

### *2.8.1 Foils and Wires*[19]

The assembly illustrated in Fig. 2.17[38] attaches directly to the cryostat assemply and is pumped out as part of the complete vacuum system. The metal in the form of a filament or foil is fixed across two water cooled copper electrodes supplying A.C. current (80 ~ 150

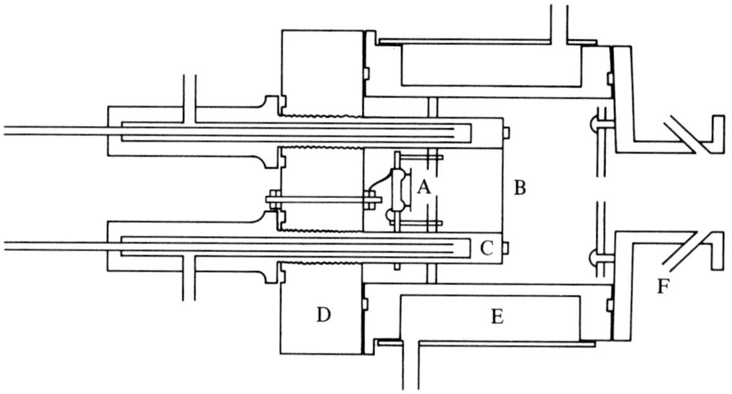


**Fig. 2.17.** Metal source for matrix isolation experiments. *A* Quartz crystal assembly; *B* metal flilament; *C* water cooled copper electrodes; *D* perspex flange; *E* water jacket; *F* gas inlet (Kundig, E. P., Moskovits, M., Ozin, G. A.: J. Mol. Struct *14*, 137 (1972); with permission of Elsevier S.P.Co.)

Amp at 1.5 volts). Typical filament dimensions are 4.2 x 0.92 x 0.0196 $cm^3$. A quartz microbalance measures the rate of deposition of metal; the rates of evaporation towards the cryostat and the microbalance are assumed equal. Gases can be cocondensed with the metal atoms via inlets F. Typical experiments last for approximately three hours and cryostat windows of CsI at 4.2 K to 10 K may be used for infrared studies.

### *2.8.2 Knudsen Cells*

Knudsen cells are small ovens, usually heated resistively or inductively[16,39,40]. They incorporate a small outlet hole for metal vapour. The inside of a Knudsen cell may be considered as a "black body" so that the metal vapour emitted is in thermal equilibrium with the bulk metal. In the double oven Knudsen cell, the vapour is passed into a second cell which is more strongly heated. The vapour is no longer in thermal equilibrium with bulk metal and the concentration of species other than atoms is often diminished. Figures 2.18 and 2.19 illustrate resistively heated and inductively heated cells. The resistively heated cell is designed to be incorporated in the apparatus described in the previous section.

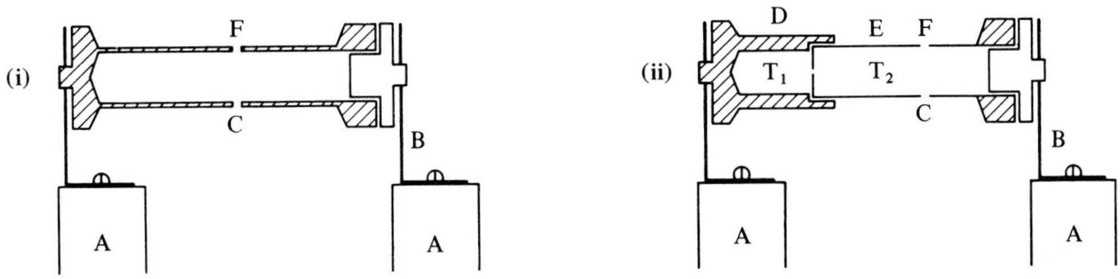


**Fig. 2.18.** Knudsen Cells. Design of **(i)** a single oven and **(ii)** a double oven tantalum Knudsen cell used for simultaneous matrix deposition and monitoring of metals. *A* water-cooled copper electrodes; *B* tantalum support clips and crews; *C* orifice directed at quartz crystal microbalance; *D* thick wall; *E* thin wall (where temperature $T_2 > T_1$); *F* orifice directed at matrix (J. Amer. Chem. Soc. *97*, 308 (1975); with permission; copyright by the American Chemical Society)

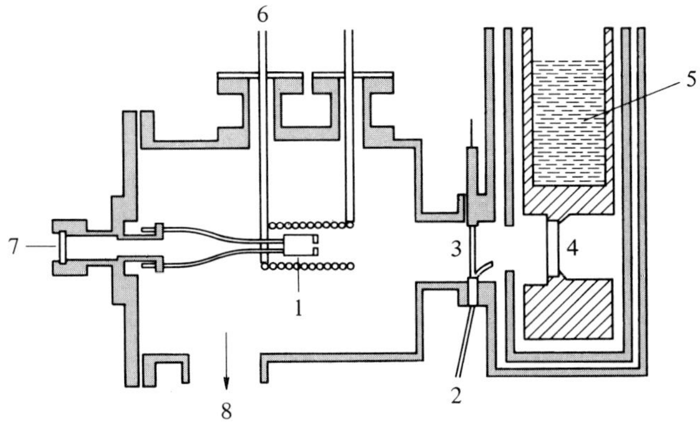


**Fig. 2.19.** Inductively heated metal atom source. Apparatus for the preparation of actinoid and lanthanoid carbonyls by matrix isolation. *1* Sample cell for metal evaporation; *2* gas inlet; *3* movable shutter; *4* cooled window; *5* liquid helium; *6* induction heater; *7* window for temperature measurement; *8* high vacuum (Angew. Chem., Int. Ed. *14*, 309 (1975))

### *2.8.3 Hollow-Cathode Sputtering Device*

A versatile hollow-cathode source of atoms or molecules for MIS studies has been described[41]. A D.C. discharge within the hollow cathode sustained by a noble gas flow sputters atoms of the cathode metal. Additions of small amount of reactive gases to the noble gas carrier can cause reactive sputtering. The cathode can be made of any machinable metal or be made in the form of a rolled foil.

## 2.9 Leak Detection

Glass vessels which develop leaks can be tested with the normal high voltage (Tesla) leak detector. Steel vessels are sprayed from a small nozzle with a gas that diffuses rapidly (He) The gas diffuses quickly into the system and is detected by its difference in thermal conductivity compared with air, by a Pirani gauge often wired to give an audio signal.

The repair of leaks in a steel vessel is a difficult job; they usually occur along weld fractures and fresh rewelding without annealing may cause further fractures. Professional help is often needed.

## 2.10 General Experimental Procedures

An MVS apparatus working continuously needs to be cleaned thoroughly and outgassed (usually overnight) before each experiment. If this is done much less maintenance of vacuum pumps is necessary and higher vacuums are achieved in much shorter times.

The pumping system must be isolable in case of sudden loss of vacuum. This is usually accomplished by a butterfly valve positioned between the cryopump and the diffusion pump; in MIS equipment the butterfly valve is often placed between the vacuum chamber and the cryopump, as the pumps are run continuously and achieve much higher vacuums ($10^{-6}$ torr). Outgassing can be achieved initially with the rotary oil pump "on ballast" then subsequently non ballasted and finally with the diffusion pump and cryopump operative. Final outgassing can be achieved by heating the source to a hundred degrees or so below the melting point of the metal with no coolant surrounding the vacuum vessel for half an hour: (This is inappropriate for chromium which sublimes.) the source power is then switched off and the coolant bath filled; the pressure should then be of the order of $10^{-5}$ torr. All reactants must be put through a freeze-pump-thaw cycle for thorough degassing before use, connected to the inlet system and pumped out again.

### *2.10.1 Experiments with a Solid Matrix*

Melting metal often outgasses vigorously and the pressure rises accordingly. Subsequently, the pressure drops sharply due to gettering by the condensing metal vapour. The reactant is then introduced:

a) Neat.
b) Diluted in an inert matrix forming solvent.
c) Separately and at the same time as a matrix forming solvent.
Continuous slow cocondensation of metal vapour and reactant (and matrix forming materials if required) then takes place. Typical flow rates for five to ten litre vessels may be: reactant: five to thirty g/h (less if a diluent also used), any matrix material: fifty to two hundred ml/h; but this also depends on metal evaporation rates (Table 2.3).

On termination (typically 1–3 hours) the pumping system is isolated, the source power turned off and an inert gas (argon, nitrogen) introduced after removal of the coolant bath. the vessel warms slowly, and the matrix melts and the resulting liquid is immediately collected and filtered to remove active metal particles. At the end of a reaction instead of introducing argon a non condensable gas may be introduced (CO) and allowed to react during the warm up period. Materials with a small but appreciable vapour pressure at $-196$ °C ($C_2H_4$, CO, $C_3H_6$ etc.) can sometimes be "gettered" into a matrix by codeposition with an inert material (toluene, hexane) but pressure fluctuations are often encountered and experimental success is never certain.

Certain experimental arrangements allow for direct investigation of volatile products. The solvent of the melted matrix is pumped into the cryopump (using the rotary oil pump only) and the remaining products are then pumped under high vacuum directly into a vacuum line for spectroscopic investigation.

### *2.10.2 Experiments in the Liquid Phase*

Metals may be evaporated directly into a rotating cryostat containing a liquid maintained at a sufficiently low vapour pressure[42]. The rotation of the vessel provides a continuously renewed liquid surface into which metal may be evaporated. The reactant is either a neat liquid or a solution. Control of vapour pressure within the vessel, whilst still maintaining the reactants in the liquid phase is critical and can be accomplished by varying the immersion depth of the cryostat in a bath of coolant material. The degassed liquid reactant can be added through the liquid inlet, and the vessel and contents brought to thermal equilibrium by adjusting the coolant bath. The vessel may be filled with argon after termination of the reaction and the liquid extracted under pressure, or the vessel disconnected and the products poured into the receptor flask (Table 2.4).

Many rotating cryostats have a capped ground glass joint on their base. The vessel can be filled with an inert gas, the cap removed and the products extracted by rotation or under pressure. This is by far the most convenient way of product recovery (Figs. 2.20, 2,21) and is particularly useful for removing metal slurries, viscous liquids and reactions on suspended solids. Reactions have been carried out between $-160$ °C and 0 °C by this method (Table 2.2). Yields of products in these liquid phase reactions are often higher than in the corresponding matrix reactions.

This method is also useful for the transportation of metal atoms onto solid supports. Silica Gel in a co-ordinating solvent (thf, toluene) will absorb iron or nickel readily. Removal of the solvent at low temperature leaves essentially metal supported on the solid. Use of olefins as metal carriers leads often to some degree of polymerisation and removal of the hydrocarbon from the support may never be complete.

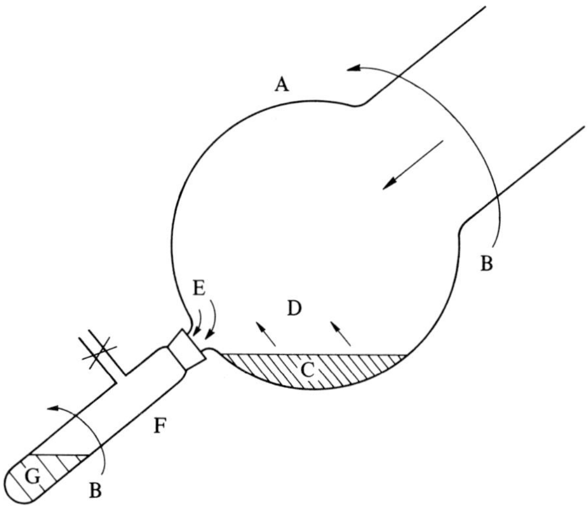


**Fig. 2.20.** *A* Reaction vessel, prefilled with inert gas; *B* rotation (*ca.* 30 r.p.m.); *C* viscous solution or suspension of product; *D* rotation drawing solution up the vessel walls; *E* gravity drain into receiving vessel; *F* receiving vessel prefilled with inert gas; *G* product

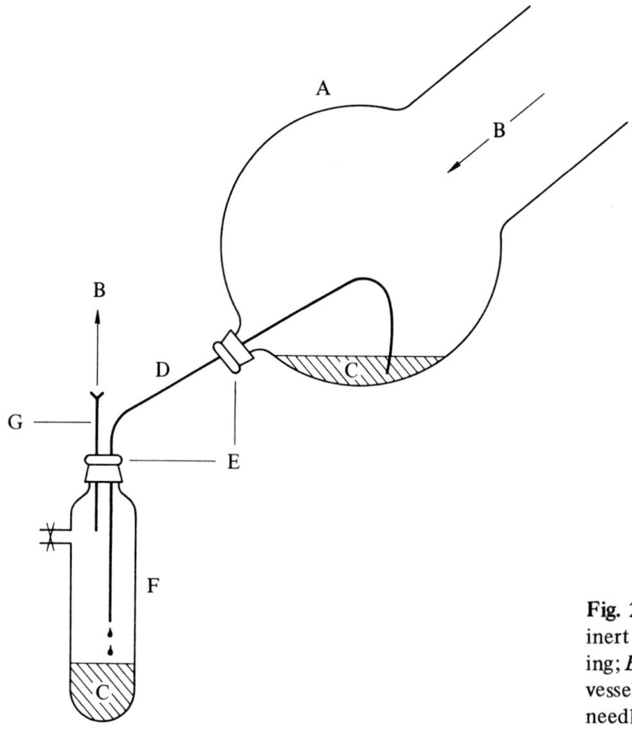


**Fig. 2.21.** *A* Reaction vessel; *B* stream of inert gas; *C* product; *D* stainless steel tubing; *E* rubber self-sealing cap; *F* receiving vessel prefilled with inert gas; *G* syringe needle for gas exit

### 2.10.3 Experiments in the Solid Phase

The direct deposition of metal on a solid support (Fig. 2.2) is relatively easy to do. Metal is evaporated into a "snow storm" of a solid support made by rapid stirring. The deposition of metal occurs only on the outer surface of the particles and for this reason may not lead to very active materials.

### 2.10.4 Experiments with Metal Slurries

Finely divided metal powders are known to be more reactive than metal shavings or chips[43,44)] but are much more easily poisoned by oxidation or aggregation. Metal slurries can be produced by evaporation of metal into a solvent. Particles so formed have structures dependent on the solvent. These slurries may then be used as normal chemical reagents. These reactions can be performed in the liquid or matrix phase.

### 2.10.5 Metal Aggregation and Catalysis

Commercially metal coated powders are prepared by a sputtering technique which may produce very thinly coated metal surfaces[49,50)]. Furthermore treatment of the powders after coating by, for example, heating (sintering) or with a heated gas ($H_2$, $O_2$, CO, $H_2O$) may lead to greatly enhanced activity of the powder as a catalyst.

It is, of course, also possible to condense two or more metals (evaporated from the same or different heating sources) onto a solid support or into a solvent or mixture of solvents. The preparation of mixed-metal coated particles and mixed-metal slurries produced in solution may lead to exciting new developments in the field of heterogeneous catalysis.

### 2.10.6 Cleaning

Scouring powders remove a lot of deposited metal on glass vessels but hydrofluoric acid in tepol is the most effective cleanser prior to washing with water and grease removers. Steel vessels can be washed with diluted hydrochloric acid or scouring powders prior to hydrocarbon solvents.

Electron guns must be kept free of deposited metal to avoid shorting. Radiation shields and inlets should be emery papered from time to time to remove deposited metal and the Penning head checked for polution fairly frequently. Optics for laser focusing and reflection must also be absolutely free of grease and dust, etc.

**Table 2.1.** Sputtering rates of metals in hydrogen after Güntherschutze[46)]

| Metal | mg/A/h | mmole/A/h | |
|---|---|---|---|
| W | 57 | 0.31 | Applied voltage 850 V |
| Mo | 56 | 0.58 | Cylindrical cathode (~ 7 $cm^2$) |
| Cr | 27 | 0.51 | $pH_2$ = 1 torr |
| Ta | 16 | 0.09 | |
| Al | 29 | 1.05 | |
| Mg | 9 | 0.38 | |
| Mn | 38 | 0.69 | |
| C | 262 | 22 | |
| Tl | 1,080 | 5.3 | |
| Pb | 400 | 1.9 | |
| Sn | 196 | 1.7 | |
| Ni | 65 | 1.11 | |
| Co | 56 | 0.95 | |
| Fe | 68 | 1.21 | |
| Au | 460 | 2.34 | |
| Ag | 740 | 6.8 | |
| Cu | 300 | 4.7 | |
| Bi | 1,470 | 7.0 | |
| Sb | 890 | 7.4 | |
| As | 1,100 | 15.3 | |
| Te | 1,200 | 9.4 | |
| Cd | 32 | 0.29 | |
| Zn | 340 | 5.2 | |

**Table 2.2.** Solvents for the rotating solution reactor (Timms, P. L.: Angew. Chem., Int. Ed. *14*, 195 (1975)

| Solvent | Operating temp.[a] | M. p. [°C] | B. p. [°C] |
|---|---|---|---|
| Propane | −185 | −187 | − 42 |
| Isopentane | −145 to −160 | −160 | + 28 |
| Methylcyclohexane | −110 to −125 | −126 | +101 |
| Pet. ether (120–160) | − 80 to − 90 | *ca.* −90 | |
| *m*-Diisopropylbenzene | − 60 to −100 | −105 | +202 |
| Bis(ethylhexyl) ether | − 10 to − 60 | *ca.* −80 | +280 |
| Light liquid paraffin | + 20 to 0 | − 20 | > +300 |
| Silicone 704 (phenylmethylsiloxane) | + 50 to − 20 | − 20 | > +350 |

[a] Vapor pressure of solution $< 10^{-3}$ torr [°C]

**Table 2.3.** Evaporation rates for metals. (From W = 200 p $(MT)^{-1/2}$. VP from Honig, R. E., Kramer, D. A.: Vapour Pressure Data for the Solid and Liquid Elements. R. C. A. Review *30,* 285 (1969))

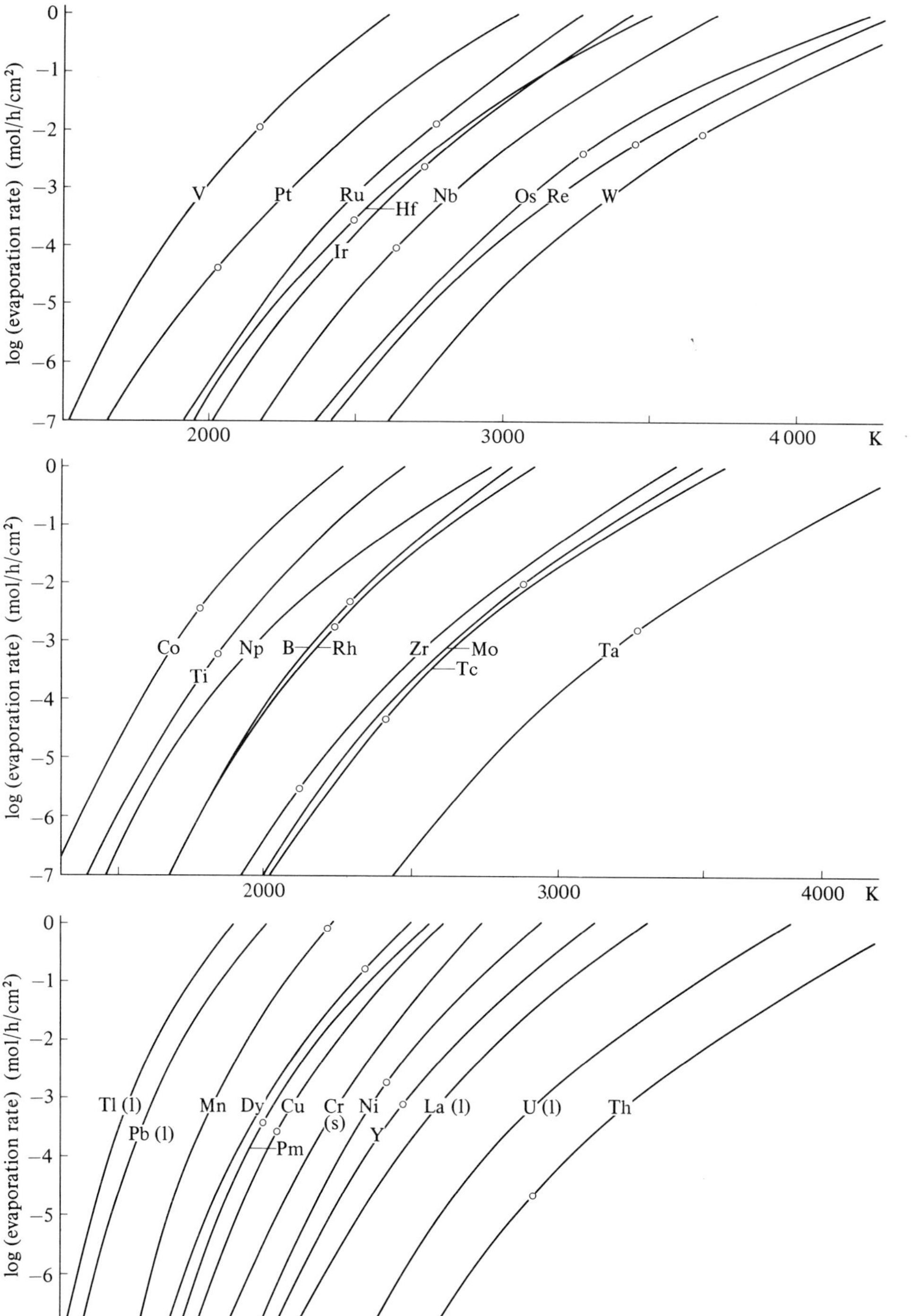

Table 2.3 (continued)

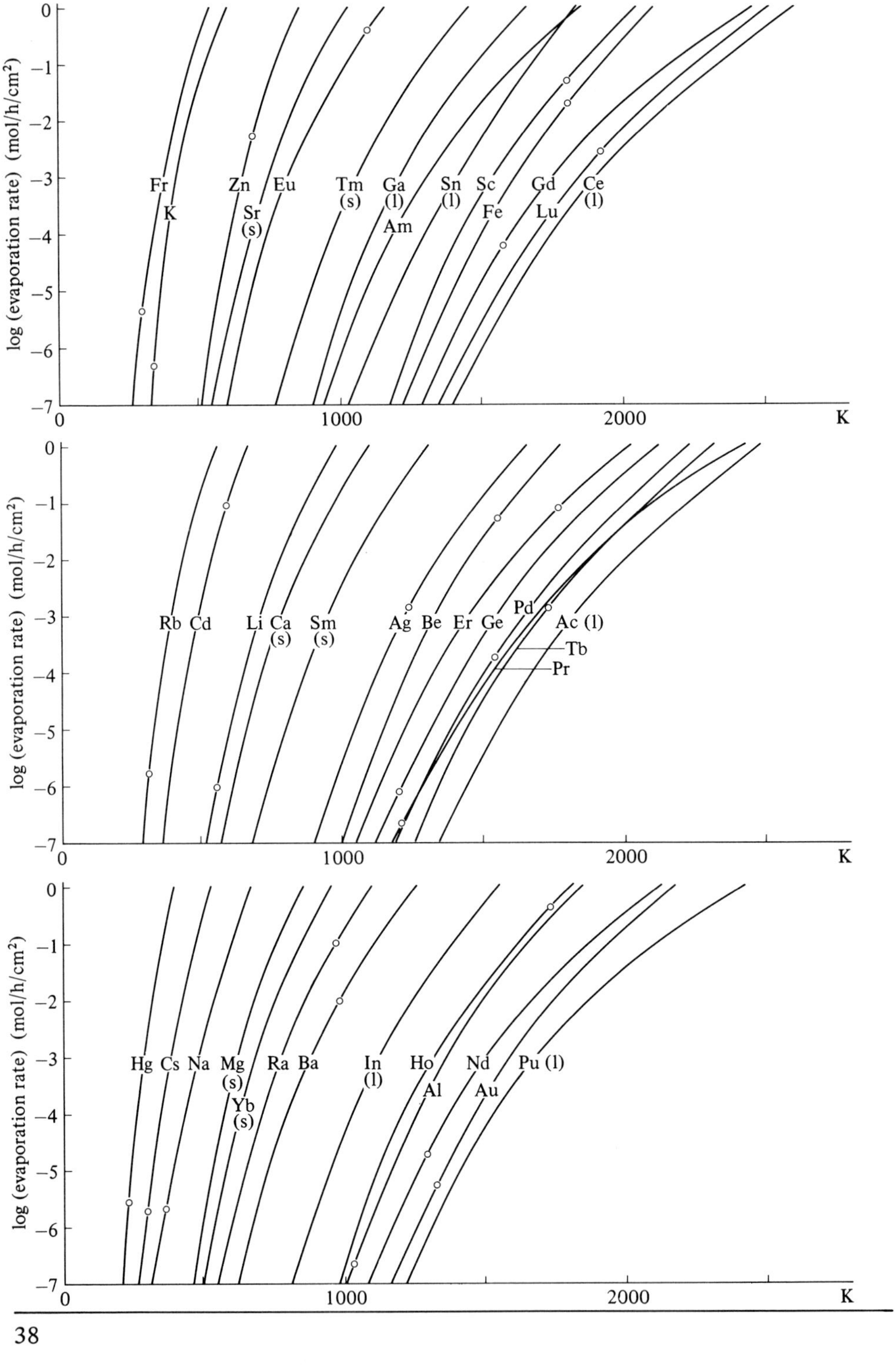

**Table 2.4.** Techniques and refractory support materials for evaporating metals, together with some useful vapor-pressure data[52,53] (from: L. Holland: Vacuum Deposition of Thin Films. London: Chapman and Hall Ltd. 1956)

| Metal | Melting Point (°C) | Evaporation Temp. (°C), Vapor Pressure = 10 microns Hg* | Resistance Heated Sources in Order of Merit | | Other Evaporation Techniques | | Remarks |
|---|---|---|---|---|---|---|---|
| | | | Filaments | Boats | Source | Method of Heating | |
| Aluminium | 660 | 996(?)<br>1,148(?) | W, Ta,<br>helical coil | ... | ... | ... | Alloys freely with refractory metals and reacts with carbon and oxide crucibles |
| Antimony ($Sb_2$) | 630 | 678 | Chromel, Ta,<br>conical basket | Mo<br>Ta | Alumina crucible | External<br>W heater | Wets Chromel<br>... |
| Arsenic | ... | ... | ... | ... | Alumina crucible | External<br>W heater | ... |
| Barium | 717 | 629 | W, Ta, Mo, Nb, Ni, Fe, Chromel, conical basket | Ta, Mo | ...<br>... | —<br>— | Freely wets without alloying with the heater metals quoted. Reacts with alumina |
| Beryllium | 1,284 | 1,246 | Ta, W, Mo,<br>conical basket | ... | Carbon crucible<br>BeO crucible | Electron bombardment<br>High frequency | Wets heater metals quoted |
| Bismuth | 271 | 698 | Chromel, Ta, W, Mo, Nb, conical basket | Fe, Mo, Ta | Alumina crucible | External<br>W heater | Wets Chromel |
| Boron | 2,000<br>~2,090 | 1,355 | — | ... | Carbon | Resistance heated | Deposits from carbon probably impure |
| Calcium | 810 | 605 | W, conical basket | ... | Alumina crucible | External<br>W heater | |
| Cadmium | 321 | 264<br>(subl.) | Chromel, Nb, Ta, Mo, W, Ni, Fe, conical basket | Mo, Ta | Alumina crucible<br>Iron crucible | External<br>W heater<br>Nichrome heater | Freely wets Chromel and Nb |

Table 2.4 (continued)

| Metal | Melting Point (°C) | Evaporation Temp. (°C), Vapor Pressure = 10 microns Hg* | Resistance Heated Sources in Order of Merit | | Other Evaporation Techniques | | Remarks |
|---|---|---|---|---|---|---|---|
| | | | Filaments | Boats | Source | Method of Heating | |
| Carbon | 3,700 ± 100 | 2,681 | . . . | . . . | Pointed carbon rods pressed together, forming high-resistance contact | Resistance heating | . . . |
| Cerium | 785 | 1,305 | . . . | . . . | . . . | . . . | . . . |
| Cesium | 29 | 153 | . . . | . . . | . . . | . . . | . . . |
| Chromium | 1,900 | 1,205 | W, conical basket | . . . | Electro-deposited Cr on W helical coil | Sublimation | |
| Cobalt | 1,478 | 1,649 | Nb, W | . . . | Alumina or BeO crucible Electroplated W spiral | Embedded W heater Resistance heater | Alloys with W, Ta, Mo, Cb, Pt. Evaporant weight must not exceed 35 per cent of that of W spiral |
| Copper | 1,083 | 1,273 | Pt, helical coil; Nb, Mo, Ta, W, concial basket | Mo, Ta | Alumina crucible | Embedded W heater | Copper alloys with Ni, Fe, Chromel. Does not wet readily Mo, W, Ta |
| Gallium | 30 | 1,093 | . . . | . . . | BeO, $SiO_2$, $Al_2O_3$ | . . . | Alloys with metals, oxides quoted resist attrack up to 1,000 °C |
| Germanium | 959 | 1,251 | Ta, Mo, W, conical basket | Mo, Ta | Alumina crucible Carbon crucible | Embedded W heater Resistance heater | Wets Ta and Mo |
| Gold | 1,063 | 1,465 | W, Mo, conical basket | Mo | . . . | . . . | Wets but reacts with Ta, possible alloy formation. Partially wets W, Mo |

| | | | | | | | |
|---|---|---|---|---|---|---|---|
| Indium | 157 | 952 | W, Fe, conical basket | Mo | . . . | . . . | . . . |
| Iridium | 2,454 | 2,556 | . . . | . . . | . . . | . . . | . . . |
| Iron | 1,535 | 1,447 | W, helical coil | . . . | Alumina or BeO crucible | Embedded W heater | Alloys with W, Ta, Mo, Cb. Evaporant must not exceed 35 per cent of the weight of W filament |
| Lead | 328 | 718 | Fe, Ni, Chromel, conical basket | Mo<br>. . . | Alumina or iron crucible<br>Carbon (? ) | External Nichrome heater<br>Resistance heater | Does not wet W, Ta, Mo, and Cb |
| Lithium | 179 | 514 | . . . | . . . | Mild steel (? ) crucible | Nichrome heater | . . . |
| Magnesium | 651 | 443 (subl) | W, Ni, Fe, Ta, Mo, Nb, Chromel, conical basket | Mo, Ta | Iron crucible<br>Carbon | External Nichrome heater<br>Resistance heater | Does not melt when volatilized from open spirals and boats |
| Manganese | 1,244 | 980 | W, Ta, Mo, Nb, conical basket | . . . | Alumina crucible | Embedded W heater | Freely wets heater metals quoted |
| Molybdenum | 2,622 | 2,533 | . . . | . . . | . . . | . . . | Refractory metal used for filament and boat-type vapor sources. Evaporates rapidly if oxidized to form $MoO_3$ |
| Nickel | 1,455 | 1,510 | W, heavy-gauge helical coil | . . . | Alumina or BeO crucible | Embedded W heater | Alloys with Mo, Ta, and W. Evaporant must not exceed 30 per cent of the weight of W filament |
| Niobium | 2,500 | (VP at MP = 1 micron Hg) | W, helical coil | . . . | . . . | . . . | Refractory metal used as a source material |
| Palladium | 1,555 | 1,566 | W, helical coil | . . . | . . . | . . . | . . . |
| Platinum | 1,774 | 2,090 | Multistrand W filament with Pt wire twisted together | . . . | Electrodeposited Pt on W spiral | Resistance heater | Alloys with Ta and partially with W. Platinum may be used as a source heater for metal oxides to prevent decomposition of charge |

Table 2.4 (continued)

| Metal | Melting Point (°C) | Evaporation Temp. (°C), Vapor Pressure = 10 microns Hg* | Resistance Heated Sources in Order of Merit: Filaments | Boats | Other Evaporation Techniques: Source | Method of Heating | Remarks |
|---|---|---|---|---|---|---|---|
| Rhodium | 1,967 | 2,149 | ... | ... | Electrodeposited Rh on W spiral<br>Resistance heated Rh foil | Sublimation | Requires very low pressure for deposition of neutral transmitting films |
| Selenium | 217 | 234 | Chromel, Fe, Mo, Nb, conical basket | Mo, Ta | Alumina crucible | Nichrome external heater or radiant heater | Very volatile, may contaminate plant. Wets filament metals quoted |
| Silicon | 1,410 | 1,343 | ... | ... | BeO crucible | Embedded W heater | Difficult to prepare Si films free from SiO contamination |
| Silver | 961 | 1,047 | Ta, Mo, Nb, Fe, Ni, Chromel, helical coil or W conical | Mo, Ta | Electrodeposited Ag on Mo helical coil | Resistance heater | Ag does not wet W. Can be kept in basket by binding fine platinum wire on outside |
| Strontium | 771 | 549 | W, Ta, Mo, Nb, conical basket | ... | Carbon | Resistance heated | Freely wets without alloying with all filament metals quoted |
| Tantalum | 2,996 | (VP = 1 micron Hg at 2,820°) | ... | ... | ... | — | Refractory metal used for source heaters |
| Tellurium | 452 | (VP = 760 mm Hg at 1,390°) | W, Ta, Mo, Nb, Ni, Fe, Chromel, conical basket | ... | Alumina crucible | External Nichrome heater | Very volatile, may lead to plant contamination. Wets without alloying all metal heaters quoted |
| Thallium | 304 | 606 | Ni, Fe, Nb, Ta, conical basket | ... | Alumina crucible | External Nichrome heater | Freely wets metal heaters quoted without alloying. Partially wets W, Ta, but not Mo |

| | | | | | | | |
|---|---|---|---|---|---|---|---|
| Thorium | 1,827 | 2,196 | W, conical basket | . . . | . . . | — | Wets W heater |
| Tin | 232 | 1,189 | Chromel, helical coil; Mo, Ta, conical basket | Mo, Ta | Alumina crucible<br>Carbon | Embedded W heater<br>Resistance heater | Wets Chromel and Mo |
| Titanium | 1,727 | 1,546 | W, Ta, conical basket or helical coil | . . . | Carbon | Resistance heated | Ti reacts with W spiral, deposit contains trace of W. Ti does not react with Ta but filament may burn out during premelting of Ti |
| Tungsten | 3,382 | 3,309 | . . . | . . . | . . . | . . . | Refractory metal used for source heaters. Evaporates more readily if surface oxidized to form volatile $WO_3$ or $WO_2$ |
| Uranium | 1,132 | 1,898 | W, conical basket | . . . | . . . | . . . | Forms oxidized deposits at lowest gas pressures |
| Vanadium | 1,697 | 1,888 | W, Mo, conical basket | . . . | . . . | . . . | Alloys with Ta and partially with W. Wets but does not alloy with Mo |
| Yttrium | 1,477 | 1,649 | . . . | . . . | . . . | . . . | . . . |
| Zinc | 419 | 343 (subl) | W, Ta, Mo, Nb, conical basket | . . . | Alumina or iron crucible<br>Carbon | External Nichrome heater<br>Resistance heater | Wets without alloy formation all filament metals quoted |
| Zirconium | 2,127 | 2,001 | W, conical basket or helical coil | . . . | . . . | . . . | Requires low pressures < 0.1 micron Hg to prevent film oxidation, Evap. characteristics similar to Ti |

* Vapor-pressure data are mainly those reported by Dushman. In many cases rapid evaporation will commence only at temperatures well above the values quoted, because of the existence of surface oxide films.

## References


1. Catalogue of High Vacuum Equipment. 5000 Köln 51: Leybold Heraeus
2. Edwards Ltd.: Catalogue of High Vacuum Equipment. Manor Royal, Crawley, Sussex, England
3. Skell, P. S., Havell, J. J., McGlinchey, M. J.: Acc. Chem. Res. *6*, 97 (1973)
4. Timms, P. L.: Adv. Inorg. Chem. Radiochem. *14*, 121 (1972); Endeavour *27*, 133 (1968)
5. Young, D., Green, M. L. H.: J. Appl. Chem. Biotechnol. *25*, 641 (1975)
6. Koerner v. Gustorf, E., Jaenicke, O., Polansky, O. E.: Angew. Chem. *84*, 547 (1972)
7. Koerner v. Gustorf, E. A., Jaenicke, O., Wolfbeis, O., Eady, C. R.: Angew. Chem., Int. Ed. *14*, 278 (1975)
8. Timms, P. L.: Angew. Chem., Int. Ed. *14*, 273 (1975)
9. Anthony, M. T., Green, M. L. H., Young, D,: J. C. S. Dalton *1975*, 1419
10. Planer, G. V.: M.V.S. Equipment. Middlesex, England: Windmill Road, Sunbury-on-Thames
11. Timms, P. L.: Adv. Inorg. Chem. Radiochem. *14*, 121 (1972)
12. Olsen, L. O., Smith, C. S., Crittenden, E. C., jr.: J. Appl. Phys. *161*, 425 (1945)
13. Bailey, J. E., Barker, H. A.: Chem. in Britain *1974*, 465
14. Balzers, A. G.: Catalogue of Vaporisation Sources and Pure Metals. Fürstentum Liechtenstein
15. Catalogue of Vaporisation Sources. Long Beach, California, U. S. A.: The R. D. Mathis Company
16. Mann, D. M., Broida, H. P.: J. Chem. Phys. *55*, 84 (1971)
17. Catalogue of Metal Sources. Surrey, England: Goodfellow Metals Ltd., Claygate, Esher
18. D'Aniello, M. J. jr., Barefield, E. K.: J. Organometal. Chem. *76*, C50 (1974)
19. Burdett, J. K., Graham, M. A., Turner, J. J.: J. C. S. Dalton *1972*, 1620 and references therein
20. Cloke, F. G. N., Green, M. L. H., Morris, G. E.: J. C. S. Chem. Comm. *1978*, 72
21. Cloke, F. G. N., Green, M. L. H., Price, D. H.: J. C. S. Chem. Comm. *1978*, 431
22. EBS 1 electron beam evaporation source supplied G. V. Planer Ltd., Windmill Road, Sunbury-on-Thames, Middlesex, England. Power continuously variable up to 1.4 kW: a more recent mode, the EBS 35/H is continuously variable up to 7 kW. See Refs. 1, 2b
23. Green, M. L. H., Young, D.: Phys. Bull. *1975*, 350; Benfield, F. W. S., Green, M. L. H., Ogden, J. S., Young, D.: J. C. S. Chem. Comm. *1973*, 866; Anthony, M. T., Green, M. L. H., Young, D.: J. C. S. Dalton *1975*, 1419; Akhmedov, V. M., Anthony, M. T., Green, M. L. H., Young, D.: J. C. S. Chem. Comm. *1974*, 777; J. C. S. Dalton *1975*, 1412
24. Reichelt, W., Dietrich, W., Hauff, A.: Metalloberfläche *20*, 474 (1966); Reichelt, W.: Angew. Chem., Int. Ed. *14*, 218 (1975)
25. Skell, P. S., Aven, P. W.: J. Am. Chem. Soc. *94*, 5434 (1972)
26. Dugdale, R. A.: Glow Discharge Material Processing M.E.5., London: Mills and Boon Monograph 1971.
27. Prengel, A. T., Dehaven, J., Johnson, E. J., Davidovits, P.: J. Appl. Phys. *48*, 3557 (1977) and references therein
28. Weber, H., Merziger, G.: Laser, Grundlagen und Anwendungen Physik, p. 207. Weinheim: Physik-Verlag GmbH 1972
29. Basov, N. G., Boiko, V. A., Krokhin, O. N., Semenov, O. G., Sklizkov, G. V.: Sov. Phys. Tech. Phys. *13*, 1581 (1969)
30. Jaenicke, O.: Dissertation, Univ. Wien 1973
31. Sheline, R. K., Slater, J. L.: Angew. Chem., Int. Ed. *14*, 309 (1975)
32. Anwärter, M.: Angew. Chem., Int. Ed. *14*, 211 (1975) and references therein
33. Bulletin 13000B. Long Beach, California, U. S. A.: The R. D. Mathis Company
34. Schiller, S., Heisig, V., Goedicke, K.: Thin Solid Films *40*, 327 (1977)
35. Haq, K. E.: Appl. Phys. Lett. *26*, 255 (1975)
36. Geraghty, K. G., Donaghy, L. F.: Thin Solid Films *40*, 327 (1977)
37. Takagi, T., Yamada I., Sasaki A.: Thin Solid Films *39*, 207 (1976)
38. Kündig, E. P., Moskovits, M., Ozin, G. A.: J. Mol. Struct. *14*, 137 (1972)
39. Huber, H., Kündig, E. P., Ozin, G. A., Pöe, A. J.: J. Amer. Chem. Soc. *97*, 308 (1975)
40. Weltner, W. jr.: Adv. High Temperature Chem., Vol. 2, p. 85 (ed. L. Eyring). New York: Academic Press 1969; Andrews, W. L. S., Pimentel, G. C.: J. Chem. Phys. *44*, 2361 (1966); Linevsky, M. J.: J. Chem. Phys. *38*, 658 (1963)

41. Carstens, D. H. W., Kozlowski, J. F., Gruen, D. M.: High Temp. Sci. *4*, 301 (1972)
42. Mackenzie, R., Timms, P. L.: Chem. Comm. *1974*, 650
43. Klabunde, K. J.: Acc. Chem. Res. *8*, 393 (1975)
44. Rieke, R. D.: Acc. Chem. Res. *10*, 301 (1977)
45. Middleton, R.: Ph. D. Thesis, Univ. Bristol 1974
46. Güntherschutze, A.: Z. Phys. *36*, 563 (1926)
47. Herzberg, G.: Atomic Structure and Atomic Spectra, and references therein. New York: Dover 1944
48. Timms, P. L.: J. Chem. Educ. *49*, 782 (1972)
49. I. C. I.: Belgium Pat. 841–924 (1977)
50. See Chap. 3, p. 63
51. Mile, B.: Angew. Chem., Int. Ed. *7*, 507 (1968); Bennett, J. E., Mile, B., Thomas, A., Ward, B.: Adv. Phys. Org. Chem. *8*, 1 (1970); Young, D.: D. Phil. Thesis, Univ. Oxford 1974
52. Dushman, S., Lafferty, J. M.: Scientific Foundations of Vacuum Technique, 2nd ed. New York: J. Wiley and Sons 1962
53. Holland, L.: The Vacuum Deposition of Thin Films. London: Chapman and Hall 1956

# 3 Properties of Metal Atoms and Interactions with Atoms and Molecules in Condensed Phases

This chapter is concerned with metal atoms and clusters; some of their physical properties, their spectroscopic investigation and their interaction in matrices. The understanding of these factors has been mainly achieved with the aid of Matrix Isolation Spectroscopy; this is an elegant technique but nevertheless a non-synthetic one. The chapter is therefore concluded with a section on how and when metal atoms react with cocondensed species on a synthetic scale.

## 3.1 Introduction

The essentials of the MVS technique have been described in the introductory chapter and it is now important to try to understand the behaviour of metal atoms with respect to their interactions in inert matrices with other metal atoms, matrix atoms and with cocondensed substrates.

Early studies of metal vapours were concerned, on the whole, with the development of atomic spectroscopy[1)] and later, quantum theory[2)]. The ground and excited state electronic energies for all the metal atoms and some polyatomic species are known[1,3)] and one may distinquish with some degree of certainty reactions involving atoms in the ground or excited electronic states[4–6)]. It is probable that MVS experiments involving transition metals involve only atoms in the electronic ground state[1]. Reactions of non-transition elements in electronically excited states with substrates in condensed phases have been found to yield different products to those obtained from the same reaction into the metal atoms in the electronic ground state[5)]. Electronic relaxation will, in general, be much faster the more states are available to the electrons, hence some elements in the first and second short periods may have excited states life times of ~ 1 sec, whereas, heavier elements such as transition elements would be expected to have excited state life times of $10^{-8}$ sec or less.

1 Cu atoms produced by resistive heating do not react with $SiCl_4$ but Cu atoms produced from electron beam vaporisation form perchlorosilanes with $SiCl_4$. It has been claimed that this may be due to the reaction of electronic excited Cu[55)]. However the first order rate constant for Cu atom relaxation $^2D_{5/2} \rightarrow {}^2S_{1/2}$ in nitrogen gas at low pressure is $4.9 \times 10^{-16}$/sec, which may mean that configurations of higher energy than $3d^9 4s$ may be involved in the $SiCl_4$ reaction[56)], or the electron beam is cracking $SiCl_4$.

Metal atoms produced at a source temperature of 3,000 °C have a translational kinetic energy of approximately 0.25 eV (24 kJ)[7]. This is much less than their vaporisation energy which usually lies between 3 and 8 eV (290 and 775 kJ).

Studies of the composition of the vapours of transition metals have been confined to mass, u.v. and visible spectroscopic investigations[8–11]. Usually over 99% of the vapour is monatomic and it is to be expected that initial condensation of metal vapours with substrates involves monatomic species. There is, however, considerable uncertainty as to the state of a matrix surface under conditions of continuous condensation[12] and it is therefore falacious to assume that all the metal atoms maintain their integrety during cocondensation.

The development of Matrix Isolation Spectroscopy over the last two decades[13,14] has led to a considerable insight into the behaviour of metal atoms and metal containing molecular fragments and their modes of chemical combination with other isolated species. The work of Pimentel[15], Andrews[16], Turner[17], Ozin[18] and Gruen[19] has contributed greatly in this respect.

The existence of monatomic metal species in inert gas matrices has been established by a comparison of their U.V.-Visible Spectra in matrices and the gas phase; these correlate reasonably well (although some early matrix isolation studies may be suspect due to metal aggregation). Early work showed that red and blue wavelength shifts were dependant on the matrix gas, temperature changes and the atomic configuration of the metal: within a given configuration transitions of higher energies were shifted more than those of lower energies and shifts were inversely proportional to matrix atom size. Atoms with odd numbers of d electrons showed smaller shifts than those with even numbers of d electrons. No satisfactory quantitative explanation was forthcoming although qualitatively the facts could be explained in terms of dispersive and covalent forces between the metal atom and the matrix material[19,20].

The quantitative estimation of diatomic metal species by U.V.-Visible Spectroscopy is more difficult than for monatomics because all the evidence points to much lower oscillator strengths thus detection and estimation in the presence of large quantities of codeposited monomer may be difficult[21]. Raman analysis of matrices is difficult because of heating effects and Mössbauer analysis although successful in detecting $Fe_2$ and $Sn_2$ is not generally applicable.

The warming of matrices containing trapped atoms to well above the Tammen temperature often causes the disappearance of the spectra of monotonic species: this is attributed to metal aggregation due to bulk diffusion[15]. Metal aggregates although often intensely active in a catalytic sense do not readily form isolable or characterisable compounds: the control and understanding of the mechanisms of aggregation is therfore most important. The spectroscopic investigation of non crystalline aggregates is really just begining[22,23].

## 3.2 Metal Atoms in Matrices

### *3.2.1 Absorbtion Frequency Shifts in Matrix Isolated Metal Atom Spectra*

The weak forces which are responsible for holding non-polar atoms and molecules together in condensed phases are London dispersion forces resulting from induced dipole – induced dipole interactions. Such interactions would be expected to exist between metal atoms and their matrix host atoms. Heats of vaporisation of the noble gases vary between 0.7 (He) and 13.7 (Xe) kJ/mol and one would expect dispersive forces between metal atoms and noble gas atoms to be at least of this order of magnitude. Clearly the strength of the interaction and the symmetry of the metal atom environment will determine the perturbation on the electronic energy levels of the valence electrons in the metal atom, which in turn will effect the size and direction of the matrix induced shift in the electronic spectrum of the metal atom.

Theoretical models suggest that the main effect of the matrix is to increase the energy of the outer electrons of the metal due to electron-electron repulsion. The effect on each electronic state is quantitatively different and this difference correlates with observed frequency shifts. According to the models the appropriate wavefunctions are no longer pure metal atomic orbitals, and orbital mixing between metal atomic orbitals and both ligand orbitals and other symmetry related orbitals can occur. The mixing is small but may give rise to the altered intensity relationships observed in most matrices. Obviously as the site symmetry decreases, the number of distinct matrix shifts increases[24)].

More recent work seems to indicate that matrix spectra of metal atoms are usually blue-shifted with respect to gas phase spectra although no immediate relationship between gas phase and matrix oscillator strengths is observed[25)].

The matrix shift (relative to the gas phase) in the electronic spectrum of nickel atoms is approximately constant when expressed in terms of wavelength; more so than when expressed in terms of frequency[19,22)]. This may be explained in terms of a repulsive term R and an attractive term A, for the ground and excited states connecting the transitions. When the attractive term A is expressed as a Van der Waals interaction it becomes proportional to the polarisability and hence to the refractive index of the host matrix. It is relatively easy to show $\frac{\delta\Delta\lambda}{\delta V}$ is more constant than $\frac{\delta\Delta\nu}{\delta\nu}$ without expressing R analytically. This result may be a useful empirical guide to preliminary spectral assignment. A more comprehensive review of matrix shifts is given in ref. 19.

### *3.2.2 Site Symmetry Effects*

Matrix-metal interactions may be geometrically specific and their effect on spectral line shifts and oscillator strengths may also depend on the proximate symmetry at the metal atom. The Mössbauer spectra of $^{57}$Fe atoms for example, show the same isomer shift in argon, krypton and xenon (−0.75 ± 0.03 mm/s) and no quadrupole splitting: in $\alpha$-$N_2$ matrices $^{57}$Fe atoms show practically the same isotope shift (−0.78 ± 0.02 mm/s) but now have a quadrupole splitting of 2.70 ± 0.04 mm/s. These results indicate that the cubic symmetry of the iron atom existing in the inert gas matrices is lowered in the nitrogen matrix although the extent of interelectronic interaction (as indicated by the nuclear electron density i.e. the isomer shift) is the same[26,27)]. The interaction of matrix isolated $Cr(CO)_5$ with inert gas atoms or methane is evidenced by strong matrix sensitive absorption bonds [Ne, 624 nm; Ar 533 nm; $CH_4$, 489 nm]. The $Cr(CO)_5$ geometry is $C_{4V}$ and it appears that the sensitivity of the absorbtion maxima is caused by a specific Keesom[30a)] interaction of one matrix atom with $Cr(CO)_5$ in the region of the "hole" available at the vacant coordination site of the chromium atom[29,30]. Elegant work on photo-orientation has shown that even molecular rotation is hindered strongly by matrix interactions[35].

### *3.2.3 Reactions of Metal Atoms with Substrates*

The interaction of a metal atom with its surrounding matrix atoms represents the first "fixing" of the atom after traversing the quasi-fluid matrix surface zone where most of the kinetic energy is absorbed. The reactive substrate is either the matrix material itself or material cocondensed within the matrix. Chemical interaction between some metal atoms and substrate may be immediate or commence only after a "warm-up" period depending on the activation energy and bulk diffusion rates. To avoid metal aggregation, low metal and high substrate concentrations are ideally necessary.

Activation enthalpies of the metal substrate reaction must be low enough to allow a favourable competition with metal aggregation, and it has been pointed out that within experimental bounds (low vapour pressure of substrate and thermal stability of products) the temperature chosen for the condensation should be as high as possible[31]. Chromium atoms cocondensed with benzene in an argon matrix at 14 K form dibenzene chromium: on controlled annealing of the matrix to 45 K the infra-red absorbtions characteristic of dibenzene chromium increase. This has been attributed to bulk diffusion of chromium atoms through the matrix to a position where they have two benzene molecules as nearest neighbours[32].

Although one of the principle aims of MVS experiments is to produce interesting molecules which cannot be made by wet chemistry routes it is interesting to look at systems which for reasons of activation enthalpy barriers may be made more readily by wet chemistry methods. Ring opening of some cyclobutane or cyclopropane systems, and carbene or carbyne formation are some examples. (Ring opening polymerisation of norbornene with Mo atoms is the only known metathesis reaction in MVS[7].) These reactions may well be accomplished in the fullness of time but they raise an important point. The success of an M.V.S. experiment may depend on an initial low activation enthalpy for a metal-substrate interaction but this interaction must be severe enough to persist through a warming period if any further high activation enthalpy process is to occur.

Reactions between metal atoms and substrates in solution are reported to give higher yields (w.r.t. the metal) than the corresponding reaction in the matrix[33]. Both substrate-metal and metal-metal reactions may well, in solution, approach rates which are diffusion controlled and although competative the yields may be limited in this respect.

## 3.3 Metal-Metal Interactions: Aggregation

### *3.3.1 General Considerations*

Metal aggregation is a very important aspect of MVS and will therefore be dealt with in some detail. The majority of work concerned with metal aggregation has involved MIS and two regions of interest have become distinquishable:

(a) the exact characterisation of metal atoms and oligomers and the extent of their inaction with matrix material,
(b) the mechanism and conditions which lead to metal aggregation.

The matrix isolation of metal atoms for spectroscopic investigation involves the cocondensation of metal atoms and an inert gas on a liquid helium-cooled windows transparent to the analysing radiation[34]. Extremely high metal dilutions are often used to minimise metal-metal contact in the matrix surface layers. The movement of atoms within the surface layers may be complex, but, within the polycrystalline matrix medium, diffusion at temperatures well below the Tammen temperature is limited to vacancy migration and movement along grain boundries[12,35]. Metal atom aggregation can also occur when the matrix is annealed, i.e. as energy becomes available for diffusion.

### *3.3.2 Dynamics of Metal Atom Aggregation*

The Statistical Model[26,27]

The rare gases argon, krypton and xenon have face centred cubic structures so that each lattice site has twelve nearest neighbours. The probability $P_N$ that a metal atom has N other metal atoms as nearest neighbours is then:

$$P_N = \frac{12!\, C_M N(1 - C_M)^{12-N}}{(12-N)!\, N!} \tag{3.1}$$

where $C_M$ is the metal atom concentration related to the rare gas/metal ratio, R, as $C_M = 1/(R+1)$. $P_0$ and $P_1$ the probabilities of monomer and dimer can then be calculated for differing values of $C_M$ and compared with experiment. This model is adequate for describing the $^{57}Fe$ monomer dimer ratio in krypton, xenon and nitrogen at high dilutions but is a factor of three in error for argon.

The statistical formation of metal aggregates at higher metal concentrations (up to 10 atom per cent) has been analysed by Monte Carlo methods. The ratio of the numbers of species of the form $M_x$ to those of the form M is approximately proportional to $[M_0]^{x-1}$ where $[M_0]$ represents the overall metal g.atom fraction in the matrix. Long chains are predicted more probable than compact clusters (and bear little relation to predictions from M.O. or free energy calculations)[36,37].

Kinetic Models[38]

Kinetic models have attempted to describe the processes of diffusion in the fluid surface zone of the matrix during deposition. Aggregation is considered diffusion controlled so that temperature can be effectively ignored. A further variable, that of a reactive substrate, may be introduced. The reactions which take place in the "reaction zone" may then be represente by the network:

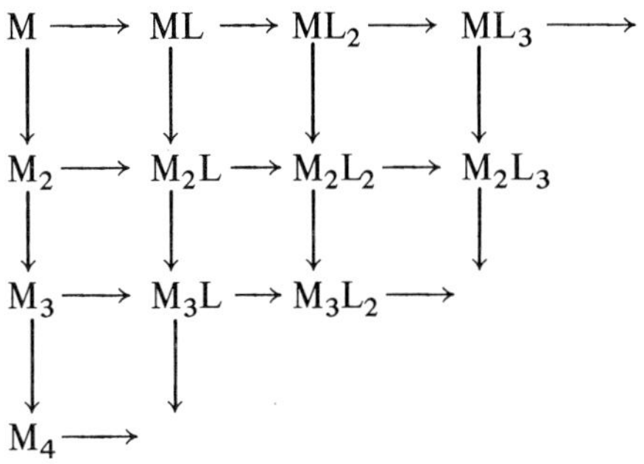


(3.2)

It is assumed that only the metal atoms M and the substrate L can diffuse so that only two rate constants $k_M$ and $k_L$ are needed to describe the system. The presence of the substrate L has the advantage that species such as $M_2L$, $M_3L$ etc. may be more easily detected spectroscopically than $M_2$ and $M_3$ due to the high oscillator strength associated with L. The system can thus be described mathematically by two models.

## The Quenched Reaction Model

This model assumes diffusion of M and L with rate constants $k_M$ and $k_L$ for a certain time $\tau_q$ after which the reaction is completely frozen out in the lower layers of the matrix.

The equations describing the system are:

$$\frac{d[L]}{dt} = - K_L [L] \{\sum_{ij} A_{ij}\} \tag{3.3}$$

$$\frac{d[M]}{dt} = - k_M A_{11} \{\sum_{ij} (1 + \delta_{ij}) A_{ij}\} \tag{3.4}$$

$$\frac{d(A_{ij})}{dt} = k_M A_{11} \{A_{i-1,j} - A_{ij}\} + k_L [L] \{A_{i,j-1} - A_{ij}\} \tag{3.5}$$

where $A_{ij}$ = concentration of species $M_i L_{j-1}$ and $A_{0,j} = A_{i,0} = 0$; $\delta_{ij}$ arises because in forming $M_2$ ($2\,M \rightarrow M_2$) the rate constant must be doubled. The Eqs. (3.3) to (3.5) solved numerically yield the relationship:

$$[M_x L_y] [ML]^{-1} = [[\text{Total metal}]/[\text{Argon}]]^m = [M_0]^m \tag{3.6}$$

Where m lies between x – 2 and x – 1 and close to the latter and provided $[M_0] \leqslant 1$ per cent.

## The Steady State Model

The concurrent deposition and freezing out of matrix material is considered to give rise to steady state conditions such that

$$\frac{d[L]}{dt} = \frac{d[M]}{dt} = \frac{dA_{ij}}{dt} = 0. \tag{3.7}$$

Terms must be added to Eqs. (3.3) to (3.5) to represent the constant rate of addition of reactants and the freezing out of products.

$$\text{Eq. (3.3) becomes } R_L - r[L] - k_L[L]\,\{\sum_{ij} A_{ij}\} = 0 \tag{3.8}$$

$$\text{Eq. (3.4) becomes } R_M - rA_{11} - k_M A_{11}\,\{\sum_{ij} (1 + \delta_{ij})\,A_{ij}\} = 0 \tag{3.9}$$

$$\text{Eq. (3.5) becomes } k_M A_{11}\,\{A_{i-1,j} - A_{ij}\} + k_L[L]\,\{A_{i,j-1} - A_{ij}\} - rA_{ij} = 0 \tag{3.10}$$

where $R_M$ and $R_L$ represent the rate of deposition of metal and ligand and r is the rate constant in units time$^{-1}$ for the freezing-out process. The rate of freezing out is considered proportional to the concentration.

Equations (3.8) to (3.10) are solved iteratively by first setting all concentrations apart from M and L equal to zero then solving for new values and repeating the process till self-consistency is obtained.

The dilute matrix limit yields the same relationships (3.6) but m does not lie so close to $x - 1$ and the product concentrations are different from those predicted in the quenched reaction model. Experimentally, for the series of reactions in the dilute matrix limit

$$M \longrightarrow ML \longrightarrow ML_2 \longrightarrow ML_3 \text{ etc.}$$

with L in a large excess, the quenched reaction model gives a reasonable fit for various concentrations of M and with [L]/[argon] ratios varing between $10^{-3}$ and $10^{-1}$. The steady state model is increasingly inappropriate for increasing concentrations of M and the reactions involving high metal concentrations are readily represented by the quenched reaction model with values of m close to $x - 1$ as predicted (Fig. 3.1).

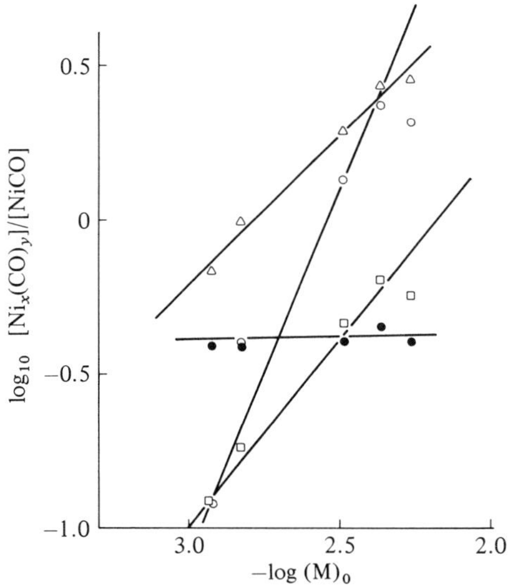


**Fig. 3.1.** Verification of the relationship $[M_xL_y][ML]^{-1} = [M_0]^{x-1}$ (J. Chem. Soc. Faraday II, *73* (4), 471 (1977)). Δ, $Ni_2CO$, $x - 1 = 1$; $Ni_3CO$, $x - 1 = 2$; •, $Ni(CO)_2$, $x - 1 = 0$; □, $Ni_2(CO)_2$, $x - 1 = 1$

### *3.3.3 The Spectroscopic Characterisation of Metal Clusters*

Dimers

All the first row transition metal dimers and many of the second and some third row dimers have been observed, and many have had their U.V.-Visible Spectra recorded in inert matrices[12]. The gas phase valence state dissociation energies of the first row dimers all lie in the range 234 ± 42 kJ/mole which ensures a significant stability for $M_2$ species. Bond dissociation energies for some second row dimers are found to be greater than the corresponding first row dimers[39] (Table 3.1).

The mixed dimer MoCr has recently been isolated in argon matrices Fig. 3.2[40]. The molecule shows similar U.V.-Visible Spectra to gas phase MoCr produced from flash photolysis of $Mo(CO)_6/Cr(CO)_6$[41]. The bond energy is calculated as 2.0 eV (194 kJ) intermediate between that for $Mo_2$ (3.4 eV (339 kJ) calc. value) and $Cr_2$ (1.56 ± 0.3 eV (151 ± 30 kJ) expt. value). Methods have been proposed for the calculation of metal dimer bond energies[42–44] which compare reasonably well with experimental values obtained from mass spectral measurements (Table 3.1).

The relationship $[M_2]\,[M_0]^{-1} \propto [M_0]$ which is the limiting condition of Eq. (3.6) in the preceeding section (in the absence of L and with x = z and m = x − 1), has been elegantly illustrated from results obtained by depositing chromium in argon matrices[21] (see Fig. 3.3).

**Table 3.1.** Dissociation energies for diatomic molecules in the gas phase (after Gingerich[42, 43])

| Molecule | $D_o^\circ$ (kJ/mole) | Molecule | $D_o^\circ$ (kJ/mole) |
|---|---|---|---|
| $Sc_2$ | 159 ± 21 | AlAg | 170.3 ± 9 |
| $Y_2$ | 156 ± 21 | AlAu | 199.2 ± 9 |
| $La_2$ | 241 ± 21 | MnAg | 112.1 ± 21 |
| $Ce_2$ | 238 ± 21 | MnAu | 181.6 ± 13 |
| $Lu_2$ | (167) | CrAu | 210.9 ± 15 |
| $Th_2$ | (285) | LuPt | 397.5 ± 33 |
| $Ti_2$ | 134 ± 21 | CePd | 318.5 ± 17 |
| $V_2$ | 238 ± 17 | TiRh | 387.0 ± 15 |
| $Cr_2$ | 151 ± 29 | LaRh | 524.5 ± 17 |
| $Mn_2$ | 42 ± 29 | ThRh | 544.0 ± 42 |
| $Fe_2$ | 121 ± 21 | ThPt | 555.5 ± 42 |
| $Co_2$ | 163 ± 25 | ThIr | 579.0 ± 42 |
| $Ni_2$ | 219.7 ± 21 | ZrPt | 627.5 ± 42 |
| $Pd_2$ | 105 ± 21 | MoCr[a, b] | 193.0 |
| $Cu_2$ | 188.3 ± 6.3 | ZrRu[a] | 602.5 |
| $Ag_2$ | 159.0 ± 6.3 | NbRu[a] | 648.5 |
| $Au_2$ | 221.8 ± 6.3 | NbOs[a] | 640.0 |
| $Rh_2$ | 274.0 | ThOs[a] | 648.5 |
| $Mo_2$[a, b] | 328.0 | | |

[a] Calculated values
[b] Ref. 40.

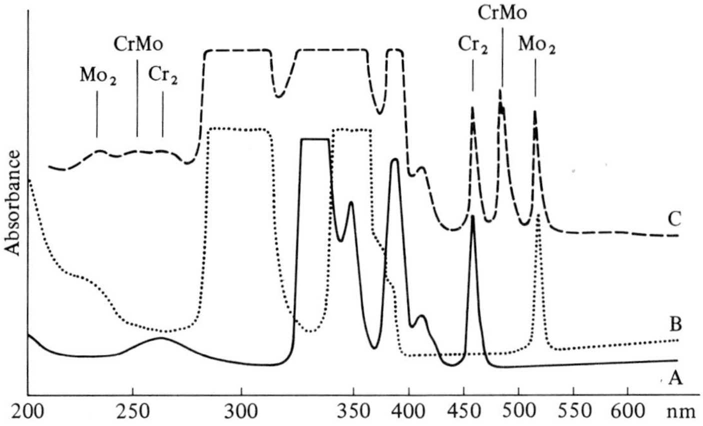


**Fig. 3.2.** The UV-visible absorption spectrum of the products of the matrix deposition. *A* Cr/Ar $\simeq$ 1/1,000; *B* Mo/Ar $\simeq$ 1/1,000, and *C* Cr/Mo/Ar $\simeq$ 1/1/1,000 at 10 K showing the bands of $Cr_2$, $Mo_2$, and CrMo molecules in the presence of Cr and Mo atoms (Inorg. Chem. *16,* 2871 (1977) with permission; copyright by the American Chemical Society)

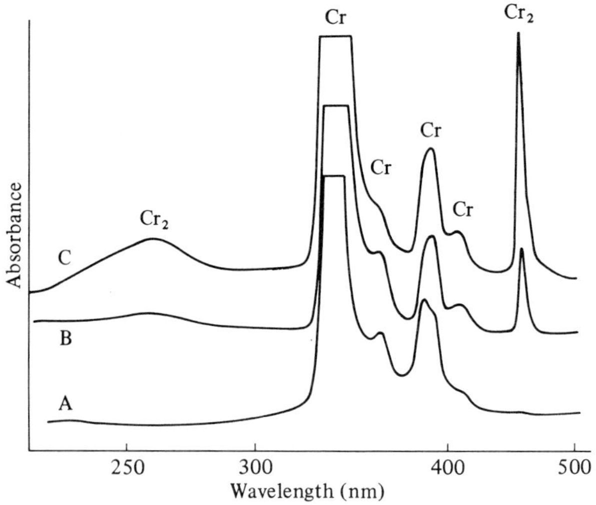


**Fig. 3.3.** A portion of the spectrum of chromium atoms and dichromium molecules in argon. *A* Low metal flow; *B* and *C* progressively higher metal flow (Kundig, et al.: Nature *254,* 503 (1975))

A similar study with vanadium atoms in inert gas matrices gave similar results with $[M_0]$ varied between $10^{-5}$ and $10^{-2}$. Higher clusters also form[25] (see Fig. 3.4).

The reaction $2\ Cu \rightarrow Cu_2$ can occur by bulk diffusion and is discussed in Sect. 4.

The deposition of metals in varying concentrations also enables the extinction coefficient ratios $\epsilon_m/\epsilon_{m_2}$ to be determined:

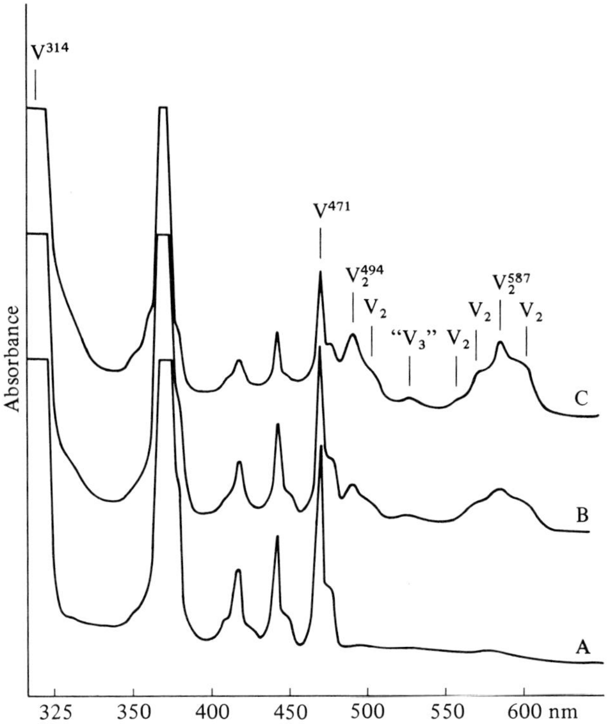


**Fig. 3.4.** The uv-visible spectrum of V in Ar matrices at 6–10 K, *A* at low V concentrations showing isolated V atoms and *B* and *C* at progressively higher V concentrations, showing both V atoms and the growth of $V_2$ molecules (J. Chem. Phys. *66*, 524 (1977))

representative values are:

$$\epsilon_V/\epsilon_{V_2} \sim 3; \quad \epsilon_{Cr}/\epsilon_{Cr_2} \sim 4; \quad \epsilon_{Rh}/\epsilon_{Rh_2} \sim 5; \quad \epsilon_{Cu}/\epsilon_{Cu_2} \sim 13$$

These ratios illustrate the difficulty in recognising $M_2$ absorbtions at low metal concentrations[25].

## Electron Configuration of $M_2$ Species

There is considerable discussion concerning the bond multiplicity and electron configuration in metal dimers and their ions[45,46]. The electron configuration of the iron dimer ($Fe_2$) in inert gas matrices as measured by the Mössbauer isomer shift was found to be $3d^{6.55}4s^{1.45}$ which is close to the value found for bulk iron[26].

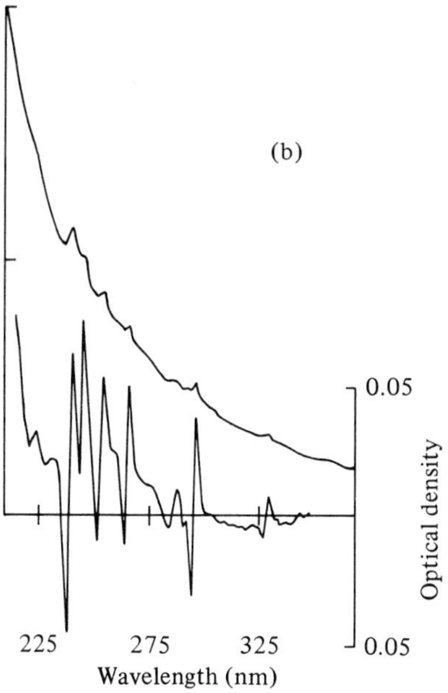


**Fig. 3.5.** The double beam absorbance spectrum (upper) and the first derivative spectrum of iron atoms in argon matrix. An illustration of the reduced effects of light scattering and ease of detection using first derivative spectra (Fe/Ar = 1/8,000) (J. Chem. Soc. Faraday II, *70*, 93 (1974))

Higher Aggregates

The electronic spectrum of $Ni_3$ has been reported and can be observed by matrix annealing of $Ni/Ni_2$ mixtures or by depositing nickel atoms at relatively high temperatures (30 K). The assignment of the $Ni_3$ spectrum is helped by the fact that $Ni_2$ forms only reluctantly by bulk diffusion. [The formation of $^3\Sigma_g^-$ $Ni_2$ from ground state ($^3F_4$) Ni atoms involves inter-system crossing at a relatively high repulsive potential and is formally spin forbidden].

Subsequent aggregation of nickel atoms results in a broad featureless spectrum attributed to higher aggregates[22)]. The non-defined structure of these higher aggregates have lead to a description of their absorbtion spectra in terms of a colloid embedded in a dielectric. The complex dielectric constant of the metal is then calculated and used to give the absorption index from which the adsorbtion A of the colloid as a function of frequency can be determined. Figure 3.6 illustrates results for nickel clusters[22,23)].

### *3.3.4 Nucleation and Free Energy of Small Clusters*

Many calculations of nucleation rates have been performed; these calculations depend on a knowledge of the free energy of the cluster[37)]. The free energy depends on the geometry of the cluster and for certain cluster sizes it is found that the minimum free energy configuration does not correspond to a lattice structure and atoms are not localised as in microcrystallites.

Practical studies of deposition of metal atoms on surfaces has led to certain rules concerning nucleation and microcrystallite growth[47)]. Initially deposited metal can grow in two

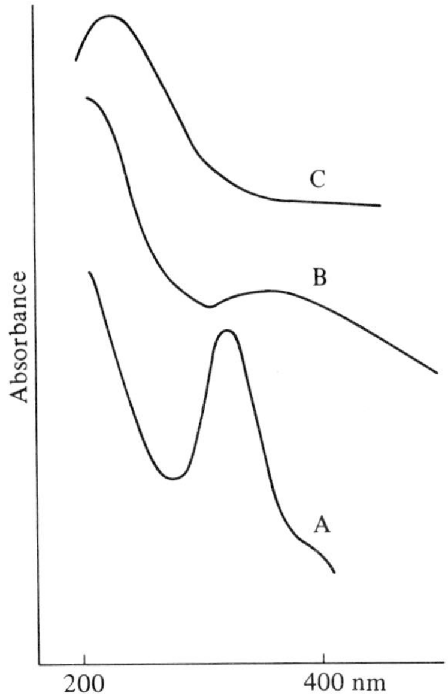


**Fig. 3.6.** *A* A portion of the raw uv-visible spectrum obtained when nickel was codeposited with argon on a very warm (35 K) plate. Only the broad feature centered at 335 nm is visible. *B* The spectrum calculated for a nickel-in-argon colloid assuming large particles. *C* The absorbance of bulk nickel (approximately what one would obtain from transmission through a thin, continuous nickel film). Ordinate is absorbance in arbitrary units and the spectra have been displaced for clarity (J. Chem. Phys. *66*, 3988 (1977))

or three dimensions on surfaces depending on the energy of interaction between the surface and the metal.

Three dimensional growth will be favoured if:

$$\Lambda_0 \geqslant 3 \cdot E_{des} \tag{3.11}$$

and two dimensional growth if:

$$\Lambda_0 \leqslant 3 \cdot (E_{des} - E_{diff}) \tag{3.12}$$

$\Lambda_0$ = metal sublimation energy; $E_{des}$ = energy of atomic desorbtion from the surface; $E_{diff}$ = energy of migration or diffusion on the surface.

The localised electrical potential of a crystal edge has been used to initiate nucleation. Absorption of negatively charged gold aggregates on kaolin crystals occurs only on crystal edges where cations in the crystal surface can overlap with the negative potential of the gold [48,49].

### *3.3.5 Theoretical Considerations*

Calculations on the bonding in metal dimers ($M_2$) have been performed using extended Hückel (EH)[50], CNDO[50] *ab initio* SCF and SCF-Xα-SW methods[51] for a variety of transition metals. Particularly interesting are the calculations on $Mo_2$, $Mo_2^{4+}$, $Nb_2$, and $V_2$ [39,51]. The exceptionally short bond distances for $Mo_2$ and $Nb_2$ suggest a quadruple bond order: the calculations for $Mo_2$ suggest a formally closed shell sextuple metal-metal bond due to the stabilisation relative to $Mo_2^{4+}$ of the 5s orbital $\sigma$-bonding combination which becomes the HOMO.

E.H. calculations[52] on the electronic properties of metal clusters of Pd, Ag, Cu, Au, Pd/Ag and Ni/Cu systems containing up to 55 atoms have been recorded.

Electronic properties important in catalysis are discussed. Molecular orbitals of the cluster contain contributions from orbitals localised throughout the cluster and account for the changes in properties with size. The valence band of the clusters expands with size and the Fermi level moves closer to the vacuum.

A quantitative M.O. description of $V_2$ using iterative E.H. calculations suggests that high spin divanadium has a strong metal-metal bond which is mainly 4s in character with only small contributions from the degenerate $d_{xz,yz}$ $\pi$ bonding set[25].

Structures and binding energies and charge distributions for two to six atom Ti, Cr, Fe and Ni clusters, and their relationship to nucleation and cluster catalysis have been investigated[53]. It was shown in the low spin approximation, titanium shows a preference for tightly packed clusters with positively charged corners, whilst nickel favours open and ring-like structures with negatively charged corner atoms. Chromium and iron prefere tightly packed configurations.

The three strong electronic transitions observed in $Ni_2$ and the two observed in $Ni_3$ can be accurately described with an M.O. theory using data obtained from gas phase ultra violet spectra of nickel atoms. Antibonding interaction of nickel atoms with the argon matrix is suggested as responsible for the nickel atom frequency shifts as compared to the gas phase[22]. Silver clusters of up to thirty nine atoms have also been investigated[54].

## 3.4 Metal Atom Interactions in Synthetic Experiments

The previous sections have been mainly concerned with metal atom properties and interactions under the stringent conditions of matrix spectroscopy. It is useful to discuss within the framework of these previous sections the behaviour of metal atoms on a synthetic scale.

There is no direct evidence to show exactly when high temperature species react with substrates. However a wealth of indirect evidence (*vide infra*) indicates that some initial reaction is very likely to occur on cocondensation. The various species are then in close contact for the first time, and, although they will have lost immediately most of their translational energy, will still form a fluid "reaction" layer for a short period as their remaining excess energy is dissipated into the bulk cocondensate. If the deposition temperature is low enough, little further reaction is likely in the rigid bulk cocondensate. Nonetheless, secondary reactions may occur on warning, to give the final isolable products.

In order to discuss in more detail the reactions that can occur during a cocondensation experiment it is convenient to divide it into a series of periods or zones and to discuss these in turn. They are:

Vapour Zone
Cocondensation Zone
Rigid Matrix Period
Fluid Matrix Period
Solution Period

It should be remembered that the "classical" cocondensation method to form a solid codeposit is not a prerequisite nor is a very low temperature. It has been shown that metal atoms can be condensed into renewed liquid surfaces to form complexes and at ambient co-condensation temperatures[57,58].

### *3.4.1 Vapour Zone*

Reaction between metal atoms and substrate molecules is most unlikely because the mean free path of the atoms is usually much larger than the dimensions of the vapour synthesis apparatus. In addition, in some apparatus, the metal and substrate vapour phases are kept separate[59].

### *3.4.2 Cocondensation Zone*

This embraces the surface where the vapours are cocondensing and the semi-fluid zone just below it where the various species are still in the process of losing their excess energy. The metal atoms and substrate molecules are still relatively mobile, in intimate contact for the first time and still have sufficient energy to overcome small activation energies to form bonds: many reactions are thought to occur or begin here. The fact that many matrices are coloured from the moment they are laid down, often with the colour of the final product lends support to this: for example[60]. The period of surface motion of a particular species is not expected to be long and will become shorter (and the cocondensation zone thinner) as the temperature of the cold surface is decreased. The temperature of the cocondensation zone will be higher than that of the bulk matrix owing to the energies of the impinging species and furnace radiation. The magnitude of this temperature difference is difficult to ascertain but there is evidence that it is relatively large. For example cocondensation of Ti atoms with a large excess of $N_2$ at 6–8 K gives only isolated Ti atoms, identified by electron spectroscopy. Only by heating this matrix to 30 K is a complex (identified as $Ti(N_2)_6$) formed. However, $Ti(N_2)_6$ can also be formed by carrying out the cocondensation experiment at 16 K, indicating an effective cocondensation zone temperature of 30 K[61].

Although reaction between a substrate and a metal atom normally has a low activation energy, it is almost always higher than that of metal-metal agglomeration. Thus unless aggregation is desired, or the temperature is so low as to prohibit atom diffusion, the cocondensation zone temperature must be high enough to allow atom-substrate reactions to compete with atom-atom reactions. Figure 1.3 shows how the difference in the rate of two reactions is reduced with increasing temperature. Thus the highly reactive species $SiF_2$ does not react with benzene at 20 K – the only product being formed is polymeric $(SiF_2)_n$ – while the same cocondensation at 77 K forms species such as *3.1*[62]. In addition, a large excess of substrate can assist the atom substrate reaction as a metal atom will then be in contact with a large number of substrate molecules before meeting another metal atom or cluster.

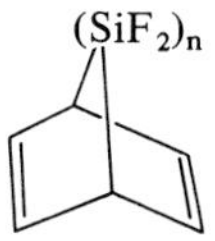


(n = 2–4) *3.1*

### *3.4.3 Rigid Matrix Zone or Period*

Once any reaction product or unreacted species has lost all its excess energy it will be "frozen" into the underlying bulk cocondensate or rigid matrix zone. The convenient liquid nitrogen coolant bath used in most metal evaporation syntheses in order to quench the substrate vapour pressure, also cools the bulk cocondensate to a temperature sufficiently low to reduce species mobility within the matrix to a negligible amount. Individual species are unlikely to meet reaction partners, thus very little further reaction can take place. No metal vapour synthesis offers direct evidence of this, but examples abound in the matrix isolation experiments. Indeed, these experiments are designed to produce rigid matrices in order to allow lengthy spectroscopic studies of e.g., species labile to polymerisation. Only by judiciously warming the matrix can bulk diffusion occur (as shown by further complexation or aggregation) while recooling to the original temperature stops the process. One particular example is the reaction between Cr atoms and benzene in an argon matrix at 14 K[32]. Some bis ($\eta$-benzene)chromium, detected by ir spectroscopy, is initially formed, presumably in the fluid cocondensation zone, after which the matrix is indefinitely stable. Further formation of $(\eta\text{-PhH})_2Cr$ can be induced by annealing the matrix at 25 K for two minutes. At this temperature some of the unreacted Cr atoms can presumably migrate to find and react with free benzene molecules. For further reaction, the matrix must now be annealed for 2 minutes at 35 K, implying that remaining Cr atoms must travel further in the available annealing time to find any unreacted benzene molecules. This process can be repeated at least once more at 45 K.

For a given matrix, bulk diffusion effectively stops below the Tammen temperature (about half the matrix mp). To a good approximation, species trapped in a host matrix lose their mobility at the same temperature as the host material, though very small, light species such as atoms and diatomic molecules retain some mobility[63–65]. Table 3.2 gives half the mp (0.5 $T_m$) of a series of substrates that have been used in metal vapour syntheses at 77 K and shows that, except for a few cases, the majority have 0.5 $T_m$ at or above 77 K. Unreacted metal atoms may still be able to diffuse but because of the high substrate : metal ratio normally used, they have every chance to react with substrate, providing the activa-

**Table 3.2.** 0.5 $T_m$ of substrates commonly used in metal vapour synthesis (Adapted from: CRC Handbook of Chemistry and Physics, 56th ed. Cleveland: CRC Press 1975/76)

| Substrate | 0.5 $T_m$ (K) | Substrate | 0.5 $T_m$ (K) |
|---|---|---|---|
| Propene | 44 | Trimethylphosphine | 94 |
| But-1-ene | 44 | Acetylene | 96 |
| Nitric Oxide | 55 | Cycloheptatriene | 97 |
| Trfluorophosphine | 61 | Ammonia | 98 |
| Allyl Chloride | 69 | Cycloocta-1,5-diene | 102 |
| Bromoethane | 77 | But-2-yne | 121 |
| 2-Chloropropane | 78 | Bromobenzene | 122 |
| Butadiene | 82 | Cyclooctetetraene | 134 |
| Propyne | 86 | Hexafluorobenzene | 139 |
| Cyclopentadiene | 88 | Benzene | 140 |
| Toluene | 89 | | |

**Table 3.3.** 0.5 $T_m$ and 0.25 $T_m$ of common matrix material in matrix isolation studies (Adapted from: CRC Handbook of Chemistry and Physics, 56th ed. Cleveland: CRC Press 1975/76)

| Material | 0.5 $T_m$ (K) | 0.25 $T_m$ (K) |
|---|---|---|
| Neon | 12 | 6 |
| Dinitrogen | 32 | 16 |
| Carbon Monoxide | 37 | 19 |
| Argon | 42 | 21 |
| Dioxygen | 45 | 23 |
| Methane | 45 | 23 |
| Ethylene | 52 | 26 |
| Krypton | 58 | 29 |
| Xenon | 76 | 38 |
| Acetylene | 86 | 43 |
| Sulphur Hexafluoride | 111 | 56 |

tion energy does not exceed the available energy: otherwise the very low activation process of agglomeration to metal clusters and crystallites will occur.

The situation is similar for matrix isolation studies, except that because inert host matrices doped with reactive substates are often used, the operating temperature (as a function of the host matrix $T_m$) must be lower in order to carefully control metal atom migration and avoid agglomeration. 0.33 $T_m$ or less is required to effectively obviate diffusion of V atoms isolated in a series or normal, branched or cyclic alkanes[66)], while for very light atoms such as those of Li, in rare gas matrices, 0.2 $T_m$ may be necessary[63–65)]. Unfortunately at temperatures as low as 0.2 $T_m$ the matrix becomes "frosty" reducing spectral quality. Therefore a compromise of *ca.* 0.25 $T_m$ is expedient. Table 3.3 gives 0.5 $T_m$ and 0.25 $T_m$ of a series of commonly used matrix materials. Appreciable diffusion of Cu atoms in 1,000 fold excess of Ar begins at 30 K[67)]. In addition, by measuring the $^{63}Cu$ atom concentration with time at 35 K by esr it was possible to measure the rate of dimerisation (Eq. (3.13)). The diffusion coefficient $D_{Cu}$ was then estimated at $4 \times 10^{-17}$ $cm^2/s$ using the formula for the dimerisation rate shown in Eq. (3.14) where $\eta_{Cu}$ is the number of Cu atoms per unit volume, the reaction radius R taken as 3 Å (the "intersite" distance in the fcc lattice of Ar) and matrix ratio Cu : Ar = 1 : 1,000

$$Cu(a) + Cu(a) \rightarrow Cu_2 \tag{3.13}$$

$$-\,d\eta_{Cu}/dt = 8\,\pi\, RD_{Cu}\eta_{Cu}^2 \left(1 + \frac{R}{(2\,\pi\, D_{Cu}t)}\ \frac{1}{2}\right) \tag{3.14}$$

Factors other than temperature, dilution, reactivity and rigidity can effect diffusion. The most notable are mass and polarisability. However they are not as large as those previously mentioned and are unlikely to become important except at the very low temperatures of the matrix isolation experiment. As may be anticipated, diffusion is inversely proportional to mass, though the exact relation is not known. Thus, light atoms such as those of H, Li, Be, and B can diffuse through quite rigid matrices[12)]. Conversely, the annealing of some matrices containing atoms of heavy elements can have little effect.

In addition, at such low temperatures the van der Waals interaction between the large, polarisable atoms and the matrix species may be higher than kT; in which case the diffusing species would be very sluggish van der Waals molecules. Furthermore the polarisability of the matrix species can be expected to have a similar effect on diffusion. Indeed the van der Waals molecule $XeAg(CO)_2$ has been postulated in xenon matrices containing Ag atoms and CO at 12 K[68].

### *3.4.4 Fluid Matrix Period*

This covers the period after deposition when the cocondensate is allowed to warm from the bulk deposit temperature, a state of rigidity, to its mp. This range can be short or long depending on the substrate mp and the bulk deposition temperature of typically 77 K.

Initially the warming will have similar effects as that of annealing; i.e. diffusion of the more mobile species such as atoms. However, as the Tammen temperature is reached, the less mobile species such as substrate and product molecules will begin to diffuse. Thus any species which owed their existence to effective isolation in the rigid deposit will be converted to new, more stable products (which may be decomposition products). In addition a whole series of new reactions with previously prohibitively high activation energies may occur with rising temperature. Such reactions may also occur in the solution period.

### *3.4.5 Solution Period*

Although much chemistry may have occured on cocondensation or on warming the cocondensate further chemistry is likely when the cocondensate melts. All species (except any metal agglomerates) are now mobile and as the temperature rises, new, more complex, reactions become possible. For example, warming the $C_6F_5COCl$/Pd cocondensate to −50 °C releases CO with formation of a $C_6F_5PdCl$ complex trapped by $PEt_3$[69] (Eq. (3.15)). An acylpalladiumhalide complex can be isolated from the reaction of n-$C_3F_7COCl$ and Pd atoms if trapped at low temperatures (Eq. (3.16)). A general decomposition scheme of acylpalladium halides with temperature can therefore be written (Eq. (3.17))

$$\mathrm{Pd(a) + C_6F_5COCl \xrightarrow{i,\,ii} \textit{trans}\text{-}C_6F_5PdCl(PEt_3)_2} \qquad (3.15)$$

$$\mathrm{Pd(a) + n\text{-}C_3F_7COCl \xrightarrow{i} [n\text{-}C_3F_7COPdCl]} \begin{cases} \xrightarrow{iii} \mathrm{\textit{trans}\text{-}n\text{-}C_3F_7COPdCl(PEt_3)_2} \\ \xrightarrow{iv} \mathrm{\textit{trans}\text{-}PdCl_2(PEt_3)_2} \end{cases} \qquad (3.16)$$

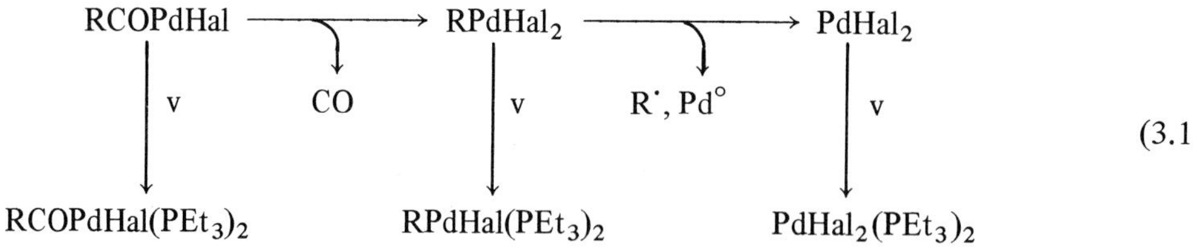


(3.17)

(i) cocondense, 77 K; (ii) $PEt_3$, 25 °C; (iii) $PEt_3$, −78 °C; (iv) $PEt_3$, 0 °C; (v) $PEt_3$

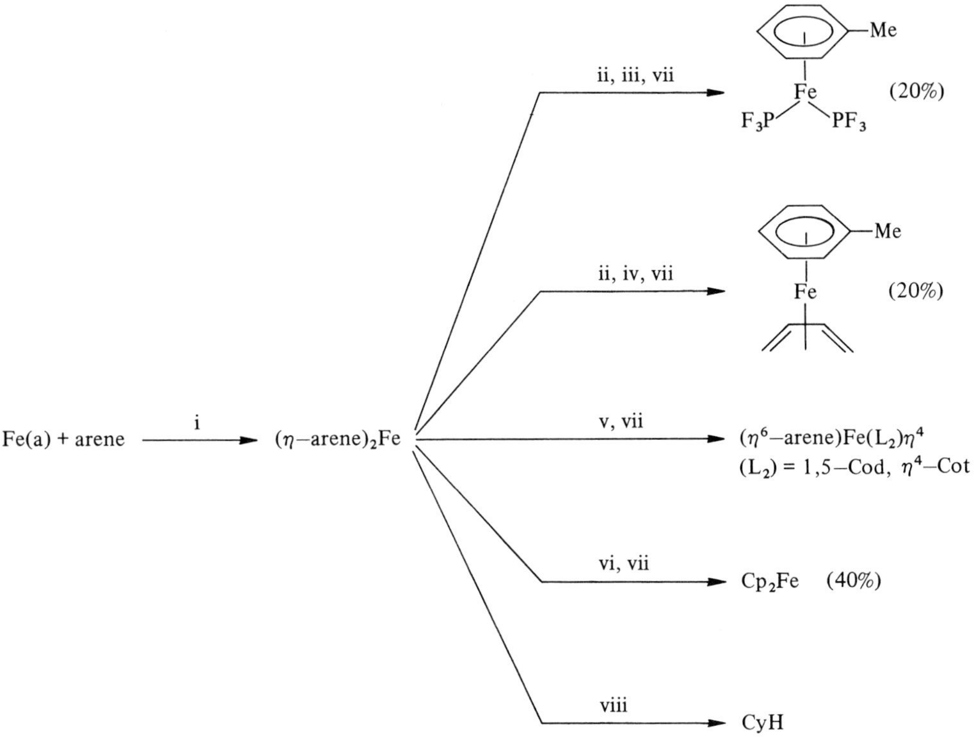


**Scheme 3.1.** (i) cocondense, 77 K; (ii) stir, $-94\ ^\circ C$; (iii) $PF_3$, 77 K; (iv) Bd, 77 K; (v) $(L_2)$, 77 K; (vi) CpH, 77 K; (vii) $25\ ^\circ C$; (viii) warm in atmosphere of $H_2$

The technique of trapping by adding further ligand precursors such as phosphines has proved a very useful method of isolating derivatives of compounds which may themselve be to sensitive for isolation. Cocondensation of Fe atoms and benzene forms a solid which explodes on warming to $-50\ ^\circ C$[70)]. However, this highly reactive complex, which spectroscopic observations at low temperatures indicate as dibenzeneiron[71)], and the related Fe/toluene cocondensate can be trapped to make a number of derivatives[70,72–75)] (Scheme 3.1).

Trapping can cause reorganisation of sensitive species. The labile $(1,5\text{-Cod})_2Fe$ complex can be trapped by number of ligand precursors to form $(1,5\text{-Cod})FeL_3$ (L = CO, $PF_3$, $Bu^tNC$), but rearrangement occurs when trapped by trialkyl phosphites to give $(1,3\text{-Cod})FeP(OR)_3$[57,77,78)].

The products that are isolated from a metal vapour cocondensation reaction may not necessarily be formed from reaction betweens metal *atoms* and the substrate. It may be that at low temperatures, reaction between metal atoms and such a substrate is not favoured, perhaps because of too high an activation energy, and instead the metal agglomerates. At higher temperatures, when more energy is available, this finely divided solid but active metal then reacts with the substrate to form the isolated products. There is no certain example of this but it has been shown that finely divided bulk metal formed by metal vapour synthesis can react with substrates added after cocondensation[79–82)] (Eqs. (3.18) to (3.20)).

$$Mg(a) + thf \xrightarrow{i, ii} Mg(s)/thf \xrightarrow{iii} C_6F_5MgBr\ (80\%) \quad (3.18)$$

$$Ni(a) + thf \xrightarrow{i, ii} Ni(s)/thf \xrightarrow{iv} PhCH_2CH_2Ph \quad (3.19)$$

$$Zn(a) + diglyme \xrightarrow{i, ii} Zn(s)/diglyme \xrightarrow{v} Et_2Zn \quad (3.20)$$

(i) cocondense, 77 K; (ii) 25 °C; (iii) $C_6F_5Br$, – 30 °C; (iv) $PhCH_2Cl$, 40 °C; (v) EtBr reflux

## References


1. Herzberg, G.: Atomic Spectra and Atomic Structure, and references therein. New York: Dover Publications, inc. 1944
2. Millman, S., Kusch, P.: Phys. Rev. *60*, 91 (1941) and references therein
3. Condon, E. U., Shortley, G. H.: The Theory of Atomic Spectra. Cambridge University Press 1935
4. Ref. 1, p. 51
5. Skell, P. S., Harvel, J. J., McGlinchey, M. J.: Acc. Chem. Res. *6*, 97 (1973); Girard, J. E.: Ph. D. Thesis, Penn. State Pa. USA (1971)
6. Stern, O., Gerlach, W.: Z. Phys. *1923*, 349
7. Skell, P. S., McGlinchey, M. J.: Angew. Chem., Int. Ed. *14*, 195 (1975)
8. Siegel, B.: Quart. Rev. *19*, 77 (1965)
9. Travis, D. N., Barrow, R. F.: Proc. Chem. Soc. *1962*, 64
10. Lin, S. S., Kant, A.: J. Phys. Chem. *73*, 2450 (1969); Kant, A., Strauss, B.: J. Chem. Phys. *41*, 3806 (1964)
11. Schissel, P.: J. Chem. Phys. *26*, 1276 (1957)
12. Kündig, E. P., Moskovits, M., Ozin, G. A.: Angew. Chem. Int. Ed. *14*, 292 (1975)
13. Meyer, B.: Low Temperature Spectroscopy. New York: Elsevier 1971
14. Hallam, H.: Vibrational Spectroscopy of Trapped Species. New York: Wiley 1972
15. Pimentel, G. C.: Angew. Chem., Int. Ed. *14*, 199 (1975)
16. Andrews, L.: Infra-red Spectra of Free Radicals and Chemical Intermediates in Inert Metrices. Ann. Rev. Phys. Chem. *22*, 109 (1971)
17. Turner, J. J.: Angew. Chem., Int. Ed. *14*, 304 (1975) and references therein
18. Ozin, G. A.: in Ref. 14
19. Gruen, D. M.: in: Cryochemistry. Moskovits, M., Ozin, G. A. (eds.). New York: Wiley-Interscience 1976
20. Brewer, L., King, B. A., Wang, J. L., Meyer, B., Moore, G. F.: J. Chem. Phys. *49*, 5209 (1968); McCarty, M., Robinson, G. W.: Mol. Phys. *2*, 415 (1959)
21. Kündig, E. P., Moskovits, M., Ozin, G. A.: Nature *254*, 503 (1975)
22. Moskovits, M., Hulse, J. E.: J. Chem. Phys. *66*, 3988 (1977)
23. Dignam, M. J., Moskovits, M.: J. Chem. Soc. Faraday Trans. 2, *69*, 65 (1973)
24. De Vore, T. C.: J. Chem. Phys. *62*, 520 (1975)
25. Ford, T. A., Huber, H., Klotzbücher, W., Kündig, E. P., Moskovits, M., Ozin, G. A.: J. Chem. Phys. *66*, 524 (1977)
26. McNab, T. K., Micklitz, H., Barrett, P. H.: Phys. Rev. *B 4*, 3787 (1971)
27. Barrett, P. H., Montano, P. A.: J. Chem. Soc. Faraday 2, *73*, 378 (1977)
28. Perutz, R. N., Turner, J. J.: Inorg. Chem. *14*, 262 (1975)
29. Perutz, R. N., Turner, J. J.: J. Amer. Chem. Soc. *97*, 4791 (1975)
30. Burdett, J. K., Perutz, R. N., Poliakoff, M., Turner, J. J.: Chem. Comm. *1975*, 157

30a. Smalley, R. E., Wharton, L., Levy, D. H.: J. Chem. Phys. *68*, 671 (1978)

31. Bassler, J. M., Timms, P. L., Margrave, J. L.: Inorg. Chem. *5*, 729 (1966)
32. Boyd, J. W., Lavoie, J. M., Gruen, D. M.: J. Chem. Phys. *60*, 4068 (1974)
33. Mackenzie, R., Timms, P. L.: Chem. Comm. *1974*, 650
34. Ozin, G. A., Vander Voet, A.: Acc. Chem. Res. *6*, 313 (1973)

35. Bass, A. M., Broida, H. P.: Formation and Trapping of Free Radicals, Chap. 4. New York: Academic Press 1960
36. Anderson, A. B.: J. Chem. Phys. *64*, 4046 (1976)
37. Burton, J. J.: J. Chem. Soc. Faraday II, *69*, 540 (1973) and references therein
38. Moskovits, M., Hulse, J. E.: J. Chem. Soc. Faraday II, *73*, 471 (1977)
39. Klötzbücher, W., Ozin, G. A.: Inorg. Chem. *16*, 984 (1977)
40. Klötzbücher, W., Ozin, G. A., Norman, J. G. jr., Kolari, H. J.: Inorg. Chem. *16*, 2871 (1977)
41. Efremov, Yu. M., Samoilova, A. N., Gurvich, L. V.: Chem. Phys. Lett. *44*, 108 (1976)
42. Gingerich, K. A.: Chem. Phys. Lett. *23*, 270 (1973); Gingerich, K. A.: J. Cryst. Growth *9*, 31 (1971)
43. Gingerich, K. A., Cocke, D. L.: Chem. Comm. *1972*, 536
44. Brewer, L.: Science *161*, 115 (1968)
45. Messmer, R. P., Knudson, S. K., Johnson, K. H., Diamond, J. B., Young, C. Y.: Phys. Rev. *B 13*, 1396 (1976)
46. Cotton, F. A.: Chem. Rev. *4*, 27 (1975)
47. Niedermayer, R.: Angew. Chem., Int. Ed. *14*, 212 (1975)
48. Thiessen, P. A.: Z. Elektrochem. Angew. Phys. Chem. *48*, 675 (1942)
49. Kohlschütter, H. W.: Angew. Chem., Int. Ed. *14*, 193 (1975)
50. Baetzold, R. C.: J. Chem. Phys. *55*, 4355 (1971)
51. Norman, Jr., J. G., Kolari, H. J., Gray, H. B., Trogler, W. C.: Inorg. Chem. *16*, 987 (1977)
52. Baetzold, R. C., Mack, R. E.: J. Chem. Phys. *62*, 1513 (1975)
53. Anderson, A. B.: J. Chem. Phys. *66*, 5108 (1977)
54. Baetzold, R. C.: J. Chem. Phys. *68*, 555 (1978)
55. Timms, P. L.: Adv. Inorg. Chem. Radiochem. *14*, 121 (1972)
56. Trainer, D. W.: J. Chem. Phys. *64*, 4131 (1976)
57. MacKenzie, R. A., Timms, P. L.: J. C. S. Chem. Comm. *1974*, 650
58. Francis, C. G., Timms, P. L.: J. C. S. Chem. Comm. *1977*, 446
59. Green, M. L. H., Young, D.: Phys. Bull. *1975*, 350
60. Anthony, M. T., Green, M. L. H., Young, D.: J. C. S. Dalton *1975*, 1419
61. Busby, R., Klotzbücher, W., Ozin, G. A.: Inorg. Chem. *16*, 822 (1977)
62. Bassler, J. M., Timms, P. L., Margrave, J. L.: Inorg. Chem. *5*, 729 (1966); Timms, P. L., Stump, D. D., Kent, R. A., Margrave, J. L.: J. Amer. Chem. Soc. *88*, 940 (1966)
63. Pimentel, G. C.: Radical Formation in the Solid Phase. In: Ref. 35
64. Haller, L.: Dissertation, Univ. California, Berkeley 1960
65. Andrews, L., Pimentel, G. C.: J. Chem. Phys. *49*, 5156 (1968)
66. Klotzbücher, W. E., Mitchell, S. A., Ozin, G. A.: Inorg. Chem. *16*, 3063 (1977)
67. Ozin, G. A.: Appl. Spect. *30*, 573 (1976)
68. McIntosh, D., Ozin, G. A.: J. Amer. Chem. Soc. *98*, 3167 (1976)
69. Klabunde, K. J., Low, J. Y. F.: J. Amer. Chem. Soc. *96*, 7675 (1974)
70. Timms, P. L.: Chem. Comm. *1969*, 1033
71. Leach, G.: B. Sc. Thesis, Univ. Bristol 1974
72. Williams-Smith, D. L., Wolf, L. R., Skell, P. S.: J. Amer. Chem. Soc. *94*, 4042 (1972)
73. Francis, C.: B. Sc. Thesis, Univ. Bristol 1974
74. Hill, A. H., Timms, P. L.: unpublished, reported in Ref. 76
75. MacKenzie, R. E., Timms, P. L.: unpublished, reported in Ref. 76
76. Timms, P. L., Turney, T. W.: Adv. Organometal. Chem. *15*, 53 (1977)
77. Cable, R. A., Green, M., MacKenzie, R. E., Timms, P. L., Turney, T. W.: J. C. S. Chem. Comm. *1976*, 270
78. English, A. D., Jesson, J. P., Tolman, C. A.: Inorg. Chem. *15*, 1730 (1976)
79. Klabunde, K. J.: Angew. Chem., Int. Ed. *14*, 287 (1975)
80. Klabunde, K. J., Efner, H. F., Murdock, T. O., Ropple, R.: J. Amer. Chem. Soc. *98*, 1021 (1976)
81. Murdock, T. O., Klabunde, K. J.: J. Org. Chem. *41*, 1076 (1976)
82. Report on work by Klabunde, K. J. In: Metal Atom Research Spawns New Catalysts. Chem. Eng. News, January 24, p. 23 (1977)

# 4 The Reactions of Metal Atoms

## 4.1 Types of Reaction

Metal atoms are the ultimate coordinately and electronically unsaturated neutral species and owe much of their high reactivity to this unsaturation and the ready availability of their valence electrons and orbitals. Not surprisingly, their reactions – certainly their initial reactions – are ones which increase the atoms' electron and coordination numbers.

The most numerous and almost certainly the first process in a metal atom-reactive substrate reaction is *addition*, the bonding of a ligand to a metal moiety through electron rearrangement. This can range from the donor-acceptor bonds of e.g. the Dewar-Chatt-Duncanson type[1)] (Fig. 4.1) formed by Lewis base or 2n-electron ligands such as $C_2H_4$, CO, Bd or PhH to interactions with more electron-demanding or defficient ligands such as $O_2$ or perfluoroalkenes where electron rearrangement is much higher leading to ligand rehybridisation and a large net drift of electron charge away from the metal moiety.

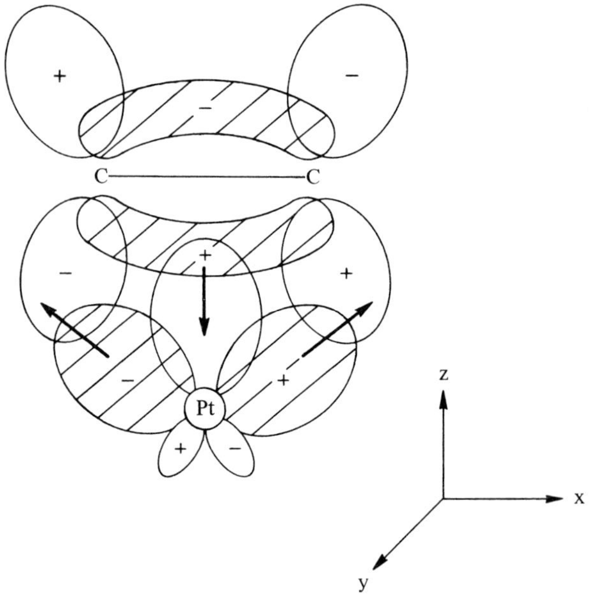


Fig. 4.1.

$$Re(a) + CO \xrightarrow{i} Re(CO)_5 \qquad (4.1)$$ [2]

$$W(a) + 3Bd \xrightarrow{ii,\ iv} (\eta^4 - Bd)_3W \qquad (4.2)$$ [3]

$$V(a) + 3PhH \xrightarrow{ii,\ iv} (\eta^6 - PhH)_2V \qquad (4.3)$$ [4]

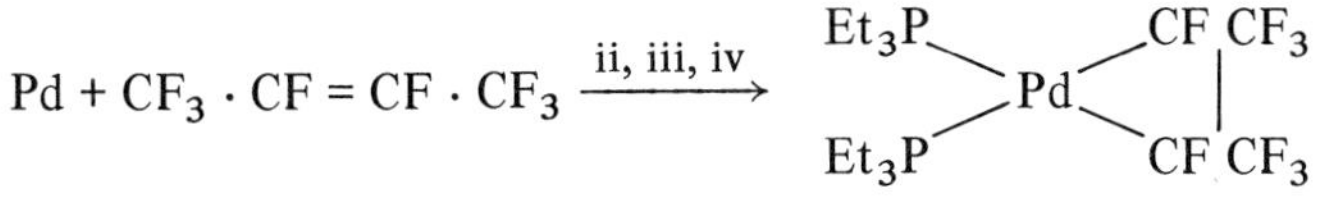


(4.4) [5]

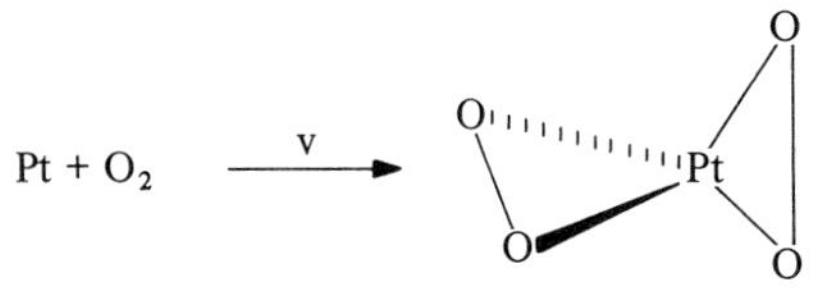


(4.5) [6]

(i) cocondense 14K; (ii) cocondense,77K;
(iii) $PEt_3$, 77K; (iv) warm; (v) cocondense 10K

Traditionally, the organometallic chemist divides addition into two areas exemplified by the two extremes, *simple addition* and *oxidative addition*, but as expected there are many "intermediate" cases. In addition a ligand can belong to either area depending on the electron-richness or availability of the metal moiety. $O_2$ adds simply to an Au atom[7] which has a first IP of 9.22 but oxidatively to a Ni atom[6] with a first IP of 7.63 eV.

After initial addition, further reaction may occur such as insertion into a substrate bond, *electron transfer* from metal to substrate, or *radical* formation (Eqs. (4.6) to (4.12)).

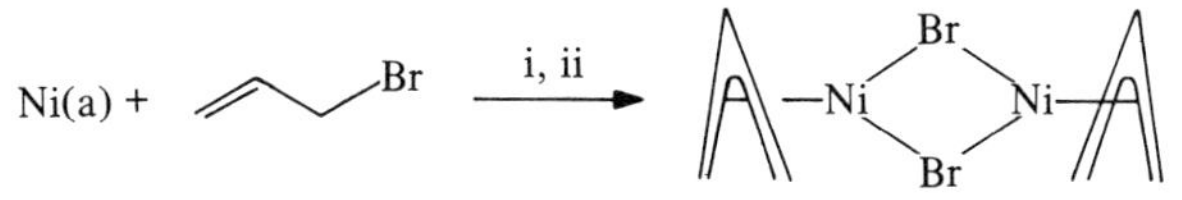


(4.6) [8]

$$Ca(a) + C_6F_6 \xrightarrow{i,\ iii} C_6F_5CaF \qquad (4.7)$$ [9]

$$Mo(a) + CpH \xrightarrow{i,\ ii} Cp_2MoH_2 \qquad (4.8)$$ [10,11]

$$Ge(a) + Me_3SiH \xrightarrow{i,\ ii} (Me_3Si)_2GeH_2 \qquad (4.9)$$ [12]

$$M(a) + TCNQ \xrightarrow{i} M^{2+}(TCNQ)^{2-} \qquad (4.10)$$ [392]

$$Na(a) + Cl_2 \xrightarrow{iv} Na^+Cl_2^- \qquad (4.11)$$ [13]

$$Au(a) + EtBr \xrightarrow{i,\ v} [Et^{\cdot} + AuBr] \xrightarrow{ii} C_4H_{10} + C_2H_4 + C_2H_6 \qquad (4.12)$$ [14]

(i) cocondense, 77 K; (ii) 25 °C; (iii) – 150 °C; (iv) cocondense, xs Ar, 15 K; (v) warm

Complexes may then undergo other reactions such as *deinsertion* or *ligand transfer* to other molecules (Eqs. (4.13) to (4.15)).

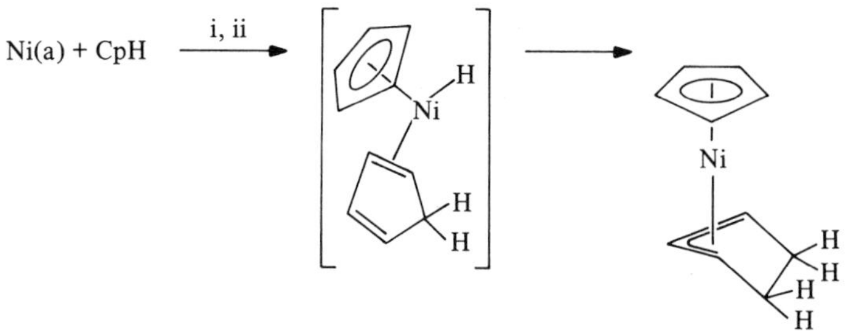


(4.13)[381]

$Pd(a) + C_6F_5COCl \xrightarrow{i, ii} [C_6F_5COPdCl]$ (4.14)[380]

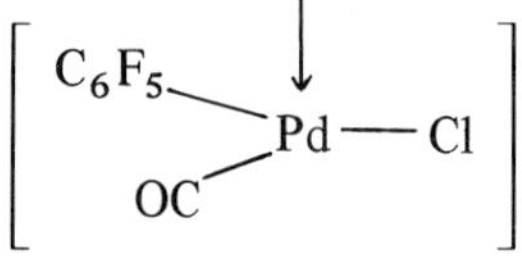


$\longrightarrow C_6F_5PdCl + CO$

$Co(a) + C_6F_5Br \xrightarrow{i, ii} (C_6F_5)_2Co + CoBr_2$ (4.15)[15]

(i) cocondense, 77 K; (ii) warm

It must be stressed that these labels are formal expressions and may not reflect the true mechanism. For example the formal insertion of singlet Mg atoms into the C-Hal bond of an organic halide may occur *via* for example insertion, $S_N2$ attack, free radicals, a tight radical pair or none of these. The actual mechanism is not yet known, but more information may be forthcoming from the use of optically active halides[16,389].

The reactions of metal atoms are now surveyed according to ligand.

## 4.2 Carbon Monoxide

CO does not freeze till 66 K and in any case has too high a vapour pressure to be used in cocondensations at 77 K. However, this problem can be circumvented by "gettering" CO with thf; i.e. CO is condensed with a large excess of thf to trap the CO in a thf matrix[17,18]. When metal atoms are cocondensed with gettered CO, metal carbonyl complexes can be prepared but in disappointingly low yield (Eqs. (4.16) to (4.18)). $Fe_3(CO)_{12}$ and $Fe_2(CO)_9$ are most likely formed *via* the intermediacy of the known reactive intermediate (thf)$Fe(CO)_4$

$Cr(a) + CO + thf \xrightarrow{i, ii} Cr(CO)_6$ (2.3%) (4.16)

$Mn(a) + CO + thf \xrightarrow{i, ii} Mn_2(CO)_{10}$ (1.0%) (4.17)

$Fe(a) + CO + thf \xrightarrow{i, ii} Fe(CO)_5$ (0.4%) + $Fe_2(CO)_9$(tr) + $Fe_3(CO)_{12}$ (0.2%) (4.18)

(i) cocondense, 77 K; (ii) warm

The formation of $Pd(CO)_4$ is infered from the reaction of Pd atoms condensed at 77 K in the presence of 1 torr pressure of CO, as addition of $PF_3$ at the end of the cocondensation results in isolation of $Pd(PF_3)_4$[19].

CO is more amenable to reactions with metal atoms in the range 4–20 K and has been a favourite ligand of the matrix isolation spectroscopist. It reacts with a wide variety of atoms at these temperatures, and usually in a simple, straightforward way (see Table 4.3). In addition the ready availability of isotopically substituted CO molecules ($^{12}C^{16}O$, $^{13}C^{16}O$ and $^{12}C^{18}O$) has, along with annealing, proved a powerful tool in determining reaction stoichiometries such as identifying CuCO, $Cu(CO)_2$ and $Cu(CO)_3$ formed in the reaction mixture of Cu atoms and CO in Ar matrices at 10 K[20,377]. This example will be discussed at length to show the way in which such techniques are used to identify species and the way in which experiments can be designed to produce such species.

Cocondensation of Cu atoms with $^{12}C^{16}O$ ($1:10^4$) at 10–15 K forms a deep purple matrix which gives an ir spectrum (Fig 4.2 (a)) showing two strong absorbances at 1,990.0 and 1,976.8 $cm^{-1}$ which are shown to be due to the same absorbing species (termed 3) by the invariability of their intensity ratio over a wide range of Cu matrix concentrations.

Condensation of Cu atoms with $^{12}C^{16}O$ and Ar as diluent ($1:125:10^4$) shows new lines in addition to those of 3 (Fig 4.2 (b)). Annealing the matrix to 35 K shows that they belong to two species (1 and 2) as the line due to 1 decreases while that due to 2 (and 3) increases (Fig. 4.2 (c)).

At very low CO concentrations, only 1 is observed (Fig. 4.2 (d)).

If an equimolar mixture of $^{12}C^{16}O$ and $^{13}C^{16}O$, with Ar diluent is condensed with Cu atoms at a concentration where 3 is not formed ($Cu:CO:Ar = 1:10^2:10^4$) 1 is split into a doublet (intensity 1:1) and 2 into a triplet (intensity *ca.* 1:2:1) as confirmed by annealing (Fig. 4.2 (f)). This immediately suggests that 1 is the monocarbonyl $Cu_xCO_1$ the doublet being due to $Cu_x(^{12}C^{16}O)$ and $Cu_x(^{13}C^{16}O)$ while 2 is the dicarbonyl $Cu_y(CO)_2$, the triplet being due to $Cu_y(^{12}C^{16}O)_2$, $Cu_y(^{12}C^{16}O)(^{13}C^{16}O)$ and $Cu_y(^{13}C^{16}O)_2$. The weak features 2′ which always appear just below a line of 2 are assigned to a multiple trapping site of 2 as they too produce an isotope pattern typical of a dicarbonyl but with all lines shifted to slightly lower frequencies than those of 2.

The identification of 1 is reinforced by Fig. 4.2 (g) and (h) which show isotope patterns of $Cu_x(^{12}C^{16}O)$ and $Cu_x(^{13}C^{16}O)$, and of $Cu_x(^{12}C^{16}O)$, $Cu_x(^{13}C^{16}O)$ and $Cu_x(^{12}C^{18}O)$ respectively.

On annealing to 30 K the matrix formed in Eq. (e) the lines due to 1 decrease, those due to 2 and 2′ stay roughly constant and new lines appear which can be analysed into each line of the original doublet in Fig. 4.2 (a) giving rise to an isotope pattern characteristic of a triangular planar tricarbonyl $Cu_z(CO)_3$ with two slightly different matrix sites in Ar (3 and 3′). Not all 8 lines for each site are observed; some accidentally overlap, some are ir inactive and some are hidden by $^{12}C^{13}O$ and $^{13}C^{16}O$ lines at *ca.* $2{,}100^{-1}$ cm.

When esr spectra are taken of a matrix at 20 K containing mainly $Cu_z(CO)_3$ ($Cu:CO:Ar = 1:500:10^4$) the resulting spectra show that the molecule has axial symmetry, $S = 1/2$, g values close to the free electron value ($g_1 = 2{,}006$, $g_2 = 2.010$) and hyperfine splitting for two Cu isotopes $^{63}Cu$ ($I = 3/2$, natural abundance 69%) and $^{65}Cu$ ($I = 3/2$, natural abundance 31%). These data are best interpreted as being due to mononuclear $Cu(CO)_3$.

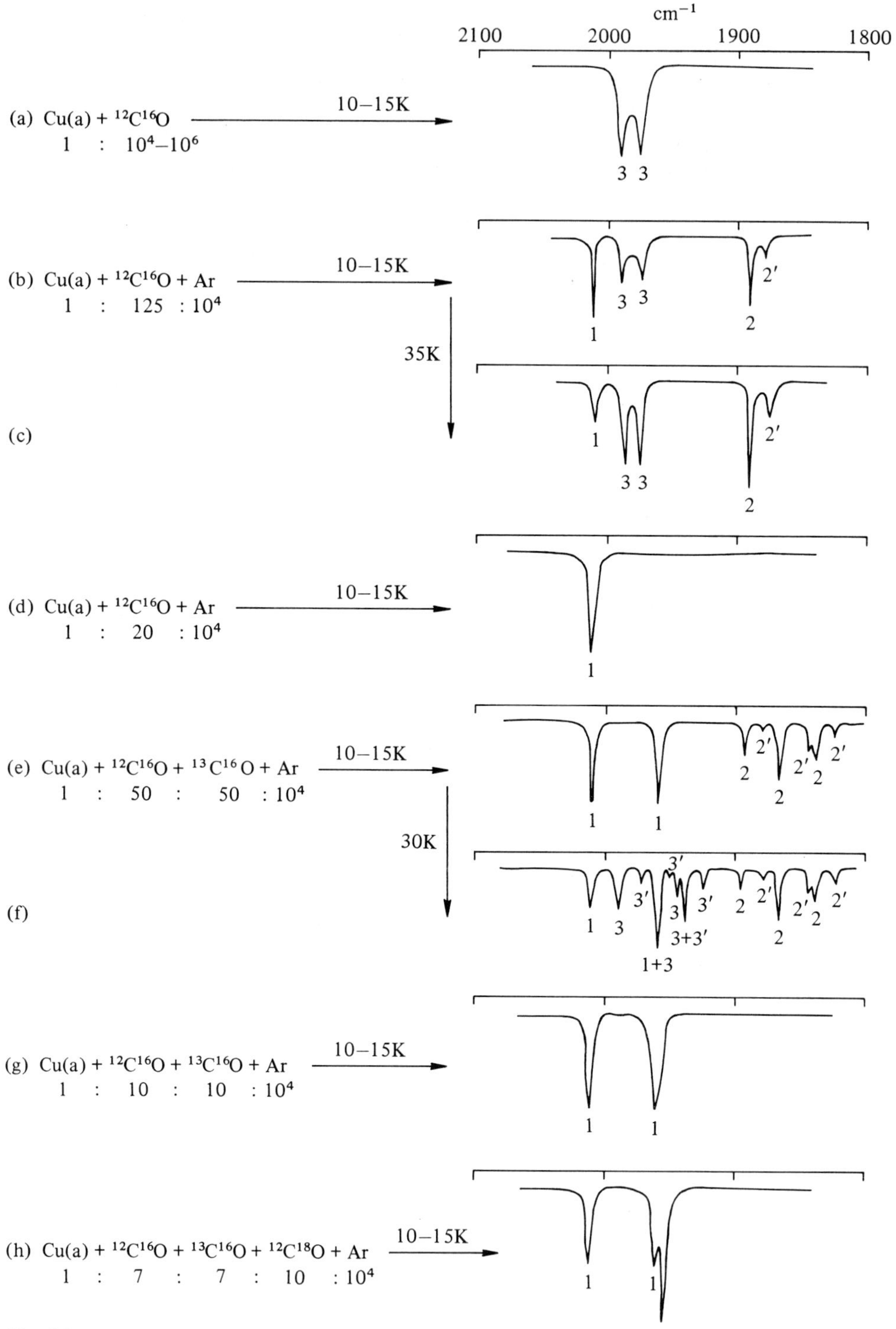
cm$^{-1}$
2100
2000
1900
1800
(a) Cu(a) + $^{12}C^{16}O$
1 : $10^4$–$10^6$
10–15K
3 3
(b) Cu(a) + $^{12}C^{16}O$ + Ar
1 : 125 : $10^4$
10–15K
35K
1
3 3
2
2′
(c)
1
3 3
2
2′
(d) Cu(a) + $^{12}C^{16}O$ + Ar
1 : 20 : $10^4$
10–15K
1
(e) Cu(a) + $^{12}C^{16}O$ + $^{13}C^{16}O$ + Ar
1 : 50 : 50 : $10^4$
10–15K
30K
1 1
2 2′ 2 2′ 2 2′
(f)
1 3 3′ 1+3 3′ 3 3+3′ 3′
2 2′ 2 2′ 2 2′
(g) Cu(a) + $^{12}C^{16}O$ + $^{13}C^{16}O$ + Ar
1 : 10 : 10 : $10^4$
10–15K
1 1
(h) Cu(a) + $^{12}C^{16}O$ + $^{13}C^{16}O$ + $^{12}C^{18}O$ + Ar
1 : 7 : 7 : 10 : $10^4$
10–15K
1 1 1

**Fig. 4.2.**

Support for this comes from the Ag/CO/Ar system. Firstly, when Ag atoms are condensed with CO at 6–15 K, a green matrix is obtained which has similar ir spectra to that of $Cu(^{12}C^{16}O)_3$ in $^{12}C^{16}O$ and to that of $Cu(^{12}C^{16}O)_n(^{13}C^{16}O)_{3-n}$ (n = 0 to 3) on using $^{12}C^{16}O/^{13}C^{16}O$ mixtures, which suggests the matrix contains $Ag(CO)_3$ analogous to $Cu(CO)_3$. Esr and other measurements support this[379)]. When approximately equimolar concentrations of Cu and Ag atoms are simultaneously cocondensed with pure CO at low metal concentrations, the resultant ir spectra are simply the superposition of those of pure $Cu(CO)_3$ and $Ag(CO)_3$. Additional lines that may be due to molecules containing both Cu and Ag atoms were not seen at any time after deposition, even on warming the matrix, supporting the monomeric nature of $Cu(CO)_3$ (and $Ag(CO)_3$).

An Eq. (4.19) has been developed[21,22,378)] to relate oligonuclear complexes such as $M_mL_p$ ($m \geqslant 1$).

$$\frac{\sum^{p}[M_mL_p]}{\sum^{q}[M_nL_q]} \propto \Big[M\Big]_0^{m-n} \tag{4.19}$$

Where $\sum^{p}[M_mL_p]$ is the sum of the concentrations of all complexes containing mM atoms, $[M]_0$ is the total concentration of M, and at concentrations of M where $\sum^{p}[M_mL_p] < \sum^{q}[M_nL_q]$ for $m > n$. Thus, the ratio of dinuclear to mononuclear species concentrations should be $[M]_0$ for example. However, it is found that the ratios $A[Cu_xCO]/A[Cu(CO)_3]$ and $A[Cu_y(CO)_2]/A[Cu(CO)_3]$ (where A[X] is the ir absorbance of a species X) are both independent of $[Cu]_0$ over a 300-fold change in $[Cu]_0$, Thus x = y = 1. Since CuCO, $Cu(CO)_2$ and $Cu(CO)_3$ are all mononuclear, Eq. (4.19) is not strictly true, but an analysis of the likely kinetics of a matrix where mononuclear carbonyls predominate indicates that the ratio $A[CuCO] : A[Cu(CO)_2] : A[Cu(CO)_3]$ is independent of $[Cu]_0$ which is of course the case.

Frequency calculations based on a Cotton-Kraihanzel force-field approximation[64)] support these assignments. The results of such an analysis for isotopic $Cu(^{12}C^{16}O)_n(^{13}C^{16}O)_{3-n}$ and $Cu(^{12}C^{16}O)_m(^{13}C^{16}O)_{2-m}$ (where n = 0 to 3, m = 0 to 2) are shown in Tables 4.1 and 4.2 (the CuCO analysis is trivial). The agreement between the observed and calculated frequencies for the $Cu(CO)_3$ and $Cu(CO)_2$ species in both their matrix sites is good, providing convincing support for $D_{3h}$ $Cu(CO)_3$ and $D_{\infty h}$ $Cu(CO)_2$ assignments.

When a Cu/CO/Ar matrix containing mainly $Cu(CO)_3$ is warmed to temperatures at which the matrix is beginning to break up and boil away, the two lines at 1,985 and 1,940.4 $cm^{-1}$ assigned to $Cu(CO)_3$ decrease in intensity rapidly while at the same time two new lines at 2,039 and 2,003 $cm^{-1}$ appear and continue to grow till the lines assigned to $Cu(CO)_3$ disappear. The two new lines conserve their relative intensity and are assigned to the same species D. Similarly, in a very dilute Cu/CO matrix ($1 : \geqslant 10^4$) at 15 K only the two lines (termed 3) due to $Cu(CO)_3$ are observed, but the ir spectra of increasingly concentrated Cu/CO matrices (as illustrated in Fig. 4.3) show the disappearance of these lines and the appearance and growth of lines due to D till in concentrated matrices ($1 : 10^2$), the two mononuclear lines are hardly observable. A graph of $A[D]/A[Cu(CO)_3]$ as a function of $[Cu]_0$ shows that this ratio is linearly dependent on $[Cu]_0$, thus D is dinuclear. The

**Table 4.1.** Observed and calculated frequencies of isotopic $Cu(^{12}C^{16}O)_n(^{13}C^{16}O)_{3-n}$ (where n = 0–3) in Ar matrices

| Site 3 | ($cm^{-1}$) | Site 3′ | ($cm^{-1}$) | Assignment | |
|---|---|---|---|---|---|
| Obs | Calc[a] | Obs | Calc[a] | | |
| 2,079.0 | 2,080.7 | 2,079.0 | 2,084.5 | $Cu(^{12}C^{16}O)_2(^{13}C^{16}O)$ | ($A_1$) |
| 2,067.0 | 2,066.3 | 2,067.0 | 2,069.9 | $Cu(^{12}C^{16}O)(^{13}C^{16}O)_2$ | ($A_1$) |
| 1,985.0 | 1,984.9 | 1,975.8 | 1,975.6 | $Cu(^{12}C^{16}O)_3$ (E) + $Cu(^{12}C^{16}O)_2(^{13}C^{16}O)$ | ($B_2$) |
| 1,966.0 | 1,965.5 | 1,956.8 | 1,956.8 | $Cu(^{12}C^{16}O)(^{13}C^{16}O)_2$ | ($A_1$) |
| 1,951.5 | 1,951.9 | –[b] | 1,943.0 | $Cu(^{12}C^{16}O)_2(^{13}C^{16}O)$ | ($A_1$) |
| 1,940.5 | 1,940.8 | 1,931.5 | 1,931.7 | $Cu(^{13}C^{16}O)_3$ (E′) + $Cu(^{12}C^{16}O)(^{13}C^{16}O)_2$ | ($B_2$) |
| | 2,091.9 | | 2,096.8 | $Cu(^{12}C^{16}O)_3$ | ($A'_1$) (Raman) |
| | 2,045.5 | | 2,050.2 | $Cu(^{13}C^{16}O)_3$ | ($A'_1$) (Raman) |

[a] The best fit CO force constants were: site 3 $K_{CO}$ = 16.51, $K_{CO\text{-}CO}$ = 0.59 mdyne/Å; site 3′ $K_{CO}$ = 16.45, $K_{CO\text{-}CO}$ = 0.66 mdyne/Å.
[b] Hidden by intense 1940.5 line of $Cu(CO)_3$ in site 3.

**Table 4.2.** Observed and calculated frequencies of $Cu(^{12}C^{16}O)_n(^{13}C^{16}O)_{n-2}$ (where n = 0–2) in Ar matrices

| Site 2[a] | ($cm^{-1}$) | Site 2′ | ($cm^{-1}$) | Assignment | |
|---|---|---|---|---|---|
| Obs | Calc[b] | Obs | Calc[b] | | |
| –[c] | 1,995.3 | –[c] | 1,994.5 | $Cu(^{12}C^{16}O)(^{13}C^{16}O)$ | ($A_1$) |
| 1,891.5 | 1,890.8 | 1,876.1 | 1,875.2 | $Cu(^{12}C^{16}O)_2$ | ($\Sigma_u^+$) |
| 1,866.0 | 1,866.0 | 1,851.0 | 1,851.0 | $Cu(^{12}C^{16}O)(^{13}C^{16}O)$ | ($A_1$) |
| 1,848.0 | 1,848.0 | 1,832.5 | 1,833.5 | $Cu(^{13}C^{16}O)_2$ | ($\Sigma_u^+$) |

[a] Intensity calculations for isotopic $Cu(^{12}C^{16}O)_n(^{13}C^{16}O)_{2-n}$ (where n = 0–2) show that $I_{1995.3}:I_{1890.8}:I_{1866.0}:I_{1848.8} = 3\cdot3:52\cdot8:100:50\cdot5$. These data support the contention that the non-observation of the out-of-phase CO stretching mode of $Cu(^{12}C^{16}O)(^{13}C^{16}O)$ relates to its low intensity compared to the respective in-phase mode.
[b] The best fit CO force constants were: site 2 $K_{CO}$ = 15.41, $K_{CO\text{-}CO}$ = 0.97 mdyne/Å; site 2′, $K_{CO}$ = 15.29, $K_{CO\text{-}CO}$ = 0.97 mdyne/Å.
[c] Not observed because of low absorbance.

most likely assignment of D is $Cu_2(CO)_6$, i.e. the dimerisation product of $Cu(CO)_3$ as it is reasonable to assume that as a matrix containing $Cu(CO)_3$ begins to break up, the bulky $Cu(CO)_3$ molecules gain some mobility and encounter other $Cu(CO)_3$ molecules to form $Cu_2(CO)_6$. Also it is normal that species in pure substrate matrices have the highest possible stoichiometry, which for a Cu dinuclear species would be $Cu_2(CO)_6$ if the 18e-rule[23,24] is to be obeyed. A simple Cu-Cu bond is suggested as no CO bridging modes are detected.

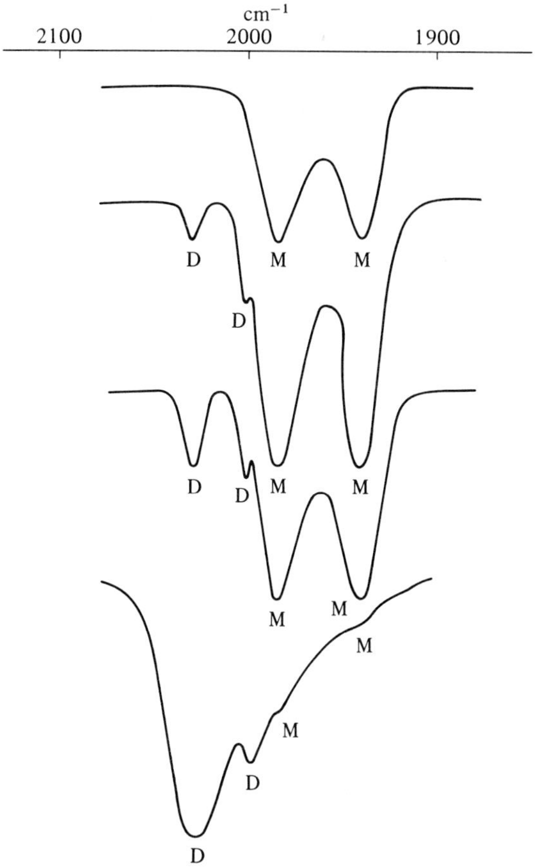


Fig. 4.3.

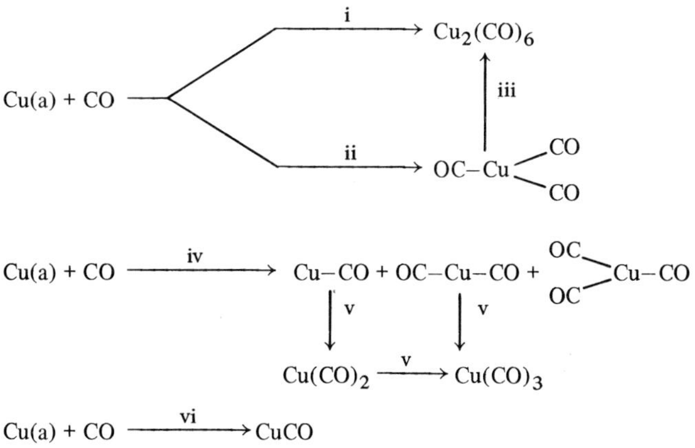


**Scheme 4.1.** (i) Cocondense, 15 K, Cu : CO = $1:10^2$; (ii) cocondense, 10–15 K, Cu : CO : Ar = $1:500:10^4$; (iii) heat till matrix begins to break up; (iv) cocondense, 10–15 K, Cu : CO : Ar = $1:125:10^4$; (v) 35 K; (vi) cocondense, 10–15 K, Cu : CO : Ar = $1:20:10^4$

The presence of the two CO stretching modes for $Cu_2(CO)_6$ in CO as well as in Ar is consistent with a $D_3$ rotameric configuration rather than the more highly symmetrical $D_{3h}$ or $D_{3d}$ configurations.

The matrix reactions of Cu atoms and CO can now be summarised in Scheme 4.1.

The chemistry of Cu atoms in CO and CO/Ar matrices nicely highlights many of the generalisations which may be made about metal atom chemistry in reactive matrices at very low temperatures.

(i) Species of highest stoichiometry (i.e. $ML_n$, $n = n_{max}$, e.g. $Ni(CO)_4$) are made in pure substrate or concentrated substrate/inert diluent matrices.

(ii) Less concentrated substrate/inert diluent matrices allow formation of the series of coordinatively unsaturated species (i.e., $ML_n$, $n \leqslant n_{max}$, e.g. $Ni(CO)_m$, $m = 1-4$).

(iii) Increased metal concentration leads to dinuclear, trinuclear and higher species.

(iv) Very low metal concentrations give only mononuclear species.

(v) Annealing allows some diffusion and therefore allows the increase of m and n in $M_mL_n$ by reaction with the mobile species M and L i.e., Eq. (4.20).

$$M_mL_n \begin{cases} \xrightarrow{\;M\;} M_{(m+1)}L_n \\ \xrightarrow{\;L\;} M_mL_{(n+1)} \end{cases} \qquad (4.20)$$

(vi) Highly rigid and polarizable inert diluent matrix gases such as Kr and Xe minimise matrix diffusion and the reactions shown in (v). Conversely, less rigid matrix gases such as Ar allow diffusion.

(vii) Light atoms tend to diffuse more than heavy atoms, e.g., Mn atoms more than Re atoms.

(viii) More complex species e.g., $ML_n$ tend to be immobile except near the break up point of the matrix.

(ix) Ir spectroscopy, including the use of isotopically substituted ligands and frequency and intensity calculations, Raman spectroscopy including depolarisation experiments, uv-visible and esr spectroscopy (in conjunction with Hückel molecular orbital calculations) are powerful tools in defining the stoichiometry of reactions.

(x) The variation of intensity ratios with metal concentration can help determine the number of metal atoms in a complex.

The reactions of metal atoms with CO in pure and diluted matrices follow this pattern and are listed in Table 4.3. Only the mononuclear species of maximum stoichiometry are given (except where anomalies occur) though it is usual that the full series $M(CO)_n$ ($n = 1$ to $n_{max}$) has been identified. Those polynuclear species which have been characterised are also included.

Many of the compounds, even those of maximum stoichiometry are too unstable outside a rigid low temperature matrix to be of any synthetic value, though obviously they are of great spectroscopic interest. They have lead to a better understanding of M-CO bonding and $M(CO)_n$ fragments[57,58] and may provide useful models to explain more complicated or less accessible systems such as chemisorption. There has been some controversy over the interpretation of the ir spectra of CO chemisorbed on transition metal surfaces, $M(CO)_{ads}$. One view[59] is that the occurance of low- and high-frequency CO stretching modes ($\nu_{CO}$) in such spectra (see Fig. 4.4) is due to there being both terminal and bridging

**Table 4.3.** Binary carbonyls formed in low temperature matrices[a]

| Metal | Species | Refs. |
|---|---|---|
| Ti | $Ti(CO)_6$ | 375) |
| V | $V(CO)_6$: $V(CO)_2$ has 3 apparent isomers<br>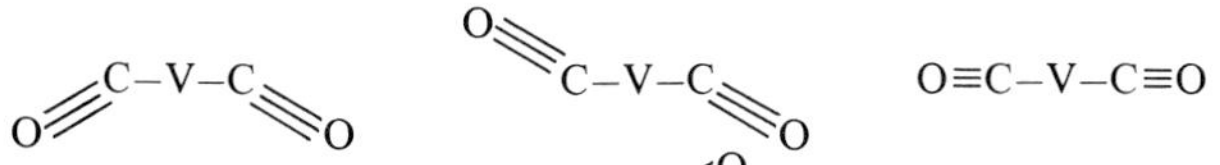<br>VCO anomalous 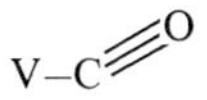<br>$(CO)_5V(\mu\text{-}CO)_2C(CO)_5$, $(CO)_6V{-}V(CO)_6$ | 20, 25–29) |
| Cr | $Cr(CO)_6$: OCCr–CrCO: $(CO)_5Cr\cdots Cr(CO)_5$? | 22, 29, 378) |
| Mn | $Mn(CO)_5$: $Mn(\mu\text{-}CO)Mn$: Mn–Mn–CO:<br>$Mn(\mu\text{-}CO)_2Mn$: $(CO)_5Mn{-}Mn(CO)_5$ | 22, 29, 378) |
| Fe | $Fe(CO)_5$ | 20, 29) |
| Co | $Co(CO)_4$: $(CO)_4Co{-}Co(CO)_4$: $(CO)_3Co(\mu\text{-}CO)_2Co(CO)_3$ | 21, 29–32, 378) |
| Rh, Ir | $M(CO)_4$: $(CO)_3M(\mu\text{-}CO)_2M(CO)_3$: $M_4(CO)_{12}$: $M_6(CO)_{16}$ | 21, 378) |
| Ni | $Ni(CO)_4$: $Ni(\mu\text{-}CO)Ni$: OCNi–NiCO:<br>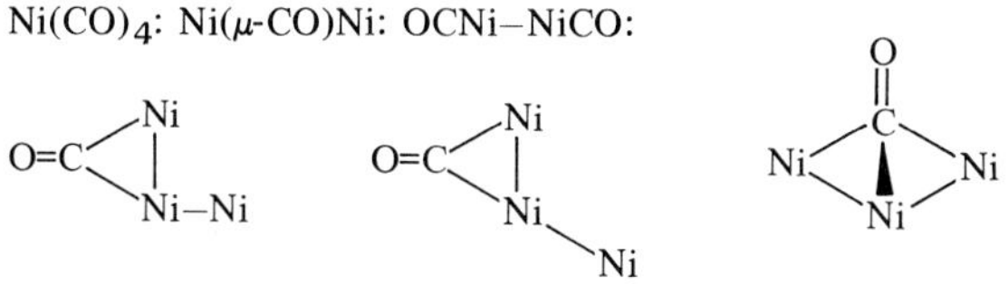<br>$(CO)_2Ni(\mu\text{-}CO)_3Ni(CO)_2$: $Ni(CO)_n(N_2)_{4-n}$ (n = 1–3) | 19, 29, 32–35, 378) |
| Pd, Pt | $M(CO)_4$ | 19, 31, 34, 36–40) |
| Cu | $Cu(CO)_3$: $Cu_2(CO)_6$: $Cu_2CO$: $Cu_3CO$: $Cu_4CO$ | 20, 29, 41, 42, 377–379) |
| Ag | $Ag(CO)_3$, $XeAg(CO)_2$ ? : $Ag_2(CO)_6$ | 41, 43, 377–379) |
| Au | $Au(CO)_2$: Au(CO)(OC) | 29, 44, 46, 55, 377–379) |
| Ln | $Ln(CO)_6$, Ln = Pr, Eu, Gd, Ho, Nd, Yb | 47–49) |
| U | $U(CO)_6$ | 49, 50) |
| Li | LiCO: LiOC: LiOCCO: LiCOOC: $Li_2OCCO$ | 51) |
| Al, Ga | $M_x(CO)_2$ X = 2? | 52, 53) |
| Si | OCSiCO: SiCO | 55) |
| Sn, Pb | $Sn(CO)_y$ | 56) |

[a] Only the species of highest known stoichiometry shown for mononuclear species except for anomalies.

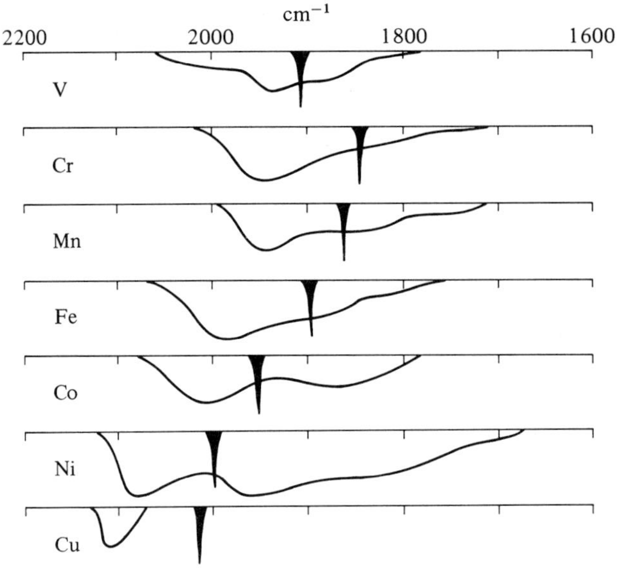


**Fig. 4.4.** A superposition of the spectra reported for CO chemisorbed on metal particles with the frequencies reported for matrix-isolated analogous monometalcarbonyls (shaded). Taken from reference[29]. Note that the frequency of chemisorbed CO can vary according to the form of the metal (powder, film, monocrystal) and the nature of any support[60]

M-CO bonds on the metal surface. However, it has also been argued[61] that they are due to terminally bonded CO on different types of metal site, *viz* corner, edge and mid-plane. The results of extended Hückel calculations for CO on a nine-atom model of a nickel surface[62] show that the degree of $\pi$-overlap between metal d-orbitals and the $\pi^*$-orbitals of terminal CO increases and thus $\nu_{CO}$ decreases in going from central to corner sites. This is a reasonable result as corner atoms have less metal near-neighbours with which to share their d electrons and thus will donate more electron charge to the $\pi(d_M - \pi^*_{CO})$ bond, weakening the $\pi(p_C - p_O)$ bond. Therefore the discrete molecule MCO, with no metal near neighbours should weaken $\nu_{CO}$ more than any of the different types of M-CO surface sites. Figure 4.4 a superposition of the ir spectra of $M(CO)_{ads}$[59,61] and of discrete MCO molecules shows that this is not so. Apart from that of Cu, the spectra of $M(CO)_{ads}$ of the first transition series from V to Ni shows an envelope of bands below $\nu_{CO}$ for MCO which implies that they are due to bridging modes. This suggests that although the thesis of different metal surface sites bonding to CO may still be true, two-centre and multi-centre bonding also takes place.

Some further chemistry of matrix isolated metal carbonyls has been observed. It can be shown in some cases that these species follow the rule[63] that warming causes their condensation with release of CO to form clusters and eventually metal lattices. A striking example is the behaviour of rhodium and iridium carbonyls on warming their matrices from 15 K to room temperature[21] (Eq. (4.21)).

$$M(a) + CO \xrightarrow{i} M(CO)_4 \xrightarrow{ii} M_2(CO)_8 \xrightarrow[\downarrow CO]{iii} M_4(CO)_{12} \xrightarrow[\downarrow CO]{iv} M_6(CO)_{16} \qquad (4.21)$$

(i) cocondense, 15 K; (ii) 50 K, CO slowly boiled away to leave solid $M_2(CO)_8$; (iii) –48 °C (M = Rh), –58 °C (Ir); (iv) room temperature (M = Rh), (M = Ir but more slowly)

Further reaction may be induced by the presence of a second substrate. CO and $O_2$ have been condensed with Ag and Au atoms in an attempt to prepare species that could be used as localised bonding models for the metal-catalysed oxidation of CO to $CO_2$[29,65,66] (Eqs. (4.22), (4.23)). Although the reactions of Ag atoms are relatively uninformative, those of Au atoms show formation of peroxyformate species which decompose in two steps to $CO_2$, suggesting a similar localised bonding description for the oxidation of CO on Au surfaces (Scheme 4.2).

$$Ag(a) + CO + O_2 \xrightarrow{i} [AgCO]^+(O_2)^- \xrightarrow{ii} \text{decomposition, no } CO_2, CO_3 \qquad (4.22)$$

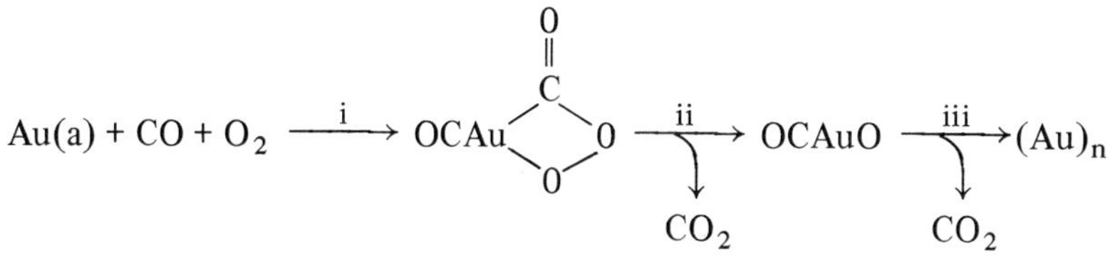
(4.23)

(i) cocondense, 10 K; (ii) 30 K; (iii) 40 K

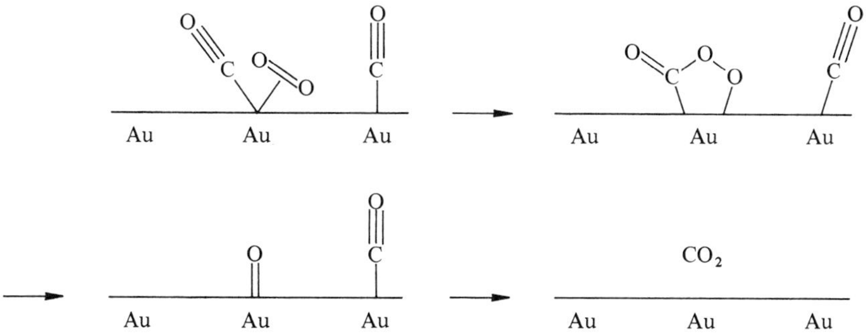

Scheme 4.2.

## 4.3 Carbon Monosulfide

Carbon monosulfide exists normally as a polymer at ambient temperatures. However, monomeric CS can be produced in an electrical discharge of $CS_2$/Ar mixtures, which if condensed at 10 K with Ni atoms forms $Ni(CS)_n$ (n = 1–4)[67]. These complexes are of interest in that the CS stretching frequencies in them are higher than that of free CS.

## 4.4 Carbon Dioxide, Carbon Oxysulfide and Carbon Disulfide

Only a few reactions have been attempted between metal atoms and $CO_2$, COS or $CS_2$, though this situation may change owing to the increased demand for CO and $CO_2$ activation[68].

In very low temperature matrices $\pi$-complexes may be formed with transition metal atoms. $CS_2$ and Ni atoms form the discrete complexes $Ni(CS_2)_n$ (n = 1–3) in matrices at 10–12 K[69]. The exact bonding in such complexes is not clear. Initial experiments show that Pd atoms behave similarly. Ni atoms form terminal-S-bonded complexes with COS[70]. An initial interpretation of the $Ni/CO_2$ matrix that it contains $CO_2$ coordinated to Ni[71] has since been shown to be wrong and in fact the resultant spectra are due to $Ni(N_2)_n$ formed from trace impurities[54]. Au atoms react with $CO_2$ in an Ar matrix to form a very weak Au-$CO_2$ coordination complex[66].

At higher temperatures, deoxygenation is an important reaction pathway. When Ni atoms and $CO_2$ are cocondensed at 77 K, CO is rapidly evolved on warming the resultant deposit to *ca.* −150 °C. $Ni(CO)_4$ can be isolated in 10% yield from the reaction mixture as can NiO[19]. Condensation of Fe or Co atoms with $CO_2$ forms no isolable organometallic product[72]. Warming the $B/CO_2$ cocondensate from 77 K to −150 °C again evolves CO but this time with emission of light. The final solid products contain $B_2O_3$, BC and BO solids. Thus some of the CO is further reduced[73].

Reaction of Na atoms with $CO_2$ or $CS_2$ forms species which are shown to be discrete ion pairs $Na^+[CO_2]^-$ and $Na^+[CS_2]^-$, where the radical ions are bent[74]. Similarly, species such as $LiCO_2$, $Li_2CO_2$ and $Li_2C_2O_4$ can be found in low temperature $Li/CO_2$ matrices[51].

## 4.5 Dinitrogen

Dinitrogen has similar bonding properties to those of isoelectronic carbon monoxide, and a growing number of transition metal complexes containing $N_2$ are known. No binary complex has been isolated at room temperature, but experiments analogous to those which form binary metal carbonyls in low temperature matrices have proved successful, and are listed in Table 4.4. Again only the mononuclear complex of highest stoichiometry is given.

The reactions roughly parallel those of CO, though the complexes are often less stable and may not reach as high a stoichiometry. If Ni atoms are cocondensed into liquid $N_2$ at 77 K and then treated with other ligands such as 1,5-Cod immediately afterwards, no nickel derivatives are isolated suggesting that $Ni(N_2)_4$, even if formed at 77 K, cannot survive at that temperature[19]. In addition $N_2$ seems to require a higher reaction temperature than does CO. Cocondensation of Ti atoms with $N_2$ ($1:10^5$) at 10–12 K leads only to monatomic titanium as shown by uv spectroscopy. Only by annealing to 30 K or depositing at 16 K can $Ti(N_2)_6$ be formed[375]. V atoms[75] and Cr atoms[78] behave similarly though the corresponding hexcarbonyl complexes are formed at 10 K or less. There is no evidence for mononuclear iron-dinitrogen complex formation at 4.2 K[80,81].

Side-on bonding of dinitrogen to a metal has been a much postulated but never proved bonding mode of $N_2$. However it has been suggested that $CoN_2$ has this structure on the basis of $^{15}N_2$, $^{15}N^{14}N$, $^{14}N_2$ isotopic substitution experiments[82].

**Table 4.4.** Binary dinitrogen complexes formed in low temperature matrices

| Metal | Species | Refs. |
|---|---|---|
| Ti | $Ti(N_2)_6$ | 375) |
| V | $V(N_2)_6$: $V_2(N_2)_x$, $V_2(N_2)_y$ (x probably 12, y probably $\leqslant$ 12) | 25, 75) |
| Nb | Very complex ir spectra (39 lines)<br>$Nb(N_2)_4$ and $NbN_2$ tentatively assigned | 76) |
| Cr | $Cr(N_2)_6$, $Cr_xN_2$ | 25, 75, 77–79) |
| Fe | $FeN_2$, $Fe_2(N_2)_x$ | 31, 80, 81) |
| Co | $Co-\begin{matrix}N\\|||\\N\end{matrix}$ | 82) |
| Rh | $Rh(N_2)_4$, $Rh_2(N_2)_x$, $Rh(N_2)_y$ | 83) |
| Ni | $Ni(N_2)_4$, $Ni(N_2)_n(CO)_{4-n}$ (n = 1–3), $Ni_x(N_2)_y$ | 26, 31, 35, 77, 79, 84–87) |
| Pd | $Pd(N_2)_3$ | 84–86) |
| Pt | $Pt(N_2)_3$, $Pt_2(N_2)_2$? , $Pt_x(N_2)_y$ (x, y $\geqslant$ 2) | 77, 85–89) |
| Cu | $Cu_x(N_2)$ | 77) |
| Li | $LiN_2$, $N_2LiN_2$ | 90) |

## 4.6 Nitrogen Oxides

Nitric oxide has a vapour pressure at 77 K of *ca.* $5 \times 10^{-3}$ torr which makes it difficult to use in cocondensation experiments at that temperature. It should be possible to getter NO but a more novel idea is to cocondense equimolar amounts of NO and $BF_3$ as they form a weak complex in the solid phase which is less volatile[91]. $BF_3$ is relatively inert towards metal atoms and should not interfere with possible reactions of NO. Fortuitously, this also eliminates an unforeseen hazard. When NO alone is condensed at 77 K with a metal vapour, the cocondensate often explodes on warming, e.g. that of Mn just above 77 K. Using $BF_3$ reduces the explosion risk but it is not clear why[91]. A possible cause of these explosions is very rapid formation of metal oxides with liberation of $N_2O$, perhaps $N_2$ and heat. Finely divided metal such as that of Co or Ni, formed after cocondensation experiments, is oxidised by NO[92]. Other metal surfaces behave similarly and the reaction is paralleled in discrete complexes.

The $PF_3$ analogues of the pseudo-nickeltetracarbonyl series $Co(CO)_3NO$, $Fe(CO)_2(NO)_2$ and $MnCO(NO)_3$ can be made in moderate yield by condensing Co, Fe and Mn atoms with mixtures of NO, $BF_3$ and $PF_3$[91] (Eqs. (4.24) to (4.26)).

$$Mn(a) + BF_3 + NO + PF_3 \xrightarrow{i,iv} Mn(NO)_3PF_3\ (5.3\%) \quad (4.24)$$

$$Fe(a) + BF_3 + NO + PF_3 \xrightarrow{ii,iv} Fe(NO)_2(PF_3)_2\ (8.3\%) + Fe(PF_3)_5 \quad (4.25)$$

$$Co(a) + BF_3 + NO + PF_3 \xrightarrow{iii,iv} CoNO(PF_3)_3\ (4.5\%) \quad (4.26)$$

(i) $BF_3$:NO:$PF_3$ = 2:2:1, 77 K; (ii) 1:1:1, 77 K; (iii) 1:1:3, 77 K; (iv) 25 °C

The new compound $Mn(NO)_3PF_3$ could not be made by conventional high pressure techniques[93]. It can be converted to $Mn(NO)_3CO$ by treatment with CO at 1 atmosphere at room temperature. If the NO pressure rises above $10^{-2}$ torr during cocondensation oxidation of Mn occurs[91].

An attempt to make the then unknown compound $Cr(NO)_4$ by cocondensing Cr atoms with NO and $BF_3$ gives a red cocondensate at low temperatures but on warming, a little $N_2$ is evolved and a red insoluble, involatile chromium-rich solid whose ir spectrum shows no NO stretching bands is formed. The chromium is apparently oxidised. $Cr(NO)_4$ can now be made by photolysis of $Cr(CO)_6$ in the presence of NO[94]. An attempt to make the plausible 18e-compound ($\eta$-PhH)$Cr(NO)_2$ from Cr atoms ended in explosion at −30 °C[91].

Cocondensation of Co atoms with either NO (with or without $BF_3$) in an attempt to make the known compound ($Co(NO)_3$ or with NO and PhH in attempt to make the unknown 18e-compound ($\eta$-PhH)CoNO gave only a black involatile cobalt-rich solid which has NO stretching bands in its ir spectrum[91]. This solid may be cobalt metal with NO chemisorbed on its surface, or perhaps even a cobalt nitrosyl cluster.

A more general route to species of the type $Fe(NO)_2L_2$ is to make use of the unstable cocondensate Fe/PhMe. For example if NO is added at 77 K and the cocondensate warmed to −70 °C, then $PPh_3$ added, $Fe(NO)_2(PPh_3)_2$ can be isolated in 10% yield. Alternatively NO may be added at low temperatures to the Fe/PhMe/$PPh_3$ cocondensate to give $Fe(NO)_2(PPh_3)_2$ in *ca.* 40% yield[95]

Condensation of alkali metal atoms with NO and a large excess of Ar at 15 K forms the discrete ion-pairs $M^+(NO)^-$ (M = Li, Na, K, Rb, Cs)[96,97]. The species $Li^+(NO)^-$ is interesting in that it shows photoisomerisation (Eq. (4.27)) while secondary reactions give $Li^+(NO)^{2-}Li^+$[96].

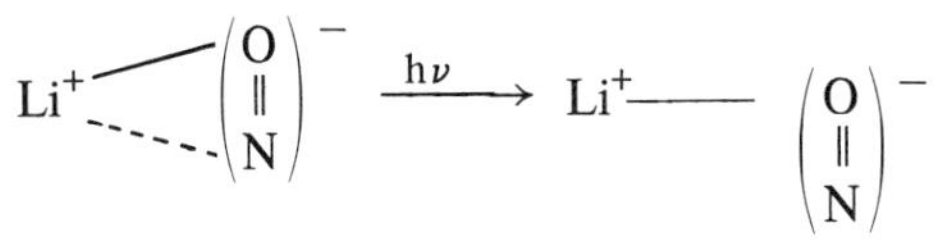

(4.27)

$N_2O$ has many similarities to the isolectronic $CO_2$, including its ability to transfer oxygen to metal atoms, presumably *via* evolution of $N_2$. As an example cocondensation of alkali metal atoms and $N_2O$ diluted by $N_2$ at 15 K produces MO (M = Li, K, Cs) and $M_2O$ (Li, K, Rb, Cs) ionic molecules[98,99]. Surprisingly, Na atoms do not react; aggregation has been suggested[99]. Li atoms, in addition to forming the regular oxide Li–O–Li, forms the isomer Li–Li–O which has either a linear or bent structure[98]. The structures of the oxides M–O–M are essentially linear except for that of Cs which is definitely bent; Cs–Cs bonding is suggested[98)].

Conversely, $NO_2$ forms nitrides N–U–N, U–N and a third, unknown nitride when condensed with U atoms in excess Ar at 14 °C[100)].

## 4.7 Phosphines and Phosphites

Trifluorophosphine has similar ligand bonding properties to CO and has a low enough vapour pressure to be cocondensed at 77 K. It reacts efficiently with transition metal atoms at 77 K to form compounds which are often analogous to those of CO. The reactions are shown in Table 4.5. This thus makes a good alternative method for the preparation of metal trifluorophosphine complexes, originally made by a high pressure process[93)].

The reaction with Ni atoms is particularly efficient. The pressure in the vacuum system during this cocondensation is always exceptionally low apparently due to gettering of the $PF_3$ by Ni atoms either in the gas phase or on the surface[101)]. Cu atoms have the same effect. Therefore it may be possible that compounds such as $Cu(PF_3)n$ (n = 1, 2 or 3) or $Cu_2(PF_3)_6$ are formed at 77 K but decompose to Cu and $PF_3$ on warming.

**Table 4.5.** Binary phosphine complexes from reaction of metal atoms

| Metal | Observations | Refs. |
|---|---|---|
| Cr | $Cr(PF_3)_6$ (65%), $(PF_3)_4Cr(PF_2)_2Cr(PF_3)_4$, $(PF_3)_4Cr(PF_2)_2Fe(PF_3)_3$ | 101, 102, 381) |
| Mo | $Mo(PF_3)_6$ (5%) | 72, 103) |
| Fe | $Fe(PF_3)_5$, $(PF_3)_3Fe(PF_2)_2Fe(PF_3)_3$, $[(PF_3)_3Fe]_3(PF)_2$ [a] $(PF_3)_3Fe(PF_2)Fe(PF_3)_3$, $(PF_3)_3Fe(PF_2)_2Cr(PF_3)_4$, $Fe[P(OMe)_3]_5$, $HFe(CH_2PMe_2)(PMe_3)_4$ | 19, 101, 102, 104, 105, 110, 381) |
| Co | $[Co(PF_3)_4]_2$ (50%), $HCo(PF_3)_4$, $Co(PMe_3)_4$ | 19, 101, 106) |
| Ni | $Ni(PF_3)_4$ (85–100%), $Ni(PF_2Cl)_4$ (32%), $Ni(PF_3)_3PH_3$, $Ni(PF_3)_2(PH_3)_2$, $Ni(PH_3)_n$ (n = 1–4), $Ni(PMe_3)_4$, $Ni(PPh_3)_4$, $(diphos)_2Ni$ | 19, 101, 106, 107, 108, 371, 381) |
| Pd | $Pd(PF_3)_4$ (70%), $Pd(PMe_3)_4$, $Pd(PPh_3)_4$ | 101, 105) |

[a] Structure:

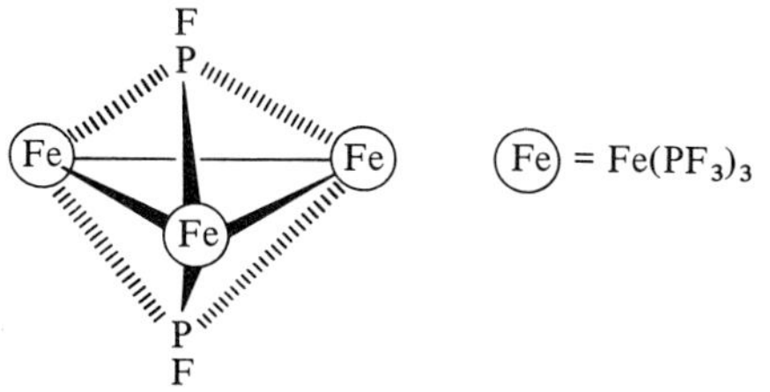

Unlike CO, $PF_3$ can have side reactions such as defluorination. This is especially so for Fe where it is found that the yield of products containing defluorinated species such as $PF_2$ and PF increases with increased metal cocondensate concentration[102)]. Similarly, $(PF_3)_4Cr(PF_2)_2Cr(PF_3)_4$ can be detected mass-spectrometrically from concentrated $Cr/PF_3$ cocondensates. One interesting corollary is that when the vapour over a Cr/Fe melt (reported to contain the species Fe-Cr) is condensed with $PF_3$, the heteronuclear complex $(PF_3)_4Cr(PF_3)_2Fe(PF_3)_3$ can be isolated[102)].

Ni atoms also react with $PF_2Cl$ to form $Ni(PF_2Cl)_4$ but in lower yields than with $PF_3$[101)]. Perhaps side-reactions such as dehalogenation are more important.

Cocondensation of Ni atoms with $PH_3$ at 77 K is accompanied by $H_2$ evolution and more $H_2$ is evolved when the cocondensate is warmed to room temperature. No volatile compound containing Ni was formed[101)]. However if the cocondensation temperature is reduced to 10 K, $Ni(PH_3)_4$ can be identified spectroscopically. In matrices diluted with Ar the species $Ni(PH_3)_n$ (n = 1–4) can be identified[106)]. $Ni(PH_3)_4$ has been previously prepared from the reaction of $(1,5\text{-Cod})_2Ni$ and $PH_3$ at –40 °C. It decomposes at –30 °C[109)]. The reaction at 77 K can be tempered by condensing Ni atoms with an equimolar mixture of $PH_3$ and $PF_3$[101)] (Eq. (4.28)) the ratio of product yields is

$$Ni(a) + PF_3 + PH_3 \xrightarrow{i,ii} Ni(PF_3)_4(tr) + Ni(PF_3)_3PH_3$$

$$(10\%) + Ni(PF_3)_2(PH_3)_2\ (15\%) \tag{4.28}$$

(i) cocondense, 77 K; (ii) 0 °C

almost exactly right for statistical production; i.e. the ligands add randomly to Ni in a diffusion-controled process and no displacement of the weaker $PH_3$ ligand by the stronger $PF_3$ occurs. The "missing" species $Ni(PF_3)(PH_3)_3$ and $Ni(PH_3)_4$ are too unstable to be isolated at room temperature.

Condensation of several metal vapours with the volatile phosphines $PMe_3$ and $P(OMe)_3$ also can form complexes (Eqs. (4.29) to (4.31)). The reaction between Fe atoms and $PMe_3$ is fascinating. The yield is higher if diluted with pentane but lower if with hexane[19)]. Cr and Mn atoms form no isolable organometallic productes when condensed with $PMe_3$[106)]. The reaction between Fe atoms and $P(OMe)_3$, whether by cocondensation or by condensing Fe atoms into liquid $P(OMe)_3$ is not clean. Cocondensation gives 5 products, the major one being $Fe[P(OMe)_3]_5$ while condensation onto $P(OMe)_3$ gives a product contaminated with $OP(OMe)_3$[110)].

$$M(a) + PMe_3 \xrightarrow{i,\ ii} M(PMe_3)_4\ (5 - 15\%)\ (M = Co, Ni, Pd) \tag{4.29}$$

$$Fe(a) + PMe_3 \xrightarrow{i,\ ii} HFe(CH_2PMe_2)(PMe_3)_4\ (6\%) \xrightarrow{iii} Fe(NO)_2(PMe_3)_2 \tag{4.30}$$

$$Fe(a) + P(OMe)_3 \xrightarrow{i,\ iii} Fe[P(OMe)_3]_5 \tag{4.31}$$

(i) cocondense, 77 K; (ii) warm; (iii) NO

Most other phosphines or phosphites are too involatile to be used in vapour cocondensation experiments, but the rotating solution reactor of Timms and MacKenzie[371] and the rotary atomiser solution diffuser of Green and Young [103, 108, 373] allow their utilisation. Condensation of Ni atoms onto a rotating solution of $PPh_3$ in silicone 704 oil at 0 °C forms $Ni(PPh_3)_4$ in good yield[371]. Silicone oil is a phenylmethoxysilane and perhaps the phenyl groups help stabilise atoms with respect to aggregation till meeting a $PPh_3$ molecule. Similarly, using the rotary atomiser solution diffuser (Fig. 2.10), Ni atoms have been condensed at 77 K with solutions of $PPh_3$ or $Ph_2PCH_2CH_2PPh_2$ (diphos) in PhMe, or Pd atoms with $PPh_3$ in PhMe to give $Ni(PPh_3)_4$, $Ni(diphos)_2$, and $Pd(PPh_3)_4$ respectively in *ca.* 100 % yield[107, 108].

Binary metal phosphine (phosphite) complexes can also be made by reactions with unstable species formed in the cocondensate at lower temperatures. $(1,5\text{-Cod})_2Pd$ has been shown by MIS to be formed at 77 K by reaction of Pd atoms and 1,5-Cod. However, it decomposes at −20 °C. Thus if Pd atoms are cocondensed with 1,5-Cod in rotating CyMe solutions at −120 °C then $P(OMe)_3$ added, warming to room temperatures leads to isolation of $Pd[P(OMe)_3]_4$. Similarly, subsequent reaction of $P(OCH_2)_3CEt$ with the unstable $(1,5\text{-Cod})_2Fe$ gives $Fe[P(OCH_2)_3CEt]_5$ in 20 % yield[110].

Phosphines and phosphites can furnish a wide variety of electronic and steric effects[112]. Thus it comes as no surprise that they, especially the volatile ones such as $PF_3$, $PMe_3$, $P(OMe)_3$, have proved such useful, effective ligands that they have been employed in many mixed ligand reactions to complete the coordination sphere, and have been added to many unstable cocondensates to trap sensitive species by converting them to more inert or stable derivatives (*vide infra*).

## 4.8 Isocyanides

Cocondensation of Cr, Fe or Ni atoms and alkyl isocyanides forms the corresponding, known isocyanide-metal complexes $Cr(CNR)_6$, $Fe(CNR)_5$ and $Ni(CNR)_4$[113].

## 4.9 Alkynes

Many atoms of the first transition series, when condensed at 77 K with alkynes, cause cyclotrimerisation to benzenes (Eqs. (4.32) to (4.35)). This well-known reaction[114] is in these cases not catalytic and no characterisable metal-containing compound has been isolated, even in the case of Cr where perhaps some ($\eta$-arene)$_2$Cr could be expected.

Cr(a) + Me•C≡C•Me —(i, ii)→ 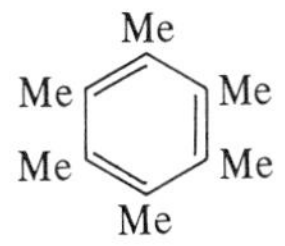

+ H•C≡C•Et (4.32)[115]

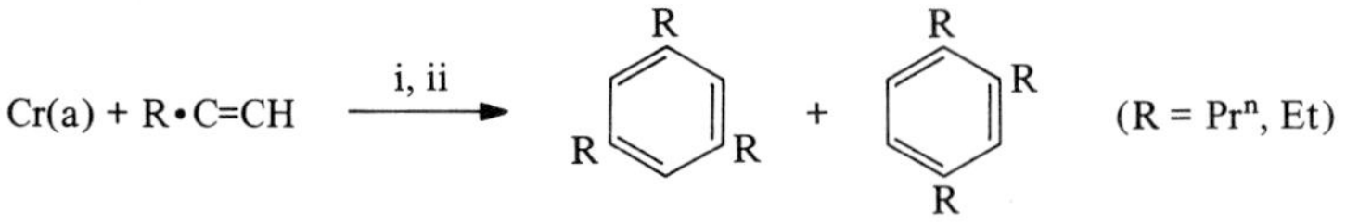


(4.33)[115]

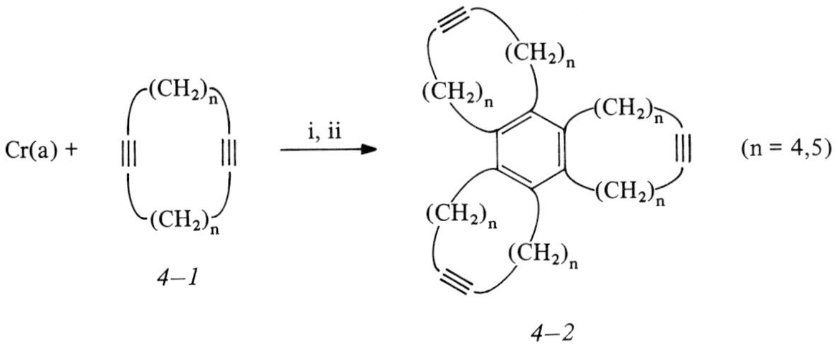


(4.34)[116]

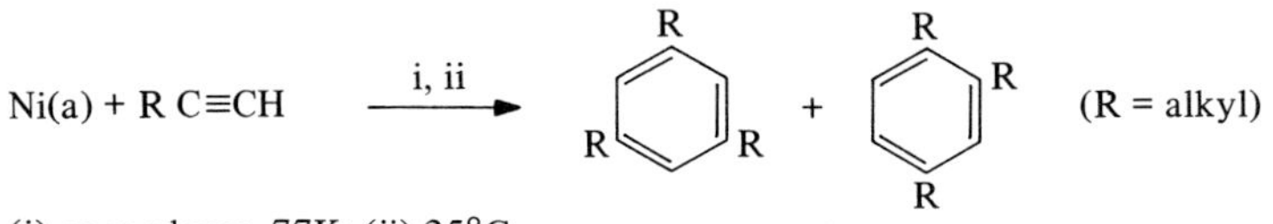


(4.35)[118]

(i) cocondense, 77K; (ii) 25°C

It has been postulated that the benzene rings are only formed one at a time at the Cr centres[115]. Less than 1 mole of arene is produced per g-atom of Cr, therefore it is possible that species responsible for forming the arene are destroyed in forming the arene. This is supported by the observation that only *4.2* is formed in Eq. (4.34). Had *4.2* remained bonded to say a metal atom there would have been the possibility of a second trimerisation to form a percyclophene-Cr complex or free percyclophane (e.g. n = 4, structures *4.3* and *4.4*).

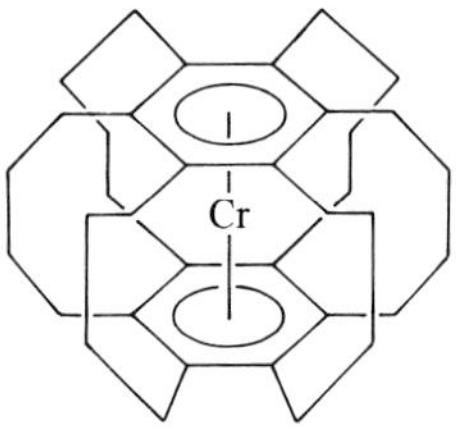


*4–3*

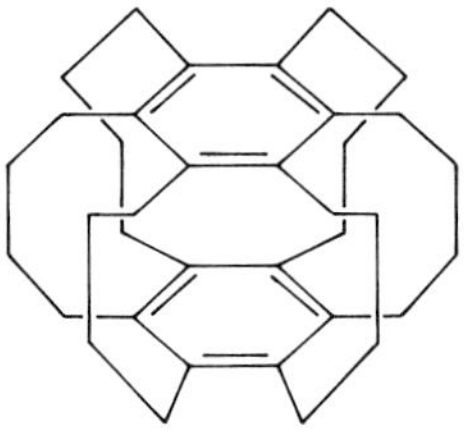

*4–4*

To avoid such reactions, metal atoms have been condensed with $Me_3Si \cdot C \equiv C \cdot SiMe_3$ which is too bulky to easily trimerise[119]. Condensation with Cr atoms gives a green involatile material of unknown structure. If CO is added, unstable species *4.5* and *4.6* can be isolated. Characterisation is hampered by decomposition to $Cr(CO)_6$ and involatiles. CO treatment of the $Fe/Me_3Si \cdot C{\equiv}CH$ cocondensate gives a species *4.7*.

Cocondensation of Fe atoms and $Me_3SiC{\equiv}CSiMe_3$ gives a solid of variable stoichiometry $Fe(R \cdot C{\equiv}C \cdot R)_n$ (n = 1.8 to 4) which on treatment with CO forms $(RC{\equiv}CR)Fe(CO)_4$ and uncharacterised compounds[119].

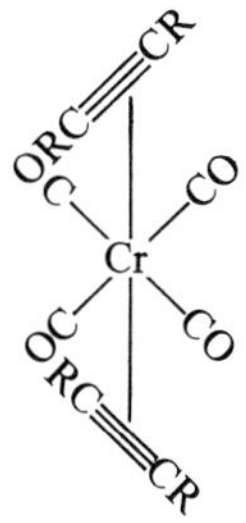

4.5

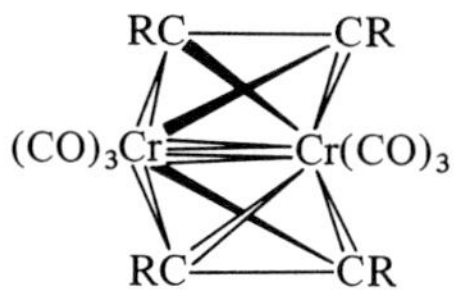

4.6

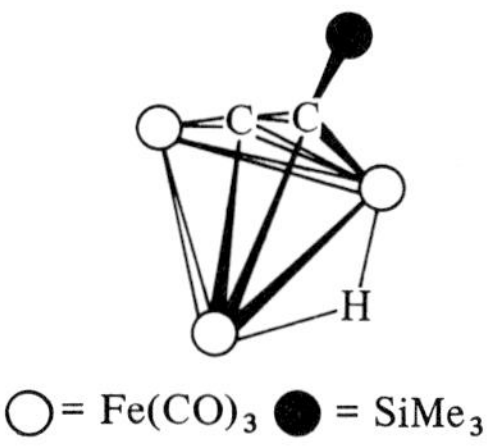


4.7

It is known that various Co species can cyclotrimerise nitriles and acetylenes to pyridines. This result is paralleled by Co atoms which when condensed with Ph · C≡CH and a nitrile form substituted benzenes and pyridines (two isomers)[119)].

Li atoms form $LiC_2H_2$ and $Li_2C_2H_2$ in very low temperature matrices[51)]. The Mg/Me · C≡C · Me/$NH_3$ cocondensate gives the stereospecific reaction (Eq. (4.36))[120)].

Mg(a) + Me·C≡C·Me + $NH_3$ $\xrightarrow{\text{i, ii}}$ 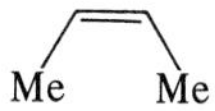 (4.36)

(i) cocondense, 77K; (ii) 25°C

Cocondensation at 77 K of Al atoms and acetylene forms on removal of excess acetylene a pale residue which on deuteriolysis gives mainly CHD=CHD. This implies that there is a strong interaction between acetylene and Al to form Al-C $\sigma$ bonds[121, 122)], similar to those between Al and olefins (*vide infra*). Other alkynes behave similarly (Eqs. (4.37), (4.38)).

$$\text{Al(a)} + \text{Me}\cdot\text{C}{\equiv}\text{CH} \xrightarrow{\text{i,ii,iii}} \text{Me}\cdot\text{CD}{=}\text{CHD} + \text{Me}\cdot\text{C}{\equiv}\text{CD}\ (+\ \text{DCH}_2\cdot\text{C}{\equiv}\text{CH}) \qquad (4.37)$$

Al(a) + Me · C≡C · Me $\xrightarrow{\text{i,ii,iii}}$  (2:1) (4.38)

(i) cocondense, 77 K; (ii) remove xs alkyne, –78 °C; (iii) $D_2O$

The presence of Me · C≡CD is most likely due to formation of $R_2Al$ · C≡C · Me species on reaction with the acidic acetylenic hydrogen.

At lower temperatures, complexation can be observed. The esr spectra of the Al/CH≡CH/Ne and Al/CD≡CD/Ne ($1:10:10^3$) matrices at 4 K identify a species *4.8*. Unlike an analogous Al/$C_2H_4$ matrix which contains a $\pi$-complex[123] (*vide infra*) the Al atoms interact strongly with the acetylene to form an Al-C $\sigma$ bond and a vinyl radical with the odd electron in an essentially p orbital[124)].

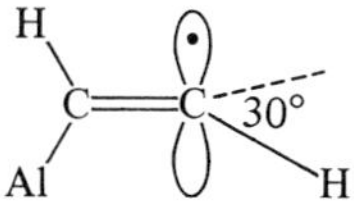

*4.8*

## 4.10 Alkenes (Monoenes)

A large number of metal atoms have been condensed with propene as a preliminary reaction in general survey of metal vapour synthesis[125]. In all cases, after removing excess substrate at −78 °C, the residue is treated with $D_2O$ which displaces or hydrolyses the organic portions of any organometallic products. The released hydrocarbons are then analysed for identity and deuterium content. The elements examined are B, Mg, Al, Si, Ti, V, Cr, Mn, Fe, Co, Ni, Cu, Zn, Zr, Mo, Pd, Ag, Sn, Tb, Pr, Er, W, Re, Pt, Au, Pb and U. Those elements underlined give no hydrocarbon hydrolysis or displacement products at −78 °C. Of the others, the products varied from element to element, but a broad general trend can be developed.

With atoms which have available but quite tightly bound d-orbitals such as those of elements at the end of the transition series (e.g. Ni and Pd) the hydrolysis products are essentially alkenes[118,126], reflecting a $\pi$-interaction between metal atom and alkene. However, if an atom at the beginning of the transition series such as Cr is cocondensed with propene deuteriolysis of the cocondensate gives propene and dideuteriopropane which is in fact the main product. The lanthanides behave similarly. This has been rationalised in terms of an increasing degree of $\sigma$-bonding in going from the end to the beginning of the transition series[115]. This is easily explained. In going from right to left the valence d-electrons are increasingly less tightly bound, thus there will be an increasing net drift of electron-charge from metal to olefin *via* the $\sigma$[spd(metal − $\pi$(alkene)] and $\pi$[d(metal) − $\pi^*$(alkene)] bonds. The main result is a build up of electron-charge in the $\pi^*$(alkene) bond which can be formally regarded as causing rehybridisation to two polar $\sigma$-bonds which on hydrolysis can be expected to give the corresponding alkane. With atoms having no available d orbitals such as Al atoms, the deuteriolysis products are dideuterioalkanes which has been interpreted as evidence of polar $\sigma$-bonding and no $\pi$-bonding between the atom and the alkene[121,127].

It may be prudent to point out that care must always be taken in basing conclusions on hydrolysis experiments, as some of the observed trends may be a result of some reaction that occurs during hydrolysis rather that as a result of chemistry in the cocondensate. Thus the preceding conclusions, while reasonable, must be regarded as hypothetical.

Co, Ni or Pd atoms readily form $\pi$-complexes M($\eta$-alkene)$_3$ when cocondensed with alkenes (e.g. M(nbe)$_3$ or M($C_2H_4$)$_3$)[105,111,128,129].

In many cases these compounds are too thermally unstable to be isolated but instead can be studied by ir spectroscopy by cocondensing the metal atoms and alkene (1 : 30) onto a CsI window kept just cold enough for cryogenic pumping, then warming the window slightly to pump away excess alkene at $10^{-4}$ torr leaving a transparent film of the complex[129]. If a complex is found not to decompose till a reasonably high temperature such as −20 °C it is then worthwhile to attempt to make and isolate the compound on a larger scale. In this way, Pd(norbornene)$_3$, first identified on the CsI window and found to decompose at 0 °C, can be made in 40% yield in a conventional metal vapour reactor[111]. Advantage can be taken of such compounds' lability to form a wide range of complexes (Scheme 4.8).

The reactions of metal atoms with ethylene in pure or diluted matrices at typically 10–12 K has been studied. A large number of complexes have been identified:

$Co(C_2H_4)_{1-3}$[29,130],
$Rh(C_2H_4)_{1-3}$[29],
$Ni(C_2H_4)_{1-3}$[29,131],
$Pd(C_2H_4)_{1-2}$[29,132],
$Pt(C_2H_4)_{2-3}$[29],
$Cu(C_2H_4)_{1-3}$[130,133],
$Cu_2(C_2H_4)_{4\ or\ 6}$[133],
$Ag(C_2H_4)$[134], and
$Au(C_2H_4)$[135].

Attempts have been made to use such species as localised bonding models of molecules chemisorbed on metal surfaces[29] and thus to test the usefulness of a localised bonding description of chemisorption[136]. For example, comparison of the ir spectra of $Pd(C_2H_4)$[132] and ethylene chemisorbed on silica-supported Pd[137] are in good agreement suggesting the same bonding mode. The ethylene in $Pd(C_2H_4)$ bonds *via* its $\pi$ and $\pi^*$ orbitals[132] which implies that ethylene is bonded to this particular surface *via* its $\pi$ and $\pi^*$ orbitals in support of other findings[137,138]. Secondly, cocondensing Ag atoms, $C_2H_4$ and $O_2$ may give species which can act as models for ethylene epoxidation by silica-supported silver. Condensation of Ag atoms with $O_2$ and $C_2H_4$ (*ca.* 1:5,000:5,000) at 10 K forms the ion pair $(C_2H_4Ag)^+(O_2)^-$ which decomposes on warming to 40–45 K without formation of an epoxide or peroxy intermediate[29]. However if Ag atoms are condensed with $O_2$ and $C_2H_4$ diluted with Ar (*ca.* 1:500:500:$10^4$) the species $Ag(C_2H_4)$, $Ag^+(O_2)^-$ and $(C_2H_4Ag)^+(O_2)^-$ are formed. If this matrix is warmed to 30 K, $Ag^+(O_2)^-$ and $C_2H_4Ag$ slowly disappear with concomitant growth of a new species which appears to have the structure *4.9*[29], not unlike the proposed surface peroxy species in the actual catalytic cycle[139] (see Scheme 4.3).

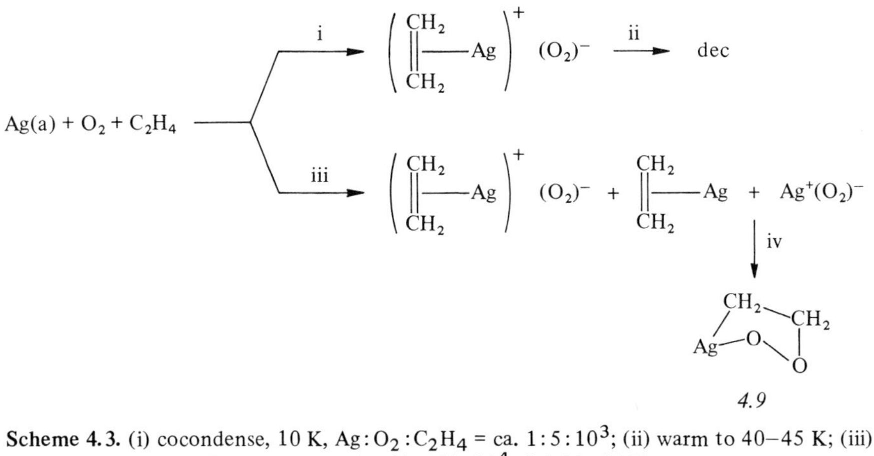


*4.9*

**Scheme 4.3.** (i) cocondense, 10 K, $Ag:O_2:C_2H_4$ = ca. 1:5:$10^3$; (ii) warm to 40–45 K; (iii) cocondense in Ar, 10 K, $Ag:O_2:C_2H_4:Ar$ = ca. 1:500:500:$10^4$; (iv) 30–35 K

The very interesting species $M(C_2H_4)$ (M = Al, Ga, In) have been formed in dilute neon matrices at *ca.* 4 K. Esr spectra of $Al(C_2H_4)$ and $Al(C_2D_4)$ complexes show the bonding is similar to that of a transition metal atom to ethylene in spite of the lack of d orbitals. It consists of a filled $sp_y$(Al) lone-pair, a filled $\sigma[sp_y(Al) - \pi(ethylene)]$ bond and a half-filled $\pi[p_x(Al) - \pi^*(ethylene)]$ bond *4.10*.

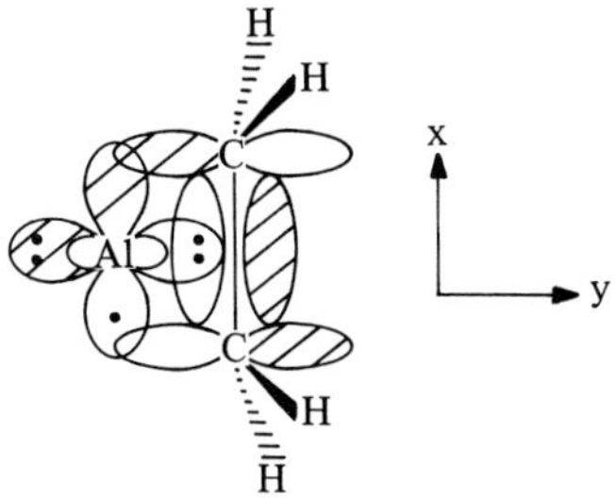


*4.10*

$Ni(C_2F_4)_n$ (n = 1–3) can be observed in matrices at 15 K[140] but these complexes and other tris (perfluoroalkene) metal complexes of Ni, Pd and Pt decompose at low temperatures[45, 141]. As an example $Pd(CF_3 \cdot CF{=}CF \cdot CF_3)$ decomposes at –30 °C but the more stable derivative $L_2Pd(CF_3 \cdot CF{=}CF \cdot CF_3)$ can be trapped by addition of $L(py, PEt_3)$ at –78 °C (Eq. (4.39)). Similarly the unstable species $Ni(styrene)_3$ forms a more stable derivative on addition of bipyridyl[142] (Eq. (4.40)).

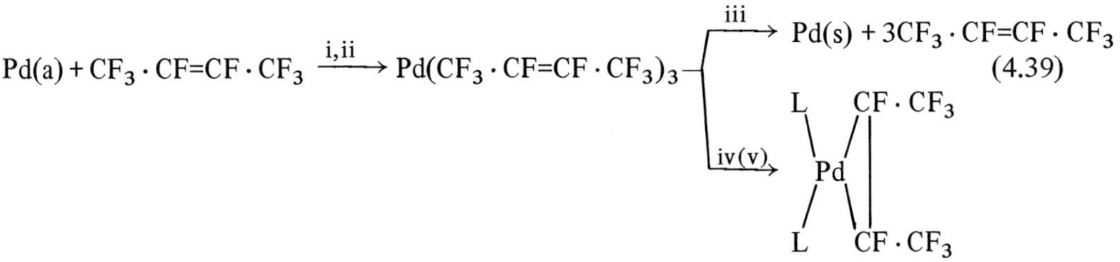


(4.39)

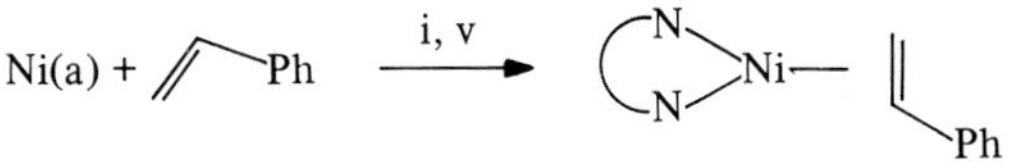


(4.40)

(i) cocondense, 77 K; (ii) remove xs substrate, –78 °C; (iii) –30 °C; (iv) L, (L = py, $PEt_3$); (v) bipy

Atoms of the first row transition series catalyse the isomerisation of the alkenes containing β-hydrogens[125] (for example Eqs. (4.41), (4,42)).

(4.41)[118]

(4.42)[115]

(i) cocondense, 77K; (ii) –40°C; (iii) –78°C

In addition cocondensation of Ni atoms and an equimolar mixture of propene and $d_6$-propene leads to scrambling[118]. These processes may proceed either via an allylmetalhydride[143,144] (Eq. (4.43)) or alkenemetalhydride[144] (Eq. (4.44)) mechanism.

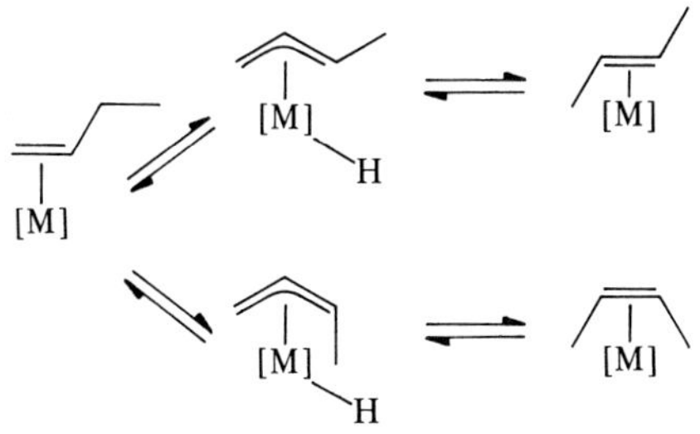


(4.43)

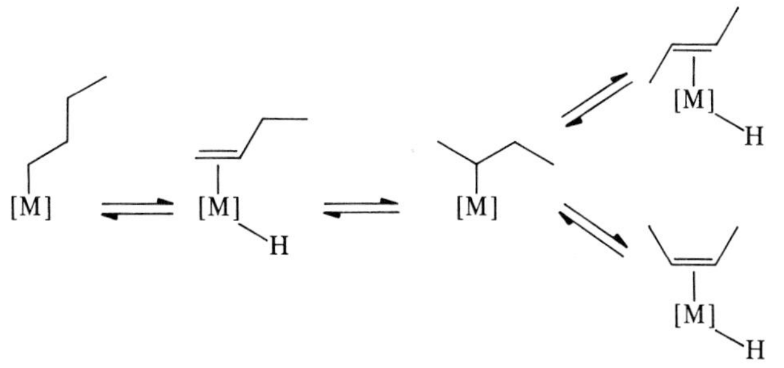


(4.44)

However, the allylmetalhydride mechanism is more likely as when Ni atoms are condensed with 2-deuteriopropene, no scrambling occurs, whereas 3-deuteriopropene gives some 1-deuteriopropene[127]. Furthermore 3-deuteriopropene leds to both undeuterated and dideuterated propenes which means that there is either intermolecular exchange occurring *via* perhaps H˙ radicals or hydride bridging, or there is more than one allyl group per Ni atom. Additional evidence comes from trapping experiments (Eqs. (4.45) to (4.47)).

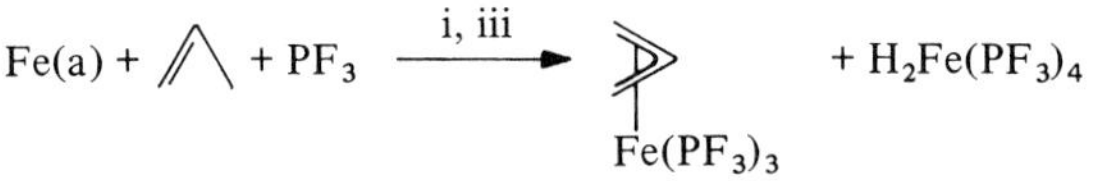


(4.45)[145,384]

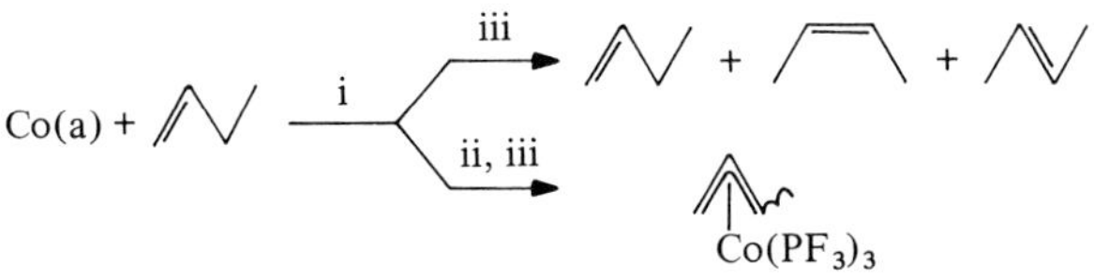


(4.46)[127]

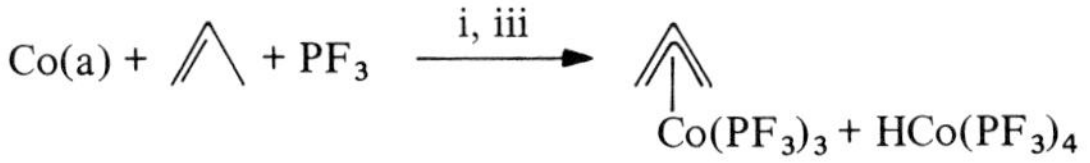


(4.47)[19]

(i) cocondense, 77K; (ii) $PF_3$, 77K; (iii) warm

Infra-red spectra of the reaction products of Co, Ni or Pd atoms with propene between 80 K and 2,000 K show only absorbances characteristic of $\pi$-bonded alkenes while similar studies of (Mn or Fe)/propene cocondensates show no identifiable organometallic product[129]. No allylmetalhydride species are observed but only a small equilibrium concentration is necessary for isomerisation.

The Pt/but-1-ene cocondensate does not isomerise but-1-ene, but addition of $H_2$ leads to catalysed conversion to butane[136]. Similarly a Co/hex-1-ene/PhMe cocondensate gives hexane[147].

In view of the ability of metal atoms to act as catalyst precursors for the oligomerisation of Bd[112] (*vide infra*), it seems likely that other olefinic systems will be subject to oligomerisation by metal-atom-derived catalytic systems. Acrylonitrile dimerisation is of commercial interest since most of the possible products can be readily hydrogenated to adiponitrile ($NC(CN_2)_4CN$) an important nylon feedstock. However, when atoms of the first row transition metals (Ti to Ni) are condensed at 77 K with acrylonitrile (An) and the products heated, glc analysis shows only a small degree of conversion to dinitriles and the major product as polymer[146]. The proportion of dinitriles can be increased by condensation of metal vapour and An with a "solvent" (see Sect. on Bd) but it is still not high. As expected, addition of $PPh_3$ reduces the dinitrile yield, presumably by blocking coordination sites. The most effective element is Fe and some of its reactions are given in Eqs. (4.48) to (4.50).

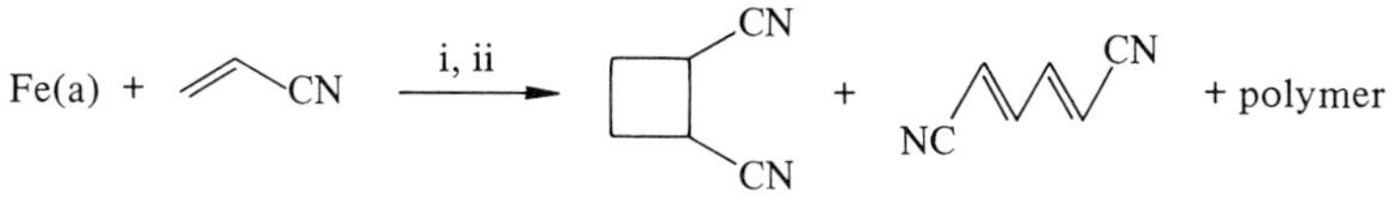

(4.48)

2.7 mmol 300 mmol 0.33 mmol 0.78 mmol 60 mmol

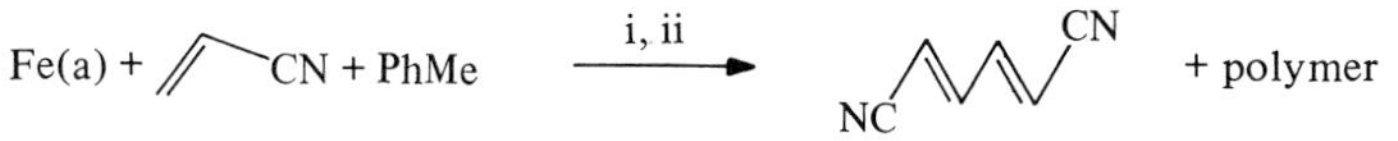

(4.49)

3.6 300 180 3.45 69

3.6 300 180 4 0.33 0.01 0.01

(4.50)

(i) cocondense, 77K; (ii) heat

A further series of experiments[146] have shown that atoms of the first transition series (Ti to Ni) do not catalyse the metathesis of *cis*-pent-2-ene in PhMe. Similar results are found in reactions where diethylaluminium chloride is cocondensed as cocatalyst. Heating the Ni/*cis*-pent-2-ene/PhMe/$Et_2AlCl$ cocondensate leads to extensive isomerisation of *cis*-pent-2-ene to *trans*-pent-2-ene (*cis* : *trans* = 23 : 77).

Cocondensation of Al atoms and propene at 77 K, followed by removal of excess propene at –78 °C gives a light yellow residue which on deuteriolysis gives, 1,2-dideuteriopropane mainly and some dideuteriohexanes[121] (Eq. (4.51)).

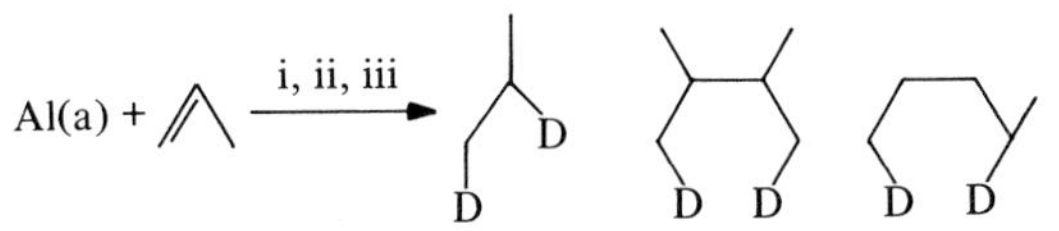

(4.51)

(i) cocondense, 77K; (ii) remove xs at −78°C; (iii) $D_2O$

This has been interpreted as meaning a σ-type interaction takes place between Al atoms and propene, and the hexanes are formed from aluminopropyl radicals[121]. Little HD was detected after deuteriolysis implying little Al-H formation. If the ir spectrum is taken of an Al/propene cocondensate at 77 K on a CsBr window there is no absorption in the 1,900–1,700 $cm^{-1}$ region (characteristic of Al-H linkages) but strong absorptions in the 700–600 $cm^{-1}$ region, indicative of Al-alkyl bonds. The ir spectrum is essentially the same after warm-up, demonstrating the reaction has gone to completion at 77 K, the deposition temperature[121].

A plausible organoaluminium intermediate which can account for the products and the observed deuterium distribution has been presented[127] (Eq. (4.52)). The most likely initial product is an aluminacyclopropane.

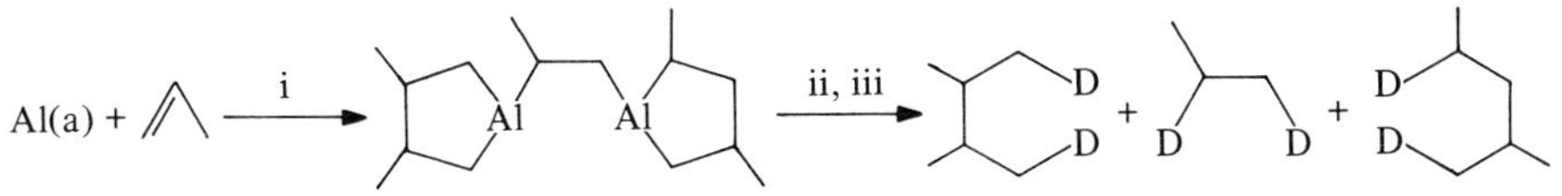

(4.52)

(i) cocondense, 77K; (ii) remove xs propene, −78°C; (iii) $D_2O$

A similar species $Al(C_2H_4)$ *4.11* has been formed in Ne matrices at *ca.* 4 K. Furthermore, Al atoms show a preference for 1,2-addition to Bd[121]. However, reaction of either *cis-* or *trans*-but-2-ene with Al atoms produces the same mixture of *meso-* and *d,l*-2,3-dideuteriobutanes[121]. This lack of stereospecificity discredits an aluminacyclopropane unless it is formed via a multi-step process involving aluminopropyl radicals or is in equilibrium with such species (Scheme 4.4). The reaction can be made stereospecific by addition of Lewis bases such as $Me_2O$ or $Me_3N$ *after* cocondensation. Deuteriolysis now gives 95% of the *cis*-adduct[127] (Eqs. (4.53) to (4.54)). This, with all the other evidence, points to Al atoms

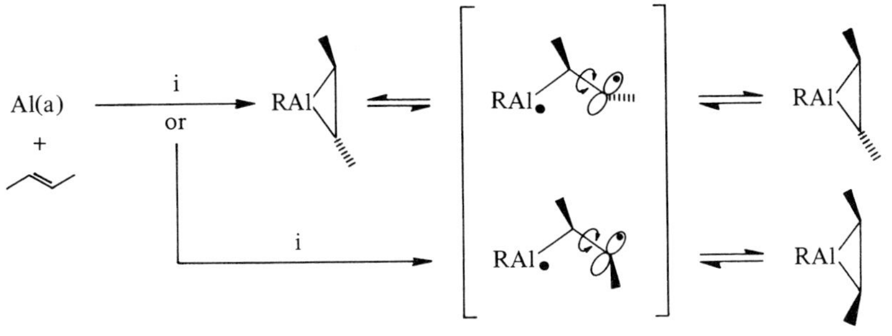


**Scheme 4.4.** (i) cocondense, 77 K

adding to olefins most probably in a concerted manner to give aluminacyclopropanes with retention of configuration and that ring-opening is much reduced by the presence of complexing agents. It is difficult to rationalise this stereochemistry if two Al atoms are involved.

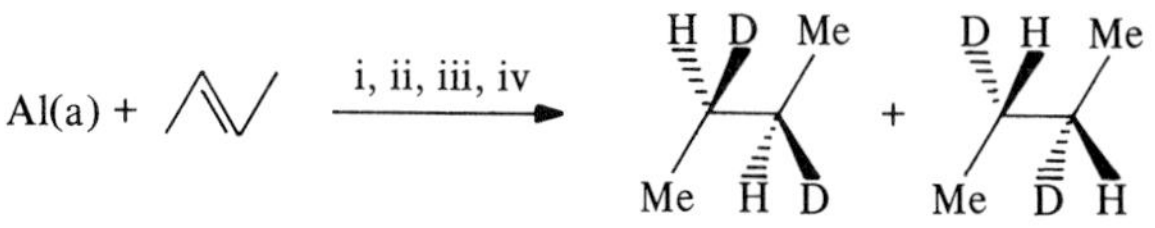


(4.53)

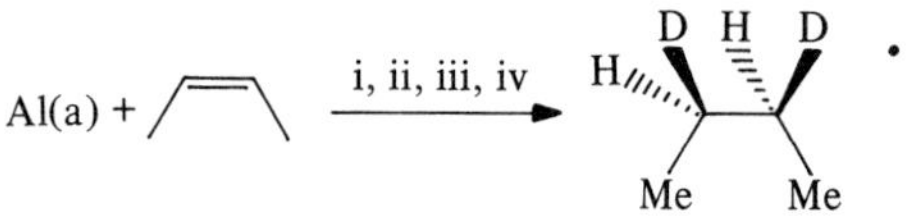


(4.54)

(i) cocondense, 77K; (ii) $Me_2O$ or $Me_3N$, 77K; (iii) remore xs substrate at −78°C; (iv) $D_2O$

It is reasonable to postulate that formation of the dideuteriohexanes if the reaction with propene is *via* a second molecule of propene adding to an aluminopropyl radical similar to the aluminobutyl radicals involved above (Scheme 4.5).

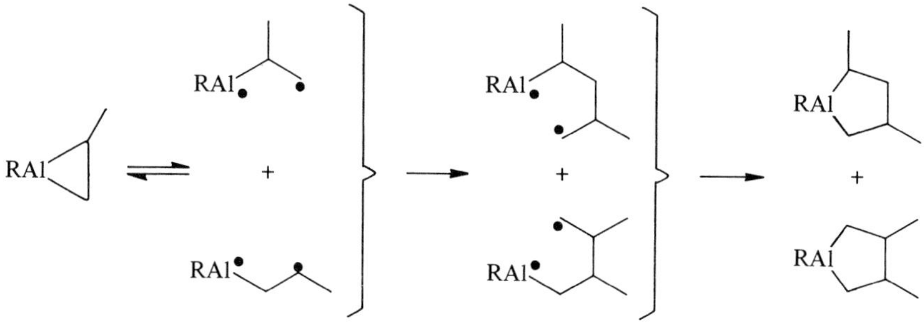


**Scheme 4.5.**

Assuming 100% reaction of the Al atoms, the total number of alkylmetal bonds per Al atom in the reaction between Al and propene is 2.7 which suggests that the other 0.3 valences per Al atom are used to form Al-Al bonds. More convincingly, the deuteriolysis products of an Al/but-2-ene cocondensate are formed in the ratio Al : $D_2$ : dideuteriobutane = 2 : 1 : 2. This strongly suggests Al-Al bonds and molecules of the type bis (2,3-dimethylaluminacyclopropane) *4.11* (Eq. (4.55)).

(4.55)

*4.11*
*inter alia*

2 : 1

(i) cocondense, 77K; (ii) $D_2O$, (iii) warm

$Bu^i_2Al$-$Albu^i_2$ has since been prepared by reaction of $Bu^i_2AlCl$ with K in hexane for 120 h[147], confirming the existence of Al-Al bonds.

Alkenes react with excess Li atoms in the vapour phase to form amorphous, heterogeneous, multicoloured solids whose composition can be infered tentatively from deuteriolysis experiments[148] (Eqs. (4.56) to (4.57)).

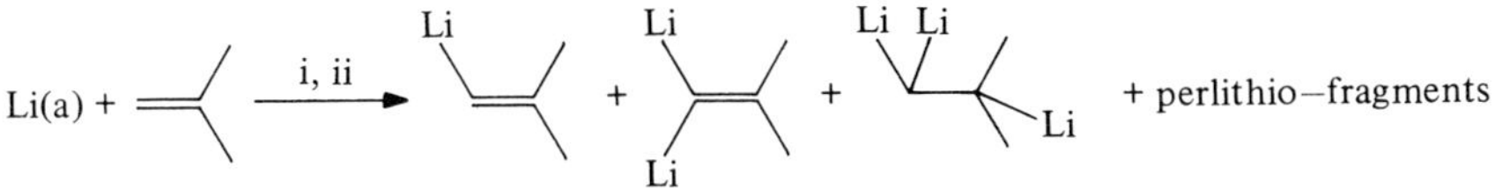


(4.56)

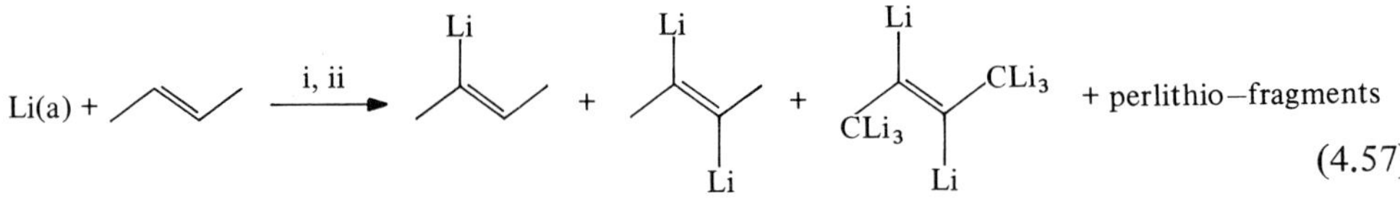


(4.57)

(i) 800°C, vapour phase, xs Li(a); (ii) collect on trap at 77K

Initial reaction is substitution of vinylic hydrogen with subsequent replacement of both vinyl and alkyl hydrogens or addition across the double bond. The reaction conditions cause alkene fragmentation leading to the perlithio-fragments $C_3Li_4$, $C_2Li_6$, $C_2Li_4$ and $C_2Li_2$ whose combined yields vary from 25 to 50 per cent. This contrasts with the behaviour of solid lithium at $-20$ °C in an ether, which, with, the aid of an electron-transfer agent such as Ph-Ph or naphthalene, reacts with ethylene to form mainly vinyllithium and lithiumhydride most probably *via* addition across the double bond to form $LiCH_2$–$CH_2Li$ which β-eliminates to $LiCH{=}CH_2$ and LiH[149].

## 4.11 Allyl

Apart from NO (and $NO_2$) the only 3e ligand precursor that has been used in MVS has been the allyl radical derived from tetraallyltin. Thus condensation of $Sn(allyl)_4$ with Ni or Cr atoms at 77 K forms $(\eta^3\text{-allyl})_2Ni$ and $Cr_2(allyl)_6$ respectively[72]. The analogous reaction with Fe atoms gives no $(\eta^3\text{-allyl})_3Fe$ but cocondensation of Fe atoms, $Sn(allyl)_4$ and $PF_3$ forms $(\eta^3\text{-allyl})Fe(PF_3)_3$, a paramagnetic purple compound stable for short periods at 20 °C. Only a trace of the 18e compound $(\eta^3\text{-allyl})_2Fe(PF_3)_2$ can be isolated[19,145,384]. A topological Hückel model for organometallic complexes, used in an attempt to indicate the electronic origins of the 18e rule has calculated that the 17e (and 16e) species of the type $(\eta^3\text{-allyl})ML_3$ have very nearly the some stability as the 18e analogue[150]. In addition $(\eta^3\text{-allyl})Fe(CO)_3$[151] and $(\eta^3\text{-allyl})Fe(CO)_3^+$[152] are known. Cocondensation of Cu atoms, $Sn(allyl)_4$ and $PF_3$ forms no isolable organometallic compounds, nor does cocondensation of Fe or Ni atoms and $N(allyl)_3$[145].

Alyllmetal derivatives can also be synthesised by H-elimination or H-addition to olefins (Eqs. (4.58) to (4.59)).

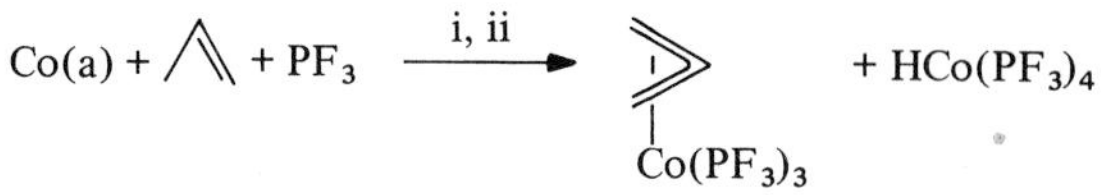


(4.58)[19]

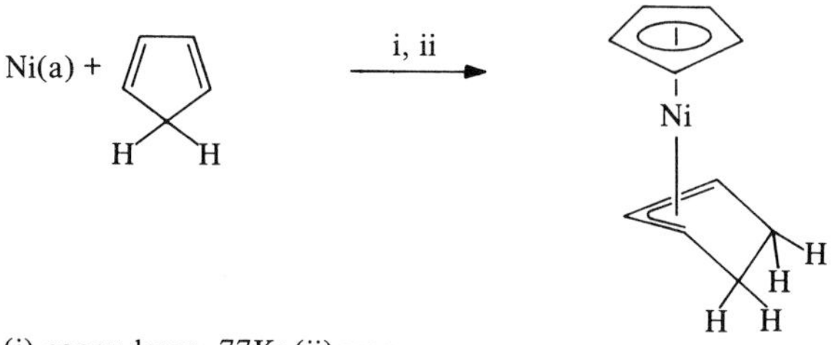


(4.59)[381]

(i) cocondense, 77K; (ii) warm

## 4.12 Alkadienes

One of the dominant interactions between metal atoms and unsaturated hydrocarbons is the insertion of the metal fragment into a C-H bond adjacent to a double bond (i.e. an active hydrogen) (Eq. (4.60)).

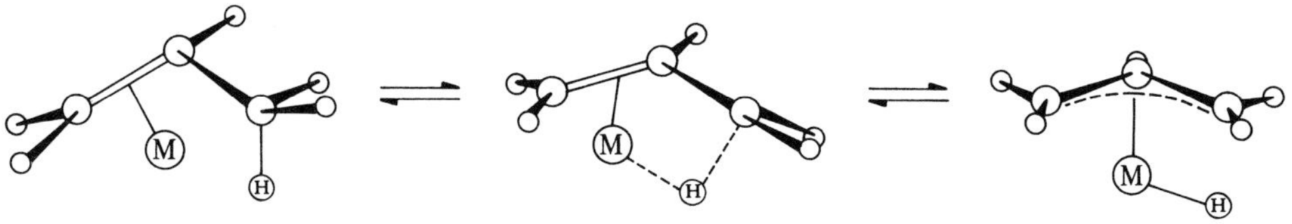


(4.60)

Alkadienes (as well as other olefins) possessing active hydrogens thus tend to have much different reactions from those possessing no active hydrogens and it is logical to treat them separately.

### *4.12.1 Butadiene*

Butadiene (Bd) is an important feedstock in the chemical industry where it is catalytically converted to materials such as artificial rubbers or nylon. Catalytic systems which oligomerise Bd to a wide variety of products have been found for all the elements of the first transition series (except V), a situation which is paralleled by the activity of metal atoms (including V) which can catalyse the conversion of Bd to a series of oligomers and polymers, whose distribution, unique for each element, can be altered by addition of suitable

**Table 4.6.** M(a) + Bd + PhH[a] —(i, ii)→

| M | Conversion of Bd/% | Conditions (θ/°C, t/h) | Product distribution[b] | | | | | | |
|---|---|---|---|---|---|---|---|---|---|
| | | | ttt-Cdt | ctt-Cdt | LT | Cod | LD | HO | Pbd |
| Ti | 0 | | | | | | | | |
| "$TiO_2$"[c] | 0 | | | | | | | | |
| V[d] | 44 | 50, 4 | 10 | 15 | | | | 75 | |
| Cr[d] | 62 | 60, 6 | 10 | | | | | 90 | |
| Mn | 65 | 60, 4 | 35 | 60 | | | 5 | | |
| Fe[e] | 90 | 50, 2 | | 30 | 63 | 7 | | | |
| "FeO"[c] | 0 | | | | | | | | |
| Co | 88 | 60, 2 | | | | 22 | 69 | 7 | |
| Ni[e] | 75 | 80, 2 | 45 | | | 55 | | | |
| "NiO"[c] | 65 | 70, 3 | 40 | 30 | | 30 | | | |
| Zr | 7 | 120, 12 | | | 14 | | 86 | | |
| Nb | 100 | 120, 2 | | | 1 | | | 99 | |
| Mo | 1 | 90, 12 | | | 100 | | | | |
| Ru | 37 | 140, 5 | | | 10 | 45 | 45[f] | | |
| Rh | 6 | 140, 6 | | | | | 100[g] | | |
| Pd | 42 | 70, 3 | | | 100 | | | | |
| Al | 0 | | | | | | | | |

(i) cocondense, 77 K; (ii) heat, θ °C.

a M(a) : Bd : PhH = 1 : 30 : 80 (approx.).
b Bd = buta-1,3-diene, Cdt = cyclododeca-1,5,9-triene, Cod = cis,cis-cycloocta-1,5-diene, LT = linear or open-chain trimers, LD = linear or open-chain dimers, HO = oligomers higher than trimers, Pbd = polybutadiene, c = *cis*, t = *trans*.
c Vapour above heated $TiO_2$, FeO, and NiO respectively.
d Analysis of the reaction mixture immediately after warming the cocondensate showed ca. 10% Bd conversion to mainly Cdt and only 10% higher oligomers.
e Aliquot portions (5 ml) of the reaction mixture were treated with Bd (ca. 10 g) in sealed ampoules at 50° (Fe) and 85 °C (Ni) giving almost quantitative conversion to the same products as obtained from the original cocondensate.
f Octatriene.
g Octatriene and methylheptatriene (7 : 3).



cocatalysts[108, 153]. Similar results have been found for second row elements, though they turn out to be less effective[154].

Cocondensation of metal atoms (Ti, V, Cr, Mn, Fe, Co, Ni, Zr, Nb, Mo, Ru, Rh, Pd), Bd and PhH leads on warming to room temperature to normally deep red-brown solutions which on heating convert the Bd to various open-chain and cyclic dimers, trimers and higher oligomers. The product distribution is unique for each element (Table 4.6) and to a certain extent reflects the catalytic activity of derivatives of these elements found by other methods (but *vide infra*).

The role of arene as solvent is important. If Ti, Mn, Fe or Ni atoms are condensed with Bd and hexane rather than an arene, Bd oligomerisation does occur with nearly exactly the same product distribution, but with much reduced activity[108, 149]. Arenes may not only act as good solvents for species in the oligomerisation cycle but also in initial "solvation" of the metal atom, even in cases where stable complexes do not survive to more ambient temperatures.

For example, it is thought that the red-brown solution formed immediately after a Ni/PhMe cocondensate has melted contains what is tantamount to solvated nickel atoms which can be trapped as $Ni[P(OMe)_3]_4$ in 39 per cent overall yield[390].

Surprisingly, Ti atoms are inactive which parallels the results obtained from ($\eta$-arene)$_2$ Ti[108], itself obtained by action of Ti atoms with arene[374]. Therefore Ti alone is not sufficient for Bd oligomerisation. However it is known that the species Arene $TiCl_2 \cdot Al_2Cl_6$ *4.12a* formed by the reaction of $TiCl_4$, Al powder, $Al_2Cl_6$ and arene[155] catalyses the conversion of Bd to *cis, trans, trans*-cyclododeca-1,5,9-triene (ctt-Cdt) and *trans*-poly-1,4-butadiene (t-1,4-Pbd)[156], and that the rate of formation of ctt-Cdt can be increased by addition of small amounts of organoaluminium compounds such as $R_2AlX$ (R = Et, $Pr^i$; X = Et, H, Cl)[157].

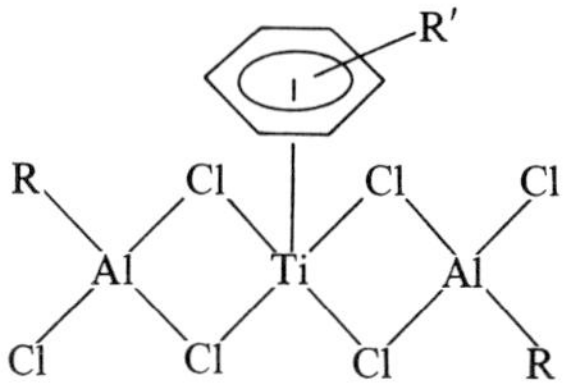

*4.12* (a) R = Cl
(b) R = Et

Therefore cocondensation of Ti atoms, Bd, arene and $Et_2AlCl$ should lead to Bd oligomerisation, which is indeed the case. Conversion to ttt-Cdt, ctt-Cdt and 1,5-Cod (15:80:5) rapidly occurs at room temperature and it is calculated that 1 mmol of Ti in this system converts at least 50 g of Bd to Cdt. $TiCl_4/Et_2AlCl$ (1:10) in PhH has a Bd conversion ratio of 20.7 g/mmol/h[158] while that of $PhR \cdot TiCl_2 \cdot Al_2Cl_6/Et_2AlCl$ (1:4) is 155 (R = H) and 202 g/mmol/h (R = Me)[159]. Cocondensation of Ti atoms, PhMe, Bd, and

**Table 4.7.** To determine the "Essential constituents" for Bd cyclotrimerisation catalysts containing Ti

| Reaction constituents[a] | | | | | Main |
|---|---|---|---|---|---|
| Ti(a) | Al(a) | EtCl | Bd | PhH | Product[b] |
| P | P | P | P | P | Cdt |
| P | | P | P | P | (–) |
| P | P | | P | P | (–) |
| P | | | P | P | (–) |
| | P | P | P | P | (–) |
| | P | | P | P | (–) |
| | | P | P | P | (–) |
| | | | P | P | (–) |

[a] P indicates the species at the head of the particular column was present in the cocondensate.
[b] Cdt refers to Cdt being present as the main observed product species, while (–) indicates no Bd oligomerisation products were observed.

$Et_3Al$ forms only Pbd. This suggest that if Al and Cl are present with Ti, Bd cyclotrimerisation may proceed. Support for this is found in a series of cocondensation experiments where the presence of the components Ti, Al and EtCl are permuted (Table 4.7)[108, 160].

**Table 4.8.** M(a) + Bd + PhH + $Et_2AlX$[a] $\xrightarrow{\text{i, ii}}$

| M | X | Conversion of Bd/% | Conditions ($\theta$/°C, t/h) | Product distribution | | | | | | |
|---|---|---|---|---|---|---|---|---|---|---|
| | | | | ttt-Cdt | ctt-Cdt | LT | Cod | LD | HO | Pbd |
| Ti | Cl | 100[b] | 20, 1 | 15 | 80 | | 5 | | | |
| Ti | Et | 85 | 20, 2 | | | | | | | 100 |
| "$TiO_2$" | Cl | 100[c] | 50, 1 | 13 | 84 | 3 | | | | |
| "$TiO_2$" | Et | 0 | | | | | | | | |
| V | Cl | 43 | 70, 3 | 20 | 30 | 30[d] | 20 | | | |
| Cr | Cl | 60[e] | 60, 3 | 60 | 35 | 2 | 3 | | | |
| Cr | Et | 55 | 60, 2 | 35 | 65 | | | | | |
| Mn | Cl | 40 | 50, 3 | 35 | 65 | | | | | |
| Fe | Cl | 100[f] | 20, 1 | | | | | | | 100 |
| Fe | Et | 100[g] | 20, 1 | | | | | | | 100 |
| Co | Cl | 10[h] | | | | | 20 | 80 | | |
| Ni | Cl | 90[j] | 60, 2 | | | | | | | 100 |
| Ni | Et | 65[k] | 20, 4 | 90 | | 3 | 7 | | | |
| "NiO" | Et | 100 | 80, 1 | 76 | | 24 | | | | |
| Zr | Cl | 90[l] | 120, 7 | | | 100 | | | | |
| Nb | Cl | 2 | 90, 0.5 | | 4 | 30 | | | 66 | |
| Mo | Cl | 1 | 200, 4 | | | 99[m] | 1 | | | |

(i) cocondense, 77 K; (ii) heat, $\theta$ °C

a M(a) : Bd : PhR : $Et_2AlX$ = ca. 1 : 30 : 80 : 1 (R = H, Me).

b Analysis of the cocondensate immediately after warming to room temperature showed 40–50% Bd conversion to ttt-Cdt (15), ctt-Cdt (80) and Cod (5%), and after 30 min at RT there was quantitative conversion. Aliquat portions (5 ml) of the reaction mixture were treated with Bd (10 g) in sealed ampoules at 60 °C giving almost quantitative conversion to Cdt.

c Aliquot portions (5 ml) of the reaction mixture were treated with Bd (ca. 10 g) in sealed ampoules at 70 °C to give almost quantitative conversion to ttt-Cdt (12), ctt-Cdt (84), and LT (3%).

d Three isomers.

e Analysis of the reaction mixture after warming the cocondensate showed ca. 12% conversion to ttt-Cdt (60), ctt-Cdt (30), Cod (3) and LT (7%).

f Analysis of the reaction mixture immediately after warming the cocondensate showed ca. 20% Bd conversion to open-chain and cyclic dimers and trimers.

g Analysis of the cocondensate immediately after warming it to RT showed ca. 10% Bd conversion to LT and Cdt.

h From analysis of the reaction mixture immediately after warming the cocondensate to RT; further heating was not undertaken.

j Analysis of the cocondensate on warming it to RT showed ca. 9% Bd conversion to mainly Cdt and small amounts of Cod.

k Aliquot portions (5 ml) of the reaction mixture were treated with Bd (10 g) in sealed ampoules at 100 °C giving almost quantitative conversion to ttt-Cdt (70), ctt-Cdt (23), and Cod and 4-Vch (7%).

l At 8% Bd conversion (90 °C/3 h) the products were ctt-Cdt (30), LT (30), and HO (8 species) (40%).

m 3 isomers.

Cocondensation of $Et_2AlCl$ (or $Et_3Al$), Bd and arene with atoms of the rest of the first transition series elements (Table 4.8) also leads to Bd oligomerisation and polymerisation. In some cases (Mn and Co), the products and their distribution are the same as when $Et_2AlCl$ is absent, but as Table 4.9 shows, in the other cases the products are different. Organoaluminiumhalides and related species are often added to transition metal salts in order to reduce them to low or zerovalent species which can then react with Bd and enter a catalytic cycle. It is normally assumed that that is the only role of the organoaluminiumhalide. However, these atom reactions show otherwise and illuminate an important question which catalysts derived from metal atoms can help to anwer.

All catalysts or catalyst precursors for Bd oligomerisation contain transition metals in the form of either discrete compounds such as $(1,5\text{-Cod})_2Ni$[161], ($\eta$-1,5,9-Cdt)Ni[161], $(\eta^3\text{-allyl})_3Co$[162] or $(\eta^3\text{-allyl})_2Cr$[161], or mixtures of a metal salt and a reducing agent such as $RhCl_3$-EtOH[163], or $Fe(acac)_2$-$Et_3Al$[164]. It is rare that the metal system initially added to Bd retains any significant equilibrium concentration during the catalytic reactions. Normally the initial compound or system reacts with the Bd, the metal enters the catalytic cycle, and some or all the original ligands around the metal are lost. ($\eta$-Cdt)Ni is one of the few well established compounds which are members of a catalytic cycle and, as such, are true catalysts rather than a catalyst precursor. Since true catalysts are highly reactive they are difficult to isolate or identify. Clearly, the nature of a catalyst precursor, whether it is a pure compound or a reaction mixture, may provide only circumstantial evidence of the nature of the true catalyst intermediates. Even the simple question of which atoms are necessary may not be easy to answer; (though it is normally an advantage of using a pure compound as catalyst precursor that the number of possible catalyst components is limited). For example since $Mn(acac)_2/PrMgCl$[165] and $Ph_2Mn$[166] both cyclotrimerise Bd to Cdt it is clear that all the components acac, Pr, Mg, Cl and Ph are not essential to the catalytic cycle. In contrast $(Cot)_2Fe$[167] and $Fe(Acac)_2$-$Et_3Al$[164] give different oligomers showing they do not give rise to the same true catalysts. Metal atom derived catalysts can be very informative in this context as clearly by judicious permuta-

**Table 4.9.** M(a) + Bd + PhH + X[a] $\xrightarrow{\text{i, ii}}$

| M | Main products | |
|---|---|---|
| | X = nothing | X = $Et_2AlCl$ |
| Ti | 0 | ctt-Cdt (80)[b] |
| V | HO (75) | Cdt (60), LT (30), LD (20) |
| Cr | HO (90) | ttt-Cdt (60), ctt-Cdt (35) |
| Fe | LT (63), ctt-Cdt (30) | Pbd (100) |
| Ni | Cod (55), ttt-Cdt (45) | Pbd (100) |
| Mn | ctt-Cdt (60), ttt-Cdt (35) | ctt-Cdt (65), ttt-Cdt (35) |
| Co | LD (69), Cod (22) | LD (80), Cod (20) |

(i) cocondense, 77 K; (ii) heat to R. T.

[a] M(a) : Bd : PhH : X = ca. 1 : 30 : 80 : 1.
[b] Percentage of oligomeric products.

tion of various additives to a metal atom cocondensate, a catalyst can be built up containing a set (or perhaps more than one set) of essential ingredients for a particular type of conversion. For example, Al and Cl are found to be one set of essential ingredients that must be found in a Ti-containing system to make it an effective Bd cyclotrimerisation catalyst[108, 160)].

Cocondensation of Al atoms and EtBr forms ethylaluminiumsesquibromide, $Et_3Al_2Br_3$[122)], thus it should be possible to form alkylaluminiumhalide cocatalysts *in situ* in the cocondensate, obviating the prior handling of these inconveniently reactive species. When samples of Al and Ti, fused to form an alloy, are coevaporated and cocondensed with Bd, PhH, and EtCl, Bd oligomerisation is indeed observed, as shown in Table 4.10. However a substantial quantity of butenylbenzenes is also formed under these conditions[108)].

It is well established that the products of Bd oligomerisation may be dramatically changed when a ligand such as a tertiary phosphine or an amine is present in the reaction

**Table 4.10.** M(a) + Al(a) + Bd + X[a] ⟶ (i, ii)

| M | X | Conversion of Bd/% | Conditions ($\theta$/°C, t/h) | Product distribution[b] | | | | | | | |
|---|---|---|---|---|---|---|---|---|---|---|---|
| | | | | ttt-Cdt | ctt-Cdt | Ddt | Cod | Vch | OO | Pbd | Bb[c] |
| Ti | EtCl | 45 | 50, 2 | 45 | | | | | | | 55 |
| Ti | – | 0 | | | | | | | | | |
| V | EtCl | 50[d] | 40, 3 | | | | | | 100[e] | | |
| Cr | EtCl | 10[f] | | 6 | 24 | | | | | | 70 |
| Mn | – | 45 | 60,2 | 30 | 70 | | | | | | |
| Fe | EtCl | 90 | 20, 6 | | | | | | | 100 | |
| Fe | $C_6Cl_6$ | 15[f] | | | 25 | 65 | | | | | 10 |
| Fe | – | 20[g] | | | 18 | 60[h] | 22 | | | | |
| Ni | EtCl | 18 | 70, 3 | 65 | | | 35 | | | | |
| Ni | – | 70 | 90, 2 | 40 | | | 45 | 15 | | | |
| – | EtCl | 5[f] | | | | | | | | | 100 |
| – | – | trace[f] | | | | | | | | | 100 |
| – | $TiCl_4$[j] | 23[f] | | | | | | | | | 100[k] |

(i) cocondense, 77 K; (ii) treat to $\theta$ °C

a M: Al: Bd: PhH: X = ca. 1:5:30:80:5.

b OO = other oligomers, Bb = butenyl benzenes.

c 2 isomers Ph·CMe:CHMe and Ph·CH:CHEt.

d Analysis of the cocondensate immediately after warming to RT ahowed ca. 7% Bd conversion to ttt-Cdt (5), ctt-Cdt (13) and Bb (82%).

e Involatile liquids which can not be analysed by glc.

f From analysis of the cocondensate immediately after being warmed to RT; further heating was not undertaken.

g An aliquot portion (10 ml) of the reaction mixture was treated with Bd (ca. 3 g) in a sealed ampoule at 55 °C for 4 h giving 40% conversion to ctt-Cdt (30), Ddt[h] (63), and Cod (7%).

h 3 isomers.

j Al: $TiCl_4$ = ca. 10:1.

k Consists of Ph·CH:CH·Me, Ph·CH:CH·Et, and o- and p-dibutenylbenzenes.

**Table 4.11.** M(a) + Bd + PhR + X[a] ——→ (i, ii)

| M | X | Conversion of Bd/% | Conditions ($\theta$/°C, t/h) | Product distribution[b] ttt-Cdt | ctt-Cdt | LT | Cod | Vch | LD | Pbd |
|---|---|---|---|---|---|---|---|---|---|---|
| Ti + $Et_2AlCl$[c] | $PPh_3$ | 100 | 40, 2 | 95 | 4 | | | | | |
| "$TiO_2$" | $PPh_3$ | 0 | | | | | | | | |
| ["$TiO_2$"][d] | $PPh_3$ | 85 | 70, 3 | 27 | 65 | 7 | | | | |
| Cr | $PPh_3$ | 100 | 20, 4 | | | | | | | 100 |
| Fe | $PPh_3$ | 37 | 50, 5 | | | | | | 100 | |
| Fe | $P(OPh)_3$ | 0 | | | | | | | | |
| Fe | $P_4$[e] | 10 | 50, 4 | 100 | | | | | | |
| Co | $PPh_3$ | 35 | 50, 5 | | | | | | 100 | |
| Co | $P(OPh)_3$ | 0 | | | | | | | | |
| Ni | $PPh_3$ | 100 | 60, 3 | 5 | | | 70 | 25 | | |
| Ni | $P(OPh)_3$ | 0 | | | | | | | | |
| Ni | $P_4$[e] | 73 | 70, 4 | 40 | 60 | | | | | |
| "NiO" | $P_4$[e] | 0 | | | | | | | | |
| Pd | $PPh_3$ | 0[f] | | | | | | | | |
| Pd | $P_4$[e] | 20 | 70, 3 | | | 100 | | | | |

(i) cocondense, 77 K; (ii) heat to $\theta$ °C

[a] M : Bd : PhR : X = ca. 1 : 30 : 80 : 1, R = H, Me.
[b] Vch = 4-vinylcyclohexene.
[c] Ti : $Et_2AlCl$ : $PPh_3$ = ca. 1 : 1 : 1.
[d] ["$TiO_2$"] is an aliquot portion (5 ml) from the reaction mixture "$TiO_2$" + $Et_2AlCl$ + Bd PhMe which is added to an equimolar amount of $PPh_3$ and Bd (5 g) in PhMe.
[e] Red phosphorus (coevaporated).
[f] Some metallic Pd was precipitated. Higher $PPh_3$ : Pd ratios lead to $Pd(PPh_3)_4$. With xs $PPh_3$ the reaction is stoichiometric with respect to Pd cocondensed.

mixture[168)]. Cocondensation of metal atoms, Bd, arene and $PPh_3$ leads to markedly altered oligomerisation products as shown in Table 4.11. For the later transition metals, there is a general trend towards dimer rather than trimer formation, indicating that $PPh_3$ is bonded to and blocking a coordination site of the true catalysts in the catalytic cycle. This result precisely parallels that found in "classic" experiments. Atoms of the earlier transition metals behave differently. Cr/Bd/PhMe cocondensates are little affected by $PPh_3$ whilst inclusion of $PPh_3$ into the Ti/Bd/PhMe/$Et_2AlCl$ cocondensate causes a pronounced increase in selectivity and essentially quantitative conversion to 95% ttt-Cdt. $P(OPh)_3$ blocks catalysis in all cases where it is used. Surprisingly addition of red phosphorus to the Fe or Ni systems, in an attempt to find a cheap substitute for $PPh_3$, causes *increased* cyclotrimerisation at the expense of open-chain and cyclic dimers[108)].

The transition metal oxides, $TiO_2$, FeO and NiO can also be used with some success as evaporant sources in generating catalytic systems. Under the conditions of evaporation the main species present in these vapours are TiO[169,170)], Fe[169)] and Ni[169)] respectively, and oxygen. Thus the vapour over NiO, when condensed with Bd and PhMe forms a catalytic system whose product distribution differs in detail from one containing Ni atoms (see Table 4.6)[108)].

**Table 4.12.** Bd oligomerisation catalysts from ($\eta$-arene)$_2$Ti and ($\eta$-arene)Ti(AlCl$_4$)$_2$

| System | Conversion of Bd/% | Conditions ($\theta$/°C, t/h) | Product distribution | | | | | |
|---|---|---|---|---|---|---|---|---|
| | | | ttt-Cdt | ctt-Cdt | Ddt | Cod | Pbd | Bb |
| (PhH)$_2$Ti | 10[a] | 80, 6 | | | | | 100 | |
| (PhH)$_2$Ti + Et$_2$AlCl[b, c] | 100 | 60, 2 | 30 | 67 | | 3 | | |
| (PhH)$_2$Ti + EtAlCl$_2$[d] | 100 | 70, 2 | 5 | 95 | | | | |
| (PhH)$_2$Ti + AlCl$_3$[d] | 100 | 60, 2 | 3 | 56 | | 2 | 30 | 10 |
| (PhMe)$_2$Ti | 20 | 20, 72 | | | | | 100 | |
| (PhMe)$_2$Ti + Et$_3$Al | 20 | 60, 3 | | | | | 100 | |
| (PhMe)$_2$Ti + Et$_2$AlCl[c] | 100 | 60, 2 | 27 | 69 | | 4 | | |
| (PhMe)$_2$Ti + EtAlCl$_2$[d] | 100 | 60, 2 | 2 | 98 | | | | |
| (PhMe)$_2$Ti + Me$_2$AlCl | 100 | 60, 2 | 50 | 43 | | 7 | | |
| (PhMe)$_2$Ti + Me$_2$AlCl + ChtH | 100 | 60, 2 | 34 | 40 | 24 | 2 | | |
| (PhH)Ti(AlCl$_4$)$_2$ | 20.3 g[a] | 50, 1 | 63[e] | | | *[f] | 37 | |
| (PhH)Ti(AlCl$_4$)$_2$ + Et$_2$AlCl[g] | 155.4 g | 50, 1 | 91 | | | | 9 | |
| (PhH)Ti(AlCl$_4$)$_2$ + Me$_2$AlCl | 58.6 g | 50, 1 | 82 | | | | 18 | |
| (PhMe)Ti(AlCl$_4$)$_2$ | 4.2 g | 50, 1 | 21 | | | | 79 | |
| (PhMe)Ti(AlCl$_4$)$_2$ + Et$_2$AlCl | 202 g | 50, 1 | 92 | | | | 8 | |
| (PhMe)Ti(AlCl$_4$)$_2$ + Me$_2$AlCl | 38.6 g | 50, 1 | 91 | | | | 9 | |

[a] Expressed in terms of the percentage of Bd converted in the given time for the (PhR)$_2$Ti systems and grams of Bd per mmol of catalyst in the hour for the (PhR)Ti(AlCl$_4$)$_2$ systems.
[b] (arene)$_2$Ti : R$_n$AlCl$_{3-n}$ = 1 : 1.
[c] Further Et$_2$AlCl causes no change.
[d] Further addition increases the proportion of Pbd.
[e] Mixture of 95–99% ctt-Cdt and 1–5% ttt-Cdt.
[f] Cod and Vch present in *ca.* 0.5% quantities.
[g] (arene)Ti(AlCl$_4$)$_2$ : R$_2$AlCl = 1 : 4.

Although it is possible by using metal atoms as the source of the metal in the catalytic reaction to learn about the necessary constituents for catalytic activity, there is little evidence for the precise nature of the members of any given catalytic cycle. Fortunately, more information can be gleaned from trapping experiments, and from more conventional experiments especially those of the monumental works of Wilke and coworkers in this area of Bd oligomerisation.

Permutation of various cocatalysts have shown that Al and Cl are one set of essential components of a Ti-containing Bd cyclotrimerisation catalyst. In addition the species ($\eta$-PhR)$_2$Ti (R = H, Me) is similarly active only in the presence of organoaluminiumhalides such as Et$_2$AlCl and EtAlCl$_2$ (see Table 4.12). Normally organoaluminiumhalides are employed as reducing agents of transition metal salts but when the metal species is already low-valent there is convincing evidence that EtAlCl$_2$, for example, can act as an oxidising agent. Addition of EtAlCl$_2$ to ($\eta$-PhH)$_2$Mo causes smooth oxidation to the cation ($\eta$-PhH)$_2$Mo$^+$ while ($\eta$-PhH)$_2$V in PhMe will reduce EtAlCl$_2$ to an Al mirror[171]: ($\eta$-PhH)$_2$Cr behaves similarly[172]. Furthermore when ($\eta$-arene)$_2$M derivatives are made by the conventional reducing Friedel-Crafts method[173] the metal salt is only reduced to the ($\eta$-arene)$_2$M$^+$ cation (M = V, Cr, Mo, V) (Eq. (4.61)).

$$3\ CrCl_3 + 2\ Al + AlCl_3 + 6\ arene \longrightarrow 3\ [(\eta\text{-arene})_2Cr]^+[AlCl_4]^- \qquad (4.61)$$

In the case of Ti, the species *4.12a* is formed[155] which acts as a very similar Bd oligomerisation agent[159] as the ($\eta$-PhMe)$_2$Ti/AlCl$_3$ systems. In addition *4.12a*, modified by addition of Et$_2$AlCl is thought to form *4.12b*[159] and this has similar catalytic properties to the system ($\eta$-PhMe)$_2$Ti/EtAlCl$_2$. Thus it is reasonable to propose that species such as *4.12* are formed on reaction of organoaluminiumhalides in ($\eta$-PhR)$_2$Ti and Ti/PhR cocondensates. The key catalytic intermediate in the formation of Cdt is thought to be Bd · TiCl$_2$ · Al$_2$Cl$_4$R$_2$, while conductivity experiments suggest that *4.12a* dissociates to form species such as (PhR)$_n$TiAlCl$_4^+$ and (PhR)$_n$Ti$^{2+}$ (n = 1 or 2) which are the precursors of the Bd polymerisation catalysts[159].

Cr atoms polymerise Bd, as do the systems Cr(acac)$_3$/Me$_3$Al or ($\eta$-allyl)$_3$Cr[160], the product being poly-1,2-butadiene. Thus it seems that Cr atoms form systems which act as monoene polymerisation catalysts. Similar results have been noted for Cr atoms with ethylene[176] and styrene[142]. Inclusion of Et$_2$AlCl in the Cr/Bd/PhR system changes the catalytic activity and essentially Cdt is formed. [($\eta$-allyl)$_2$CrCl]$_2$ in CH$_2$Cl$_2$ gives almost identical results[162]. A CrCl fragment has been postulated as the active fragment and that species such as ($\eta$-Cdt)CrCl occur in the catalytic cycle[161].

Deuteriolysis of the Cr/Bd cocondensate after removal of unreacted Bd forms essentially butenes of which the major constituent is d$_2$-but-1-ene, indicating polar Cr-Bd bonds[115] (see alkenes). If Cr atoms Bd and PF$_3$ are cocondensed the species ($\eta^4$-Bd)Cr-(PF$_3$)$_4$ can be isolated in low yields (3%)[115]. Similarly addition of CO to a Cr/Bd cocondensate at 77 K, forms on warming, ($\eta^4$-Bd)Cr(CO)$_4$ (4%)[175], a suspected catalytic intermediate in the photochemical hydrogenation of Bd by H$_2$ using Cr(CO)$_6$[176]. Its decomposition at *ca.* 0 °C and sensitivity to impurities explains its previous non-isolation and shows that 1,3-dienetetracarbonyl-chromium complexes, of which BdCr(CO)$_4$ is the first example are appreciably more labile than the relatively inert tetracarbonylchromium complexes of non-conjugated dienes[177]. Attempts have been made to rationalise this in terms of the energy ordering, symmetry and extent in space of the valence molecular orbitals of the Cr(CO)$_4$ fragment[58].

Cocondensation of Bd and either Mo or W atoms leads to formation of the stable, air-insensitive complexes ($\eta^4$-Bd)$_3$M (M = Mo, W) in good yield (50–60%)[3]. These surprising compounds are the only known examples of binuclear M-Bd complexes where Bd has kept its integrity. The molecular structure of (Bd)$_3$Mo has been determined *4.13*[127] and also holds some surprises. All the metal-carbon distances are the same (2.29 (1) Å) while the formal double bond length (1.32 (2) Å) is shorter and the intervening bond

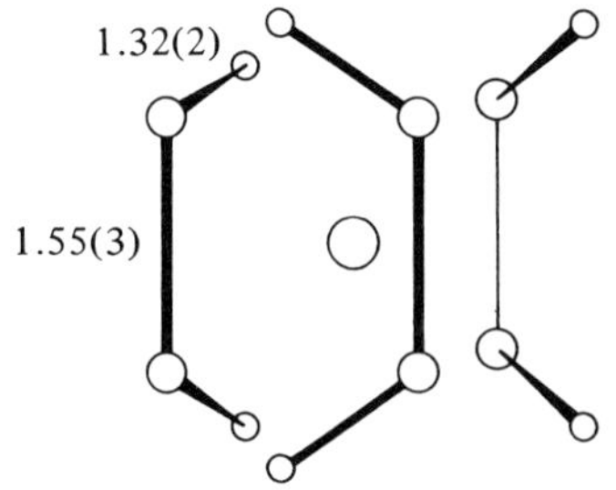


*4.13*

length (1.53 (3) Å) longer than in the free ligand. This is opposite to the normal trend of equalisation of the three ligand bond lengths due to population of the lowest lying anti-bonding orbital of Bd[23,178]. Moreover these lengths are almost exactly the C-C double and single bond lengths respectively. Thus it may be that Bd is acting as a non-conjugated diene and $(Bd)_3Mo$ is more accurately represented by $(\eta^2, \eta^2\text{-Bd})_3Mo$.

No organometallic compound has been isolated from a Mn atom/Bd cocondensate but addition of CO at 77 K leads to $(Bd)_2MnCO$ (4–5%) on warming[179].

At low temperatures (–20 °C), the Fe/Bd/PhMe cocondensate converts Bd to mainly 4-Vch (75 per cent selectivity at 35% conversion)[95]. There is little corroborative information known of species involved in the catalytic cycle except that addition of Bd to a Fe/PhMe cocondensate forms the previously unknown species $(\eta^6\text{-PhMe})(\eta^4\text{-Bd})Fe$ *4.14a* as a red volatile liquid[378]. However, it is tempting to write as the key intermediate in the catalytic cycle *4.14b* on analogy with the species *4.16* derived from the known catalyst precursor for 4-Vch formation *4.15*[161] (Eq. (4.62)), and remembering that the $(\eta^6\text{-PhMe})Fe$ and $(\eta^5\text{-Cp})Co$ fragments are isoelectronic and isolobal.

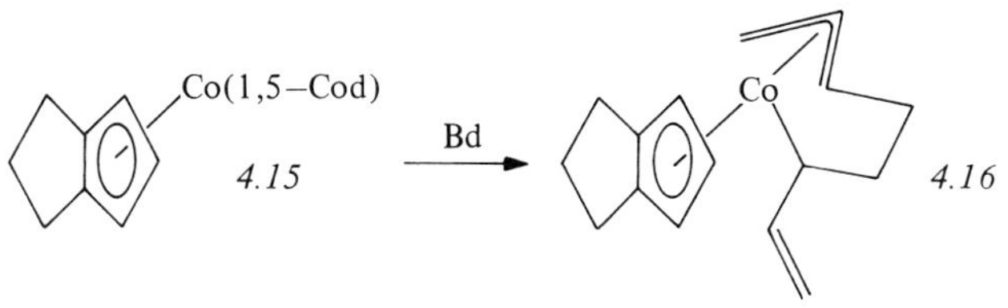


(4.62)

Furthermore $(\eta^5\text{-Cp})(\eta^3\text{-allyl})(\eta^1\text{-allyl})Co$ *4.17* is converted to $(\eta^5\text{-Cp})(\eta^4\text{-Bd})Co$ *4.18* by Bd[180] (Eq. (4.63)).

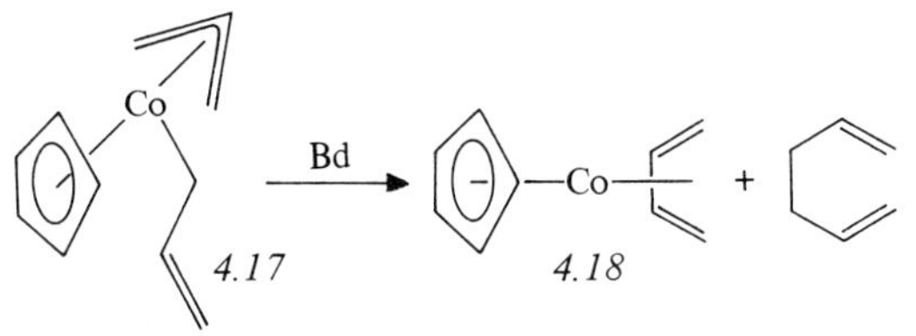


(4.63)

Therefore a plausible mechanism for the Fe reaction may be written as in Scheme 4.6. The activity and selectivity (to 100%) of the reaction can be improved by bubbling through NO. No species has been isolated, but addition of $PPh_3$ afterwards gives $Fe(NO)_2(PPh_3)_2$[95]. Such selectivity is found for similar catalyst precursors such as $Fe(NO)_2(CO)_2$ and Fe(NO)$(\eta^3\text{-allyl})(CO)_2$[181], $Na[Fe(CO)_3NO]/[Fe(NO)_2Cl]_2$[182] and $[Fe(NO)_2Cl]_2/Zn$[183]. In addition the Fe/isoprene/PhMe cocondensate gives on treatment with NO[95] the Vch derivatives, *4.19* therefore it would seem that a $Fe(NO)_2$ fragment can be involved in exclusive formation of Vch, and can replace PhMe in the preceeding reaction scheme. The

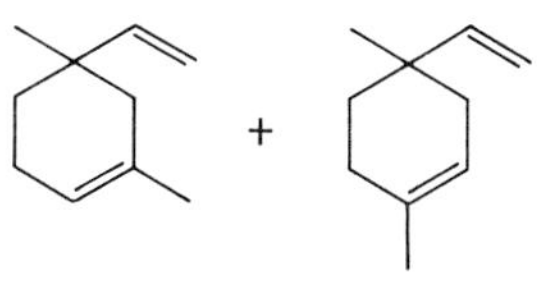


*4.19*

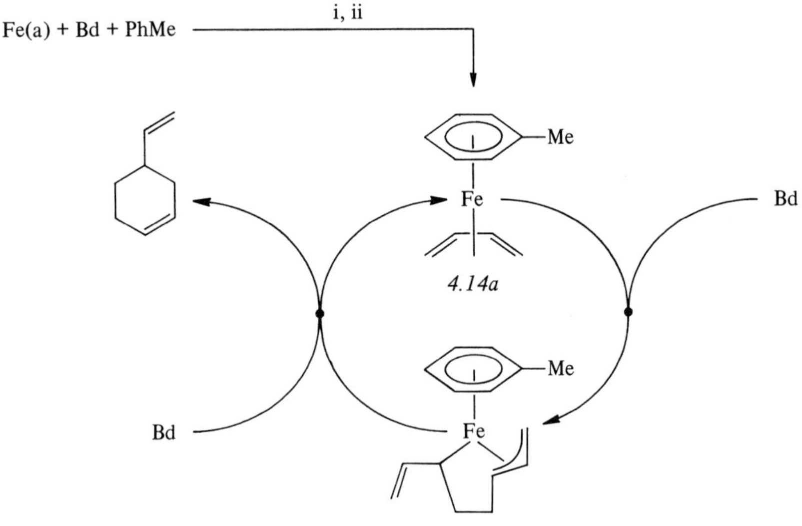


4.14a

4.14b

**Scheme 4.6.** (i) cocondense, 77 K; (ii) −20 °C

reduced selectivity of the ($\eta^6$-PhMe)Fe fragment compared with the isoelectronic $Fe(NO)_2$ fragment may be due to weaker bonding by PhMe to Fe. Certainly NO can replace PhMe in $(PhMe)Fe(PPh_3)_2$[95]. In addition the facility of $\eta^3 \leftrightarrow \eta^1$ interconversion of bonded NO is considered very important. The Fe/Bd/PhMe reaction mixture at 50 °C gives much changed products (mainly Ddt)[108]. Ddt, however, can only be formed from Bd *via* hydrogen transfer (*vide infra*) which may be facilitated by a Fe-H species; formed perhaps by insertion into the benzylic C-H bond of toluene (see arenes).

Species such as $Bd_2Fe$ or $Bd_3Fe$ have been postulated as products of the Fe atom-Bd reaction, though no tractable compound survives to room temperature[19, 117]. However, warming the Fe/Bd cocondensate under CO or $PF_3$ atmospheres forms $Bd_2FeL$ (L = CO, $PF_3$) in reasonable yield[175, 384], while warming in the presence of $P(OMe)_3$ leads to isolation of $Bd_2FeP(OMe)_3$ and $BdFe[P(OMe)_3]_3$[117]. Bis(diene)monocarbonyliron complexes at *ca.* 100 °C in the presence of $PPh_3$ are catalysts for diene dimerisation, e.g. Bd to 4-Vch and 1,5-Cod[270], and are especially suitable for the mixed dimerisation of dienes with acrylates to give heptadienoates[271]. The direction and amount of catalytic activity are strongly affected by replacing CO with phosphines (Eq. (4.64))[17].

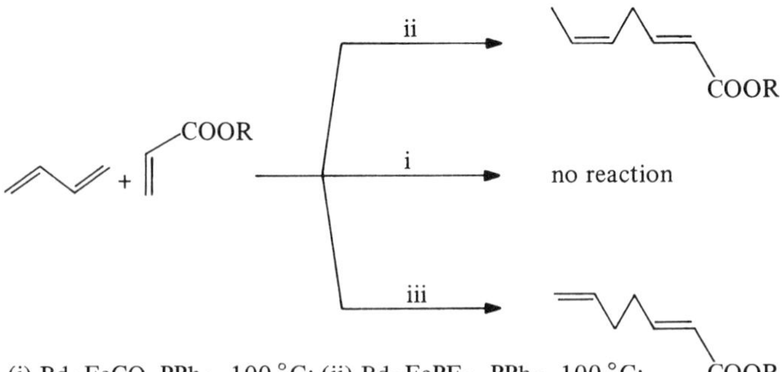


(4.64)

(i) $Bd_2FeCO$, $PPh_3$, 100 °C; (ii) $Bd_2FePF_3$, $PPh_3$, 100 °C;
(iii) $Bd_2FeP(OMe)_3$, $PPh_3$, 100 °C

This is another example of the tailoring of organometallic catalysts using metal atom reactions.

Cocondensation of Fe atoms and 2,3-dimethylbutadiene followed by warming under $N_2$ to $-20$ °C produces on filtration and evaporation of all liquid a brown residue with erratic analysis. Introduction of $P(OMe)_3$ after three days at 0 °C produces (2,3-dimethylbutadiene)tris(trimethylphosphite)iron as the only identifiable product. It therefore appears that no β-hydrogen-metal interaction to form a more stable complex can occur in this case[117)].

The main products from a Co(a)/Bd/PhMe cocondensate are the dimers methylheptatriene (Mht) *4.21a* and octatriene (Ot) *4.21b* at low conversion, while the proportion of trimers and higher oligomers increases with conversion, which precisely parallels the behaviour of the typical Co-systems $(\eta\text{-allyl})_3Co$[161)] and $Co_2(CO)_8/Et_3Al/PhH$[184)]. The action of $(\eta\text{-allyl})_3Co$ is particularly well-documented[162)]. The key intermediate is *4.20*, formed upon reaction of Bd and $(\eta\text{-allyl})_3Co$ (Eq. (4.65)).

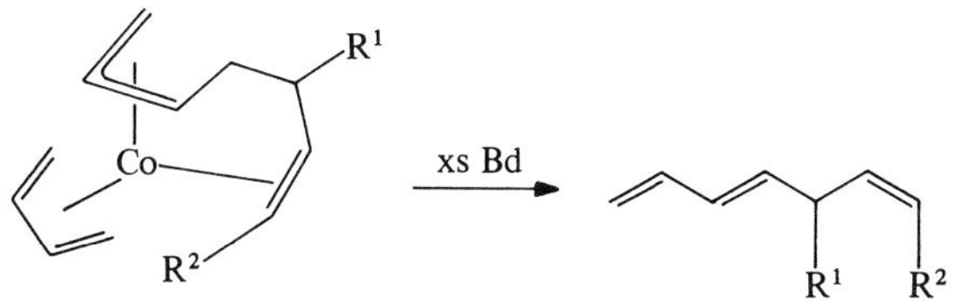


(4.65)

*4.20* *4.21* (a) $R^1$ = Me, $R^2$ = H
(b) $R^1$ = H, $R^2$ = Me

Mht is formed more abundantly than Ot (Mht : Ot = 9 : 1) and as both are terminal dienes, they can re-enter the catalytic cycle to form trimers and then higher oligomers.

Species such as *4.20* comprise Co, three Bd units and one hydrogen. Co has shown its readiness to abstract hydrogen from various substrates in many reactions[19, 127)]. Of particular relevance is the cocondensation of Co atoms, Bd and a molecule containing a readily abstractable hydrogen such as isobutane[185)] (Eq. (4.66)). Reaction of a third Bd molecule may form *4.20*.

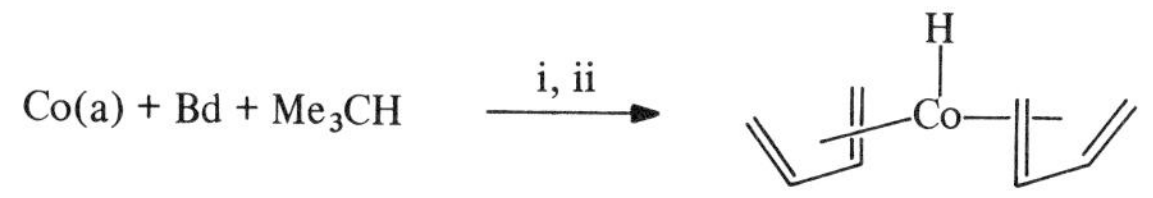


(4.66)

(i) cocondense, 77K, (ii) warm

If Ni atoms are cocondensed at 77 K with Bd and the excess Bd removed at $-78$ °C, a grey solid is formed which on deuteriolysis gives 2 moles of $d_0$-Bd per g atom of Ni[118)]. Prolonged pumping of this solid gives a yellow material in very low yield which was originally formulated as "bis(butadiene)nickel" but has since been shown to contain bis(crotyl) Ni[127)]. Treatment of this or the grey solid with excess Bd at $-20$ °C forms in good yield the blood-red species *4.22*[118)] a known true catalyst in the cyclotrimerisation of Bd to Cdt[186)]. Warming this species in the presence of further Bd or the action of ligands such as $PPh_3$ converts this species to predominantly ttt-Cdt.

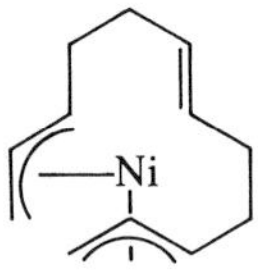


*4.22*

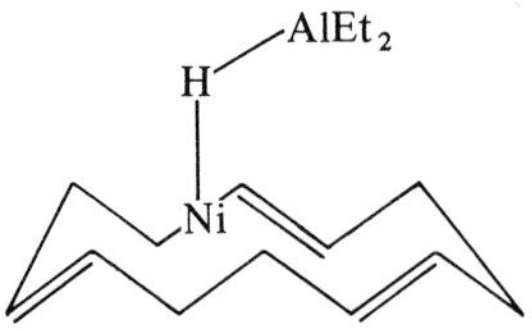


*4.23*

Heating the Ni/Bd/PhR cocondensate to 80 °C gives ttt-Cdt as expected though 1,5-Cod predominates (55% selectivity at 75% conversion). In comparison the system Ni(acac)$_2$/ Et$_2$AlOEt (1 : 2)[187] forms 81.7% ttt-Cdt at 100% and the amount of Cod produces is only 1.6%. The high proportion of Cod is difficult to explain, though Wilke and coworkers have noted that the proportion of dimers increases with temperature[187,188]. In addition the Ni/Bd/PhR/Et$_3$Al cocondensate gives ttt-Cdt with 95% selectivity at 65% conversion, which suggests involvement of the organoaluminium complex. Indeed the complex (Cdt)Ni(Et$_2$AlH) which decomposes at −60 °C is thought to have the structure *4.23* on the basis of its nmr[189] and the addition of ligands is known to accelerate the closing of the C$_{12}$-bisallyl chain in *4.22*[190].

Inclusion of PPh$_3$ in the Ni/Bd/PhR cocondensate leads to formation of the dimers 1,5-Cod and 4-Vch *via* the intermediacy of *4.24* and *4.25* for which model compounds are known[190].

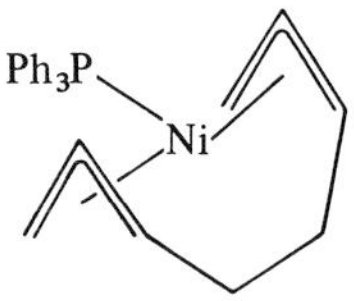


*4.24*

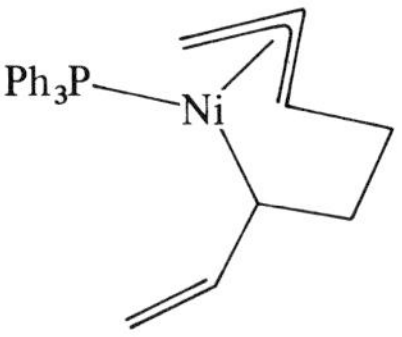


*4.25*

The Ni/Bd/PhR/Et$_2$AlCl cocondensate in contrast forms Pbd on warming which again finds a parallel in the work of Wilke and coworkers with the complex [($\eta$-allyl)Ni]$_2$($\mu$-Br)$_2$[162]. The most likely purpose of the Et$_2$AlCl is to oxidise Ni to a divalent species similar to that *4.26* formed on reaction of NiCl$_2$ and EtAlCl$_2$[163] which can then react with Bd to form a species *4.27*, similar to ($\mu$-Br)$_2$[($\eta_3$-allyl)Ni]$_2$ (Eq. (4.67)).

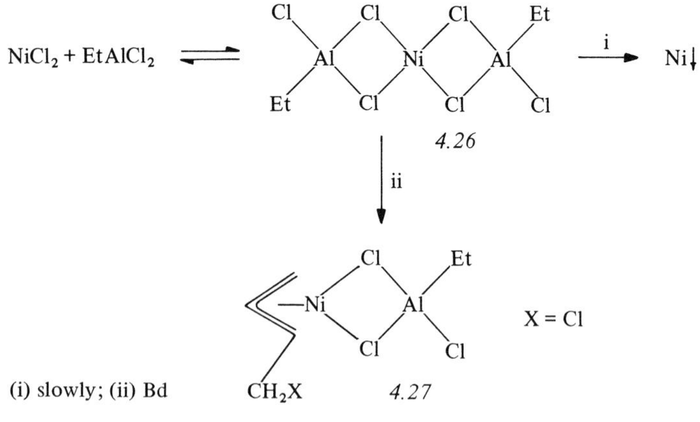


(4.67)

The oxidation of low valent complexes is not without precedent and has used to explain the catalytic properties of the Ti/$Et_2AlCl$/Bd/PhR cocondensate (*vide supra*).

As Tables 4.6 and 4.7 show, atoms of the second transition series elements can also effect Bd oligomerisation, but with reduced activity. In addition, little cyclotrimerisation occurs in contrast to the first row transition metal atoms[154]. An explanation is that trimeric Bd-derived ligands bonded to a metal, e.g. *4.22*, cannot couple to give Cdt when the metal is larger as in the case of the second row transition metals. Instead hydrogen transfer occurs to give Ddt (Eq. (4.68)).

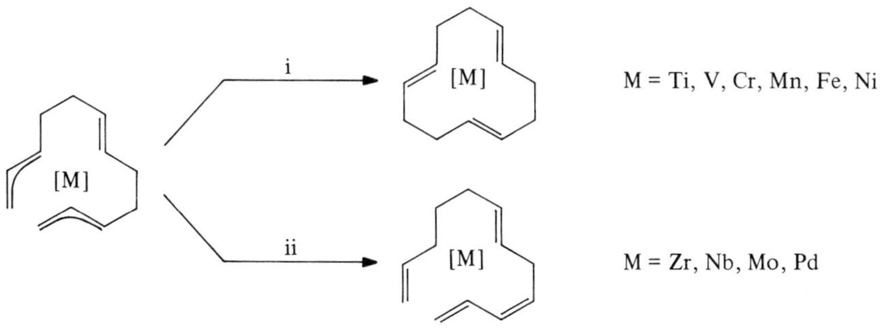


(4.68)

(i) couple; (ii) H–transfer

This distinction is nicely illustrated by Ni and Pd.

The Pt/Bd cocondensate is effective for the hydrogenation of Bd at −78 °C[126]. The major product is butane with only small amounts of butenes formed. This is the reverse of normal Pt catalytic activity[191]. In addition the Pt/but-1-ene cocondensate catalytically hydrogenates but-1-ene[126].

Bd reacts with excess Li atoms at 850 °C in the vapour phase to form a solid whose deuteriolysis products suggest that Li atoms add across the double bonds to form $Li \cdot CH_2 \cdot CH : CH \cdot CH_2Li$ (16%) and $Li \cdot CH_2 \cdot CHLi \cdot CH : CH_2$ (2%) as well as causing fragmentation to $C_3Li_4$, $C_2Li_6$, $C_2Li_4$ and $C_2Li_2$[148].

Cocondensation at 77 K of Mg atoms and Bd followed by removal of unreacted Bd at −78 °C produces a residue which on deuteriolysis gives 1,4-dideuteriobut-2-ene[127]. This indicates that polar bonds which can be represented by the various formalisms *4.28* are formed.

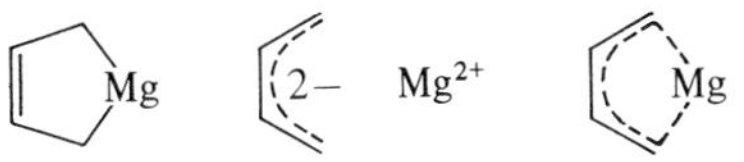


*4.28*

Similar results have been found for the reaction of activated solid magnesium with Bd in thf in the presence of Lewis acids. Deuteriolysis of this product forms $d_2$-butene and $d_2$-octadiene[192]. This may prove to be a useful way of synthesis of halide free organo-magnesium compounds which show promise of having great potential for organic and inorganic synthesis[193].

Deuteriolysis of the Al/Bd cocondensate after removal of excess Bd at −78 °C, forms mainly 3,4-dideuteriobut-1-ene which supports the postulate of Al atoms forming 1,2-addition products with species containing C=C double bonds[121,127] (Eq. (4.69)).

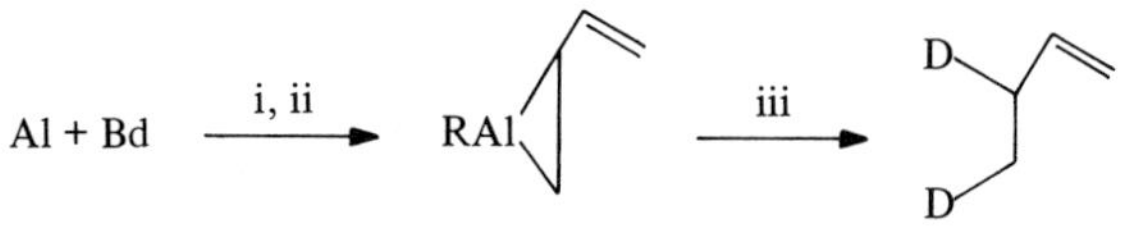


(4.69)

(i) cocondense, 77K; (ii) remove xs Bd, −78°C; (iii) $D_2O$

### *4.12.2 Alkadienes with Active Hydrogens*

Metal-hydrogen interactions are widespread in the reactions of metal atoms with alkadienes possessing active hydrogens. It is almost certain that the initial reaction is coordination through the ligand π-system and this brings the metal moiety and active hydrogen in close juxtaposition and facilitates interaction (Eq. (4.60)). This reaction manifests itself through the driving forces of increasing the coordination and electron number of the complex (Eq. (4.70)), isomerisation to a conjugated system (Eq. (4.71)), disproportionation to aromatic systems (Eq. (4.72)), disproportionation to form 18e compounds (Eq. (4.73)), and isomerisation to more stable species (Eq. (4.74)).

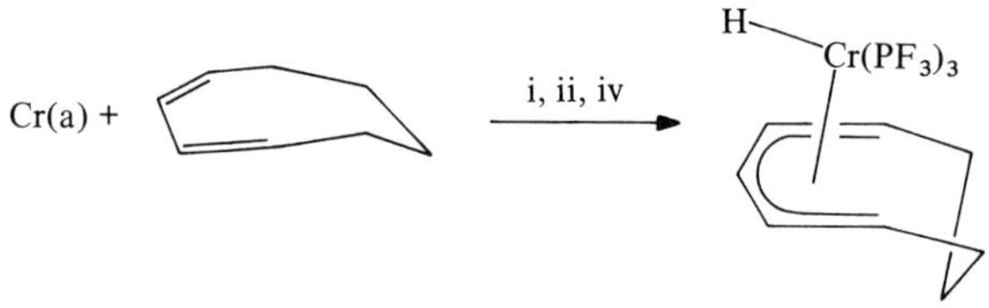


(4.70)[17]

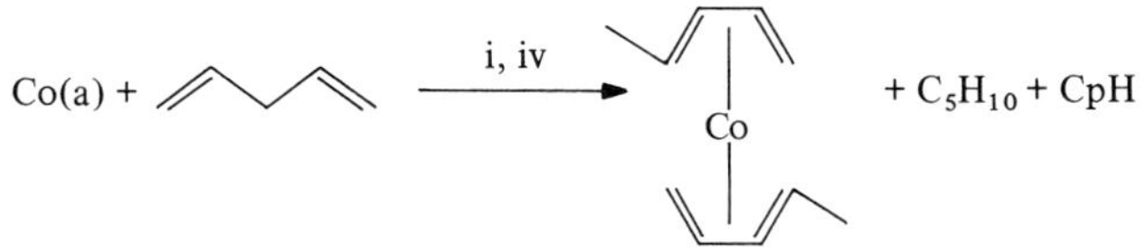


(4.71)[120,127]

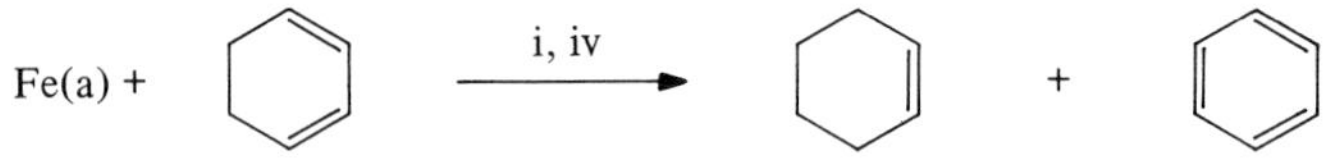


(4.72)[127]

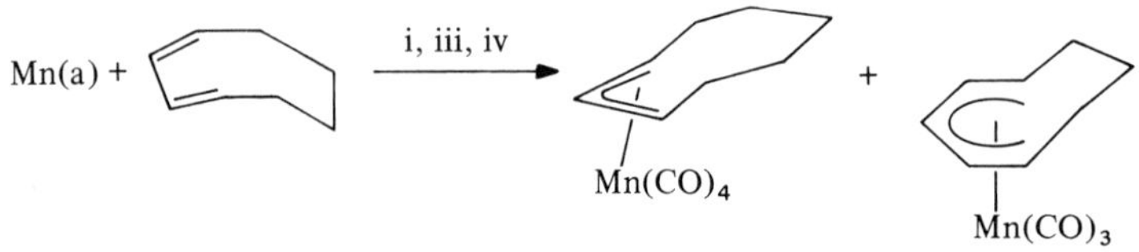


(4.73)[194]

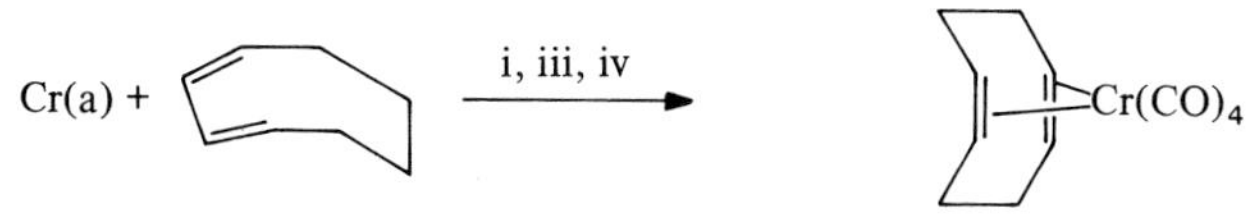


(4.74)[17]

(i) cocondense, 77K; (ii) $PF_3$, 77K; (iii) CO, 77K; (iv) warm

The reactions of metal atoms with CpH (and indene) further exemplifies this (Eqs. (4.75) to (4.80)).

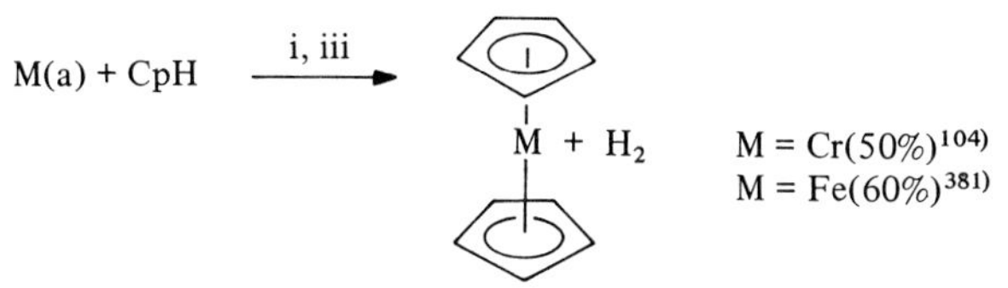


(4.75)

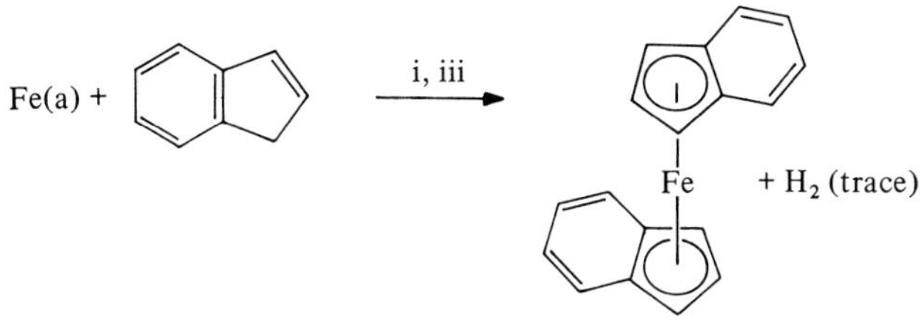


(4.76)

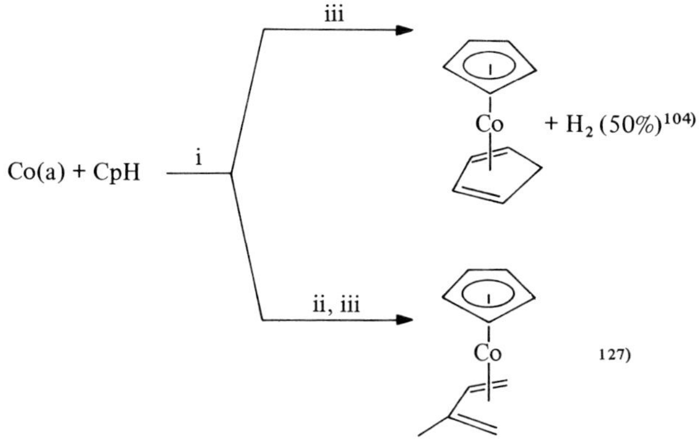


(4.77)

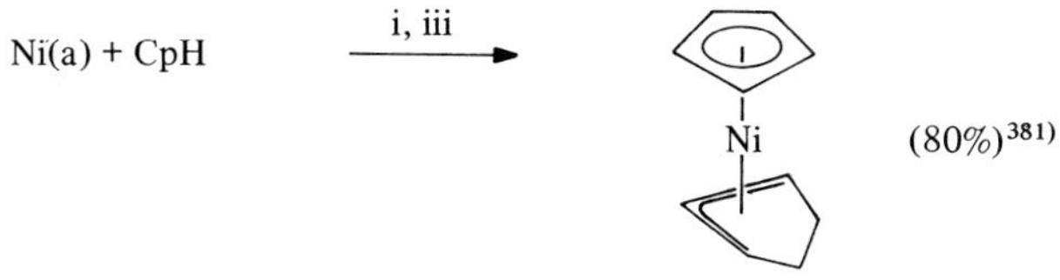


(4.78)

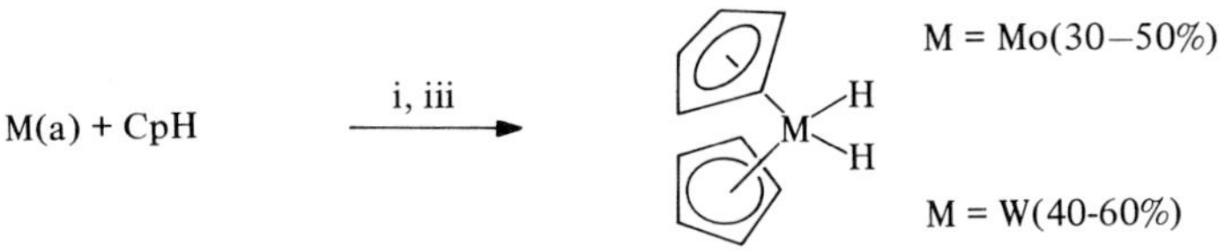


(4.79)[10,11]

(4.80)

(i) cocondense, 77K; (ii) isoprene, 77K; (iii) warm; (iv) observed by ir spectroscopy at 88K

The driving force is undoubtedly the formation of the compact, robust relatively inert pseudo aromatic cyclopentadienyl ligand which can be found in a wide variety of complexes[23,197]. The strength of this driving force can be seen in the reaction of W atoms and cyclopentene in forming $Cp_2WH_2$ (30%) and cyclopentane[11].

The mechanism is almost certainly hydrogen transfer; for example, a plausible mechanism for the reaction of Ni atoms with CpH is shown in Eq. (4.81). The model compounds *4.29*[198] and *4.30*[199], formed in Eq. (4.82) and (4.83) support this, as does the equilibrium (Eq. (4.84))[190].

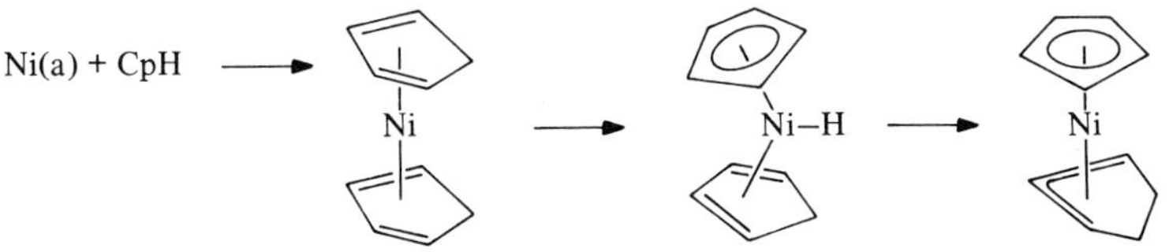


(4.81)

*4.29*

(4.82)

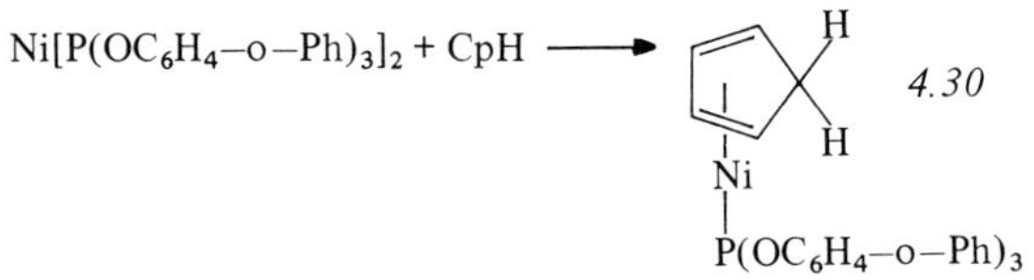


*4.30*

(4.83)

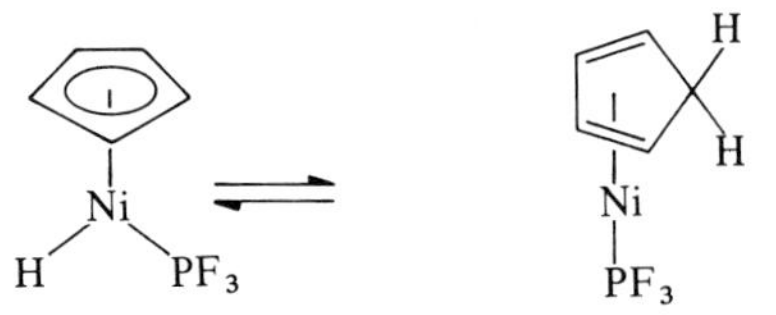


(4.84)

Hydrogen transfer can occur at very low temperatures: The ir spectrum at 77 K after codeposition at 77 K of CpH and In or Tl vapours shows CpM (M = In, Tl) is already formed[196]. Addition of CO at 77 K to the Cr/CpH cocondensate leads to isolation of ($\eta^5$-Cp)($\eta^3$-$CpH_2$)$Cr(CO)_2$ (5%)[200] which suggests a $(CpH)_2Cr$ or more likely a $Cp_2CrH_2$ intermediate.

Cocondensation of Mn atoms and CpH surprisingly does not form the known complex $Cp_2Mn$. However a mixture of CpH or CpMe and PhH, PhMe or mesitylene when condensed with Mn atoms gives in very low yields the corresponding mixed ligand complexes *4.31* (Eq. (4.85))[17,19]. $H_2$ evolution does not occur till −120 °C which implies that the cyclopentadienyl derivatives are not primary products[19]. Thus it may be that an arene-Mn complex is better able to coordinate to CpH than Mn atoms. There is some precedence in that $Cp_2Fe$ can be formed in 40% yield by addition of CpH to a Fe/PhH cocondensate[381].

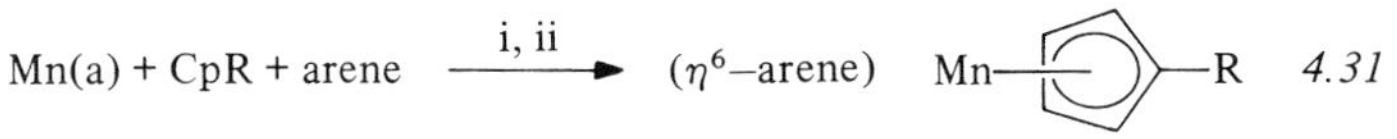


(4.85)

arene = PhH; R = H, Me
PhMe; R = H, Me
Mes; R = H

(i) cocondense, 77K; (ii) warm

Cocondensation of W atoms and mixtures of an arene and CpH also forms mixed ligand complexes (Eq. (4.86))[11]. The compounds are not separated but identified mass-spectrometrically. The results reflect a better bonding ability of PhMe over PhH in such systems. Initial experiments with the fused aromatic hydrocarbon indene, which can act as an arene or a cyclopentadiene indicates a similar reaction (Eq. (4.87))[11].

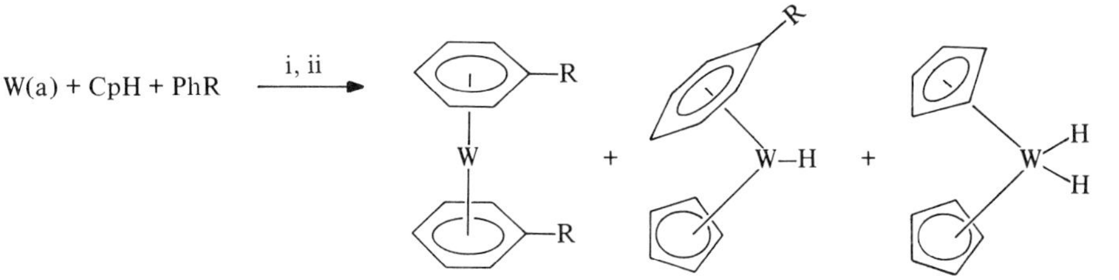


(4.86)

| | | | | |
|---|---|---|---|---|
| R = H; | CpH : PhH = 1 : 1 | 0 | 3 | 7 |
| | CpH : PhH = 1 : 5 | 0 | 4 | 6 |
| R = Me; | CpH : PhMe = 1 : 1 | 2 | 5.5 | 2.5 |

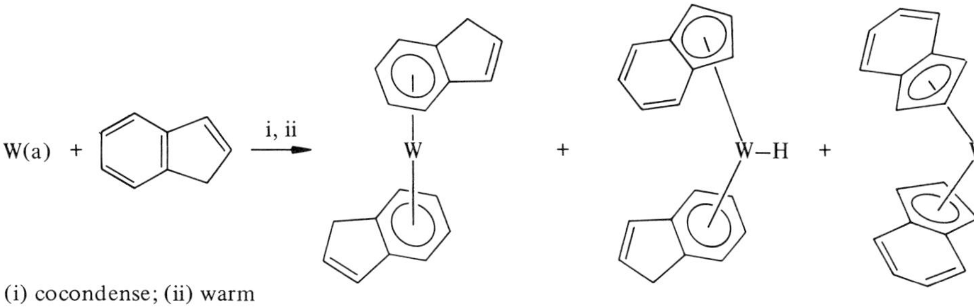


(4.87)

(i) cocondense; (ii) warm

A novel synthesis of 1,1′-alkyl-bridged ferrocenes is achieved by cocondensation of Fe atoms and dimethylfulvene and probably occurs *via* the diradical *4.32* (Eq. (4.88)[201]). This promises to be a useful synthetic method as although several routes to 1,1′-alkyl-bridged ferrocenes have been devised[202], even the most direct synthesis leads to many side products[203].

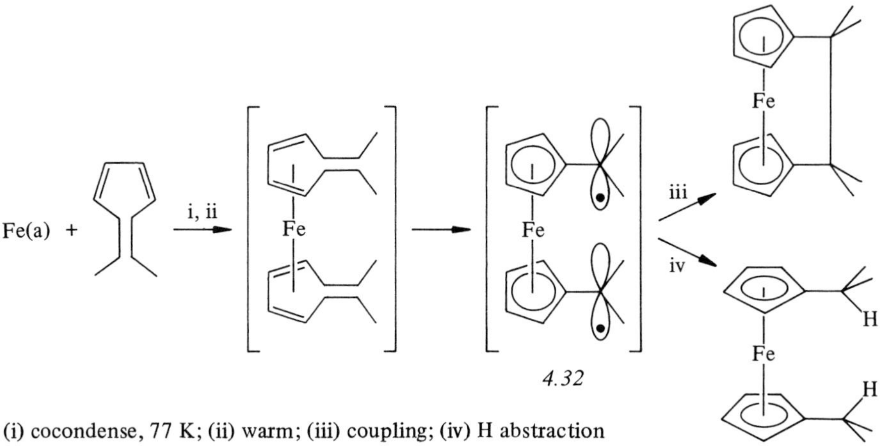


*4.32*

(4.88)

(i) cocondense, 77 K; (ii) warm; (iii) coupling; (iv) H abstraction

Metal atom/cyclohexadiene cocondensates (m = Cr, Fe, Co, Ni) catalyse the disproportionation of cyclohexadiene to benzene and cyclohexene[127]. The mechanism is most probably hydrogen transfer, which is supported by the isolation of ($\eta^6$-PhH)($\eta^4$-cyclohexadiene)Fe *4.33* from the Fe/cyclohexa-1,3-diene cocondensate (Eq. (4.89))[383].

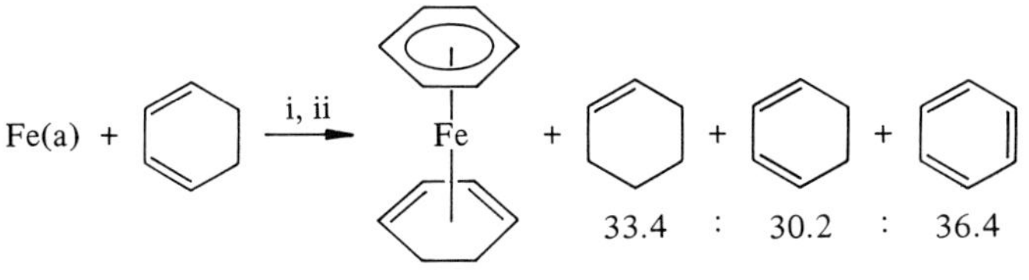


*4.33* (1.7%)

(4.89)

(i) cocondense, 77K; (ii) stir, –20°C, 1h

Further information and support can be gained from trapping experiments. Addition of $PF_3$ to the Cr/cyclohexa-1,3-diene cocondensate at 77 K and purification at temperatures below –30 °C leads to the orange-red compound *trans-anti*-bis-(cyclohexa-1,3-diene) bis(trifluorophosphine)Cr *4.34* (Eq. (4.90))[204], stable to 50 °C and analogous to the dicarbonyl compound made previously by photolysis of $Cr(CO)_6$ in the presence of cyclohexa-1,3-diene[205]. It is interesting to note that addition of $PF_3$ to a Cr/butadiene cocondensate forms only $BdCr(PF_3)_4$[115] while photolysis of $Cr(CO)_6$ in the presence of Bd similarly gives only $BdCr(CO)_4$[206]. Addition of $PF_3$ to the Cr/cyclohexa-1,4-diene cocondensate leads however to hydrogen transfer, as the sole product isolated is ($\eta^5$-cyclohexadienly)hydridotris(trifluorophosphine)chromium *4.36*[204] which rearranges in the presence of silica gel to *4.34*. It is therefore reasonable to surmise that disproportionation of cyclohexa-1,4-diene proceeds *via* species such as the 16e-complex *4.35* which are then trapped by $PF_3$ (Eq. (4.91)). However one must always bear in mind that formation of e.g. the cyclohexadienyl metal hydride may occur after complexation by $PF_3$.

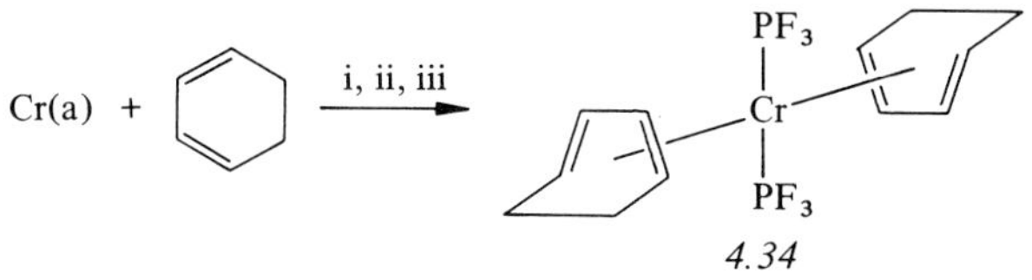


(4.90)

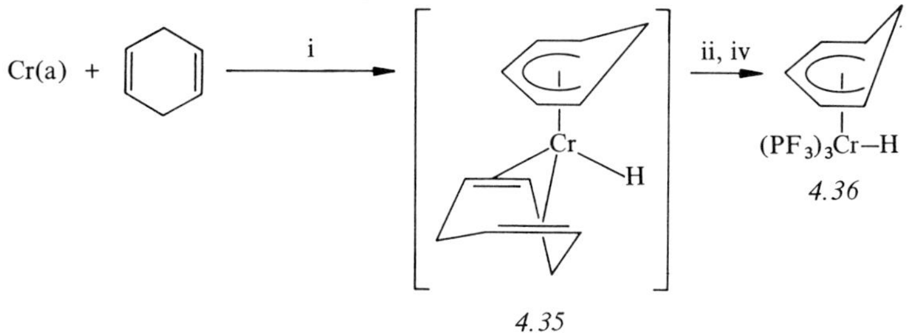


(4.91)

(i) cocondense, 77K; (ii) $PF_3$, 77K; (iii) T< –30°C; (iv) 25°C

Cocondensation of Ti atoms and 1,5-Cod forms a golden yellow deposit which decomposes at –110 °C[385]. Similarly, no organometallic products can be isolated from the Mn/1,5-Cod cocondensate[105]. Cocondensation of Cr atoms and 1,5-Cod forms a green ill-defined product[385]. Ir spectroscopy of this codeposit at low temperatures shows extensive rearrangement of the organic moiety on reaction[119]. Warming the Cr/1,3-Cod or 1,5-Cod cocondensate in an atmosphere of CO leads to the same product in both cases (1,5-Cod) Cr $(CO)_4$ *4.37a*[17]. Isomerisation from the conjugated 1,3-Cod to 1,5-Cod may seem surprising, but is in fact to be expected from the observations that non-conjugated dienes form more stable complexes with $Cr(CO)_4$ fragments than do conjugated dienes[167,175], no 1,3-Cod complex of Cr is known[137,207] and that thermal reaction of 1,3-Cod with $Mo(CO)_6$ or $W(CO)_6$ leads to 1,5-Cod complexes[208]. Attempts have been made to rationalise these effects in terms of the energy ordering, symmetry and extent in space of the free valence molecular orbitals of the $M(CO)_4$ (M = Cr, Mo, W) fragment[58]. Isomerisation to the final product may have occurred after complexation with CO, but low-temperature ir spectroscopy has indicated some Cod-rearrangement prior to attack by other ligands[129]. Addition of $PF_3$ to the Cr/1,5-Cod or 1,3-Cod cocondensates at 77 K leads to altered products[17]. No 1,5-Cod-complex is now formed from 1,3-Cod, and it is tempting to postulate that *4.38* whose structure is derived from X-ray crystallography[209] represents a model for the isomerisation occurring in the Cod/Cr/CO and Cod/$M(CO)_6$ (M = Mo, W) systems. The fortuitous trapping of this intermediate seems due to its enhanced stability over *4.37*. Heating *4.37* to 85 °C causes disproportionation to *4.38*, $Cr(PF_3)_6$ and 1,5-Cod[17]. (The reactions are summarised in Scheme 4.7.) Furthermore *4.38* rearranged in PhMe, probably *via* the intermediate diene $Cr(PF_3)_3$ species *4.39* to a complex

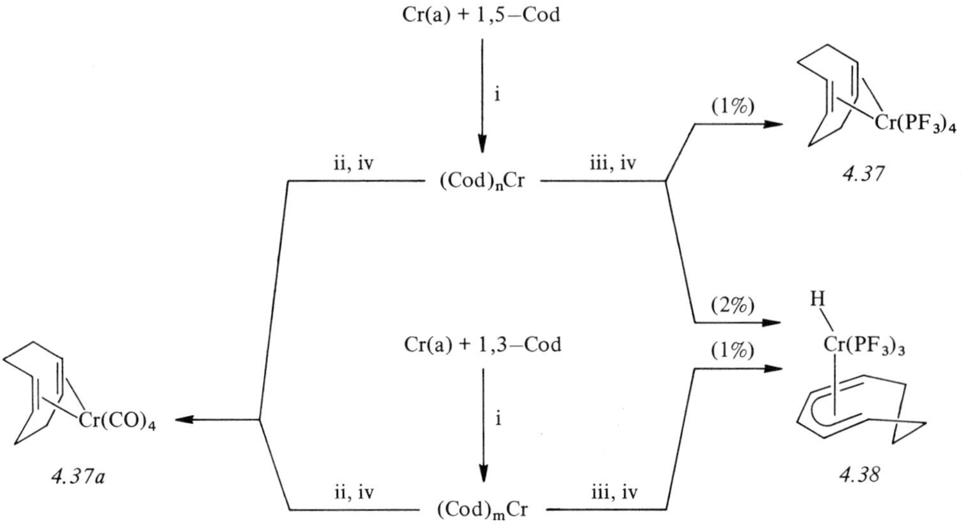


**Scheme 4.7.** (i) cocondense, 77 K; (ii) CO, 77 K; (iii) $PF_3$, 77 K; (iv) warm

*4.40* which is fluxional at room temperature (Eq. (4.92)). Furthermore, at *ca.* 180 °C in $d_{24}$-cyclododecane, all carbon atoms become equivalent and the Cr-C-σ-bond moves round the ring *via* a series of metal-to-carbon hydrogen exchanges which divides the twelve hydrogen atoms into a set of eight which remain attached to their respective carbon atoms and a set of four which exchange H-C($sp^2$) ⟷ H-Cr[119].

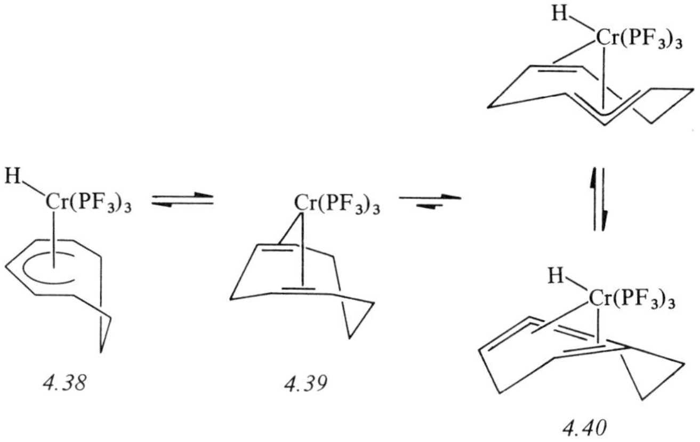


(4.92)

The low temperature ir spectrum of the Mo/1,5-Cod cocondensate shows evidence for a complex, stable below −35 °C, which contains both $\eta^2$- and $\eta^4$-1,5-Cod. At ambient temperatures the compound converts to a very air-sensitive material containing no free C=C double bonds[129].

Condensation of Mn atoms with cyclohepta-1,3-diene (1,3-Cht$H_3$) or 1,3-Cod forms no isolable organometallic products, but if CO is added to the nascent cocondensates, organometallic species can be isolated (Eqs. (4.73) and (4.93))[194].

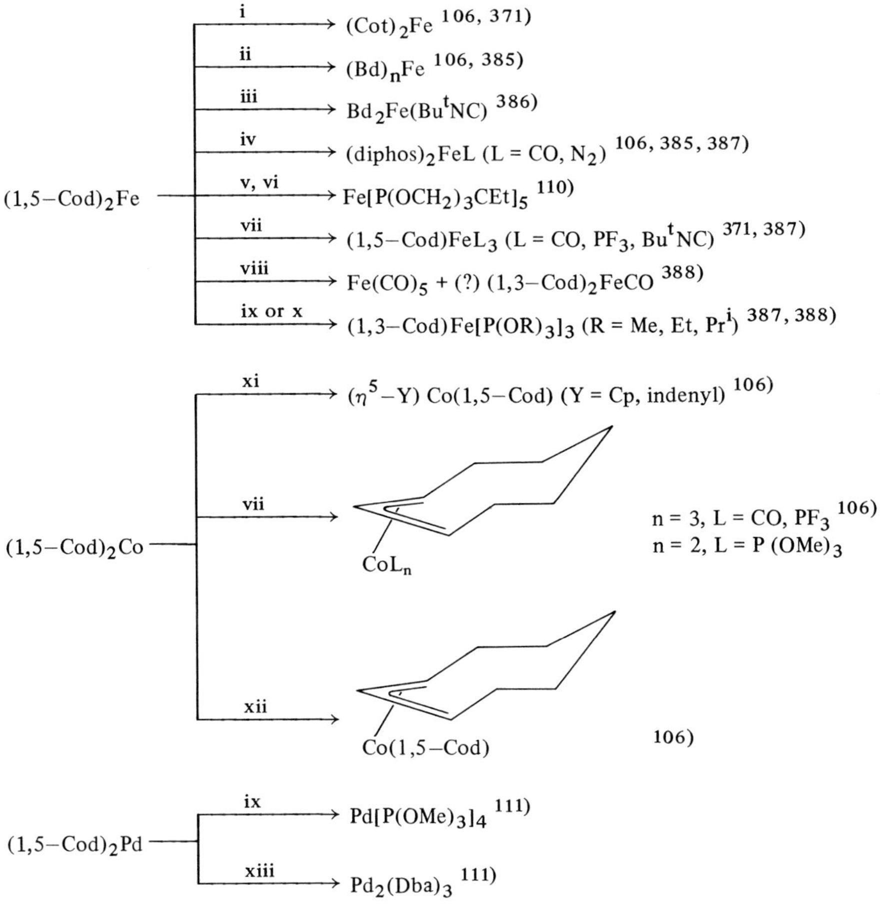


**Scheme 4.8.** (i) Cot; (ii) Bd; (iii) Bd, $Bu^tNC$; (iv) diphos, L; (v) $P(OCH_2)_3CEt$; (vi) warm; (vii) L; (viii) CO; (ix) $P(OMe)_3$; (x) $P(OR)_3$, thf; (xi) YH; (xii) $H_2$; (xiii) dibenzylidene acetone (Dba)

$$Mn(a) + 1,3\text{-}ChtH_3 \xrightarrow{i,ii,iii} (1,3\text{-}ChtH_3)_2MnCO \qquad (4.93)$$

(i) cocondense, 77 K; (ii) CO, 77 K; (iii) warm

Condensation of Fe atoms into a rotating 10% solution of 1,5-Cod in CyMe at −120 °C or with 1,5-Cod vapour at 77 K[110, 388] forms the unusual 16e-compound $(1,5\text{-Cod})_2Fe$, on the basis of analysis, displacement experiments and ir spectrum. It is the first in a series of $(1,5\text{-Cod})_2M$ complexes which can be synthesised from reaction of 1,5-Cod and atoms of the later transition metals; $(1,5\text{-Cod})_2Co$[105, 128], the known[210] compound $(1,5\text{-Cod})_2Ni$[111], $(1,5\text{-Cod})_2Pd$ (dec. −20 °C)[111], and the known[211] compound $(1,5\text{-Cod})_2Pt$[126]. The low temperature ir spectra of the new compounds of Fe, Co and Pd[129] are almost identical to, and thus presumed isostructural with $(1,5\text{-Cod})_2Ni$ whose structure[212] is *4.41*.

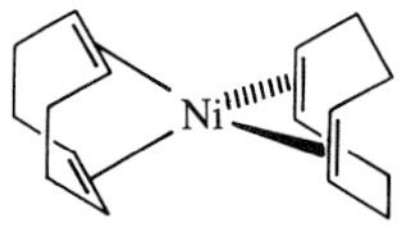


*4.41*

The ligands in these complexes are labile and can be partially or totally replaced readily at low temperatures. These reactions are shown in Scheme 4.8. In addition, reaction of Fe atoms, 1,5-Cod and $PF_3$ leads to (1,5-Cod)Fe$(PF_3)_3$[387]. Only on reaction with phosphites is any hydrogen-transfer observed for (1,5-Cod)$_2$Fe[387,388]. Thus it may be presumed that hydrogen transfer occurs on or after interaction with the incoming ligands. It is a little surprising that the other $FeL_3$ fragments do not isomerise 1,5-Cod to 1,3-Cod[213]. Molecular orbital arguments also point to the preference of $FeL_3$ for conjugated dienes[13,58].

Reaction of excess Li atoms with cyclohexa-1,5-diene in the gas phase at 850 °C leads to an amorphous, heterogeneous solid whose deuteriolysis products suggest that Li atoms have replaced 2 to 4 vinyl hydrogens (Eq. (4.94))[148]. The extreme conditions also cause fragmentation to the perlithio-fragments $C_3Li_4$, $C_2Li_6$, $C_2Li_4$ and $C_2Li_2$.

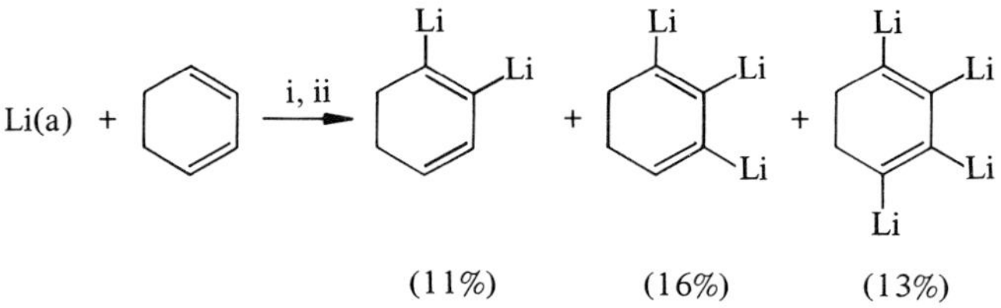


(4.94)

(i) xs Li(a), vapour phase, 850°C; (ii) collect on trap, 77K

### *4.12.3 Norbornadiene*

Norbornadiene (Nbd) is discussed separately from other dienes because of its unique properties. Although there is some evidence for through space intereaction of $\pi$-orbitals[214]. The two alkene units of Nbd can be regarded as non-conjugated, though their close juxtaposition makes Nbd a very effective chelating diene. Each double bond has adjacent $\beta$-hydrogens but they are at the bridgeheads of small fused rings thus any $\pi$-interaction with the double bonds is impossible due to negligible orbital overlap. The molecule's bicyclic nature also introduces strain into the double bonds (i.e. departure from trigonal angles) which increases the energy of the highest occupied M.O. and decreases that of the lowest unoccupied M.O. making the compound a better $\sigma$-doner, $\pi$-acceptor than a corresponding unstrained ligand precursor. In addition the strain directs many of the reactions of Nbd towards strain relief. This is just as true for its reactions with metal atoms.

No characterised binary complex has resulted as yet from metal atom reaction with Nbd. However, analysis of the organic products often shows that some interaction has occurred. Cocondensation at 77 K of Ni atoms and Nbd leads to relief of ring strain by dimerisation of Nbd to its *trans* isomers *4.42*, *4.43*, with predominance of the *exo-trans-*

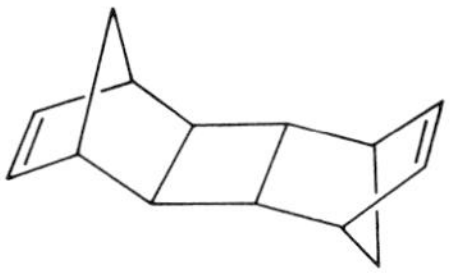

exo–trans–exo
*4.42*

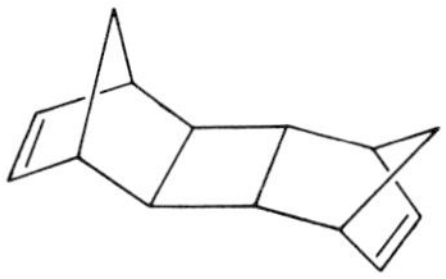

exo–trans–endo
*4.43*

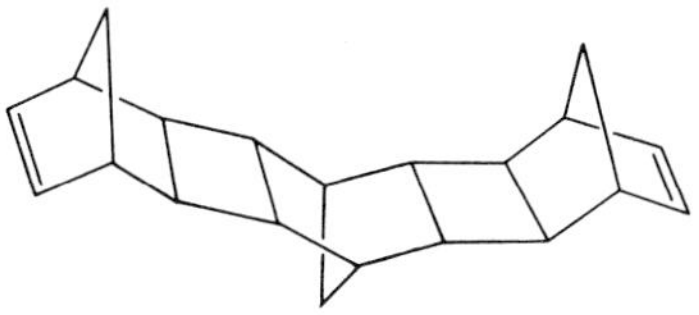

exo–trans–exo–trans–cxo
*4.44*

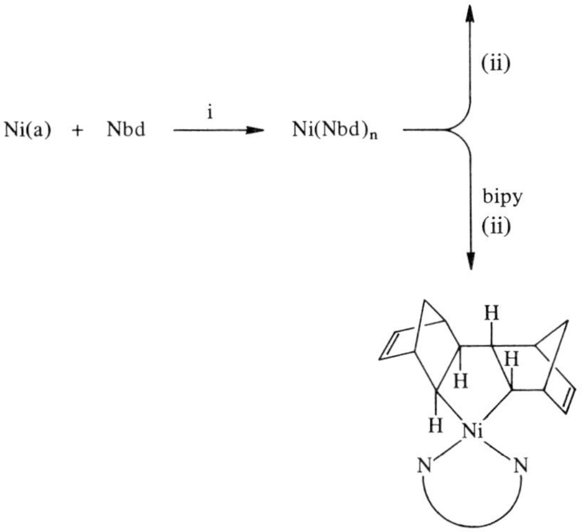


*4.45*

**Scheme 4.9.** (i) cocondense, 77 K; (ii) warm

*exo* dimer, the trimer *4.44* is also formed. If bipyridyl is added to the Ni/Nbd cocondensate at 77 K, warming gives a species shown to be α,α′-dipyridyl-*exo-trans-endo*-3-*nickela*-pentacyclo-[9,2,1,$1^{5,8}$,$0^{2,10}$,$0^{4,9}$]-pentadeca-6-12-diene *4.45*, Scheme 4.9[118,119].

A similar situation is found in the case of iron. Cocondensation of Fe atoms, Nbd and PhMe and warming to $-20\,^{\circ}C$ results in the catalytic conversion of Nbd to three dimers. However, on low temperature addition of NO, only one dimer, the *exo-trans-exo* dimer continues to be formed[95]. No complex has been isolated but addition of $PPh_3$ give $Fe(NO)_2(PPh_3)_2$ in good yield. Highly reactive species such as *4.46* and *4.47*, isoelectronic with the bipyNi-analogues seem likely. In support, $Fe(Nbe)_3(NO)_2$ has been isolated from the reduction of $[Fe(NO)_2Cl]_n$ in the presence of norbornene (Nbe)[215]. Addition of $PPh_3$ forms $Fe(NO)_2(PPh_3)_2$ and equal amounts of Nbe and the dimer $(Nbe)_2$, which suggests the structure *4.48*. Similarly, $PPh_3$ attack on $[Ir(Nbd)_3Cl]_n$ forms the *exo-trans-exo* dimer *4.42* while a derivative $Ir(Nbd)_3(Acac)$ has the molecular structure *4.49* (Eq. (4.95)).

*4.46*

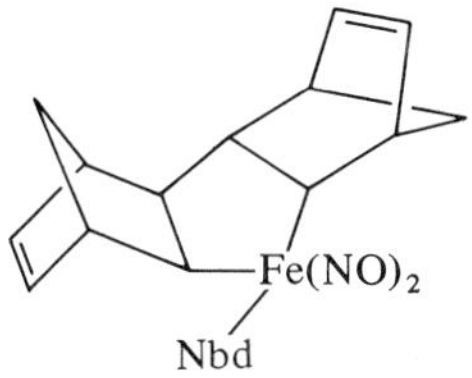


*4.47*

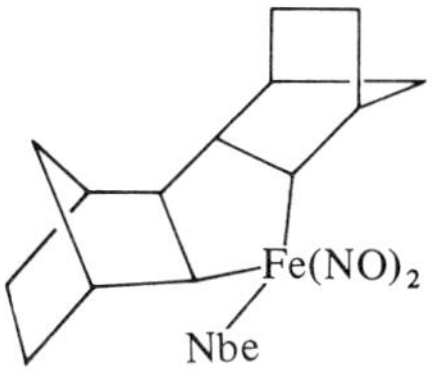


*4.48*

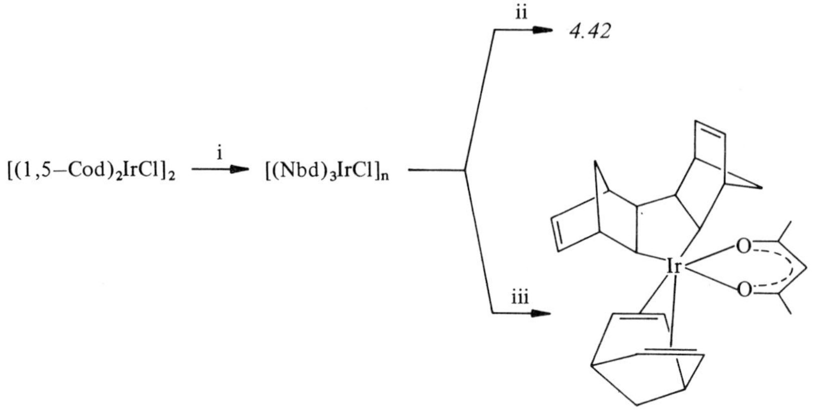


(4.95)[216)]

(i) Nbd; (ii) $\phi_3$P; (iii) acacH

4.49

The Co/Nbd/PhMe cocondensate dimerises Nbd without stereoselectivity and the situation is not improved by addition of NO[95)].

Reaction of Mo atoms and Nbd leads to ring-opened polymerisation[127)] which strongly suggests olefin metathesis, for which many Mo derivatives are well-established catalyst precursors[217)]. A reaction mechanism involving metal-carbene and metallocyclobutane intermediates has the most support[218)], and it is easy to rationalise in this that α-elimination[219)] to form a vinylmetalhydride is followed by β-H insertion; the driving force being relief of strain (Eq. (4.96)), such a mechanism has been touched upon previously[220)]. Metathesis can then proceed (Eq. (4.97)).

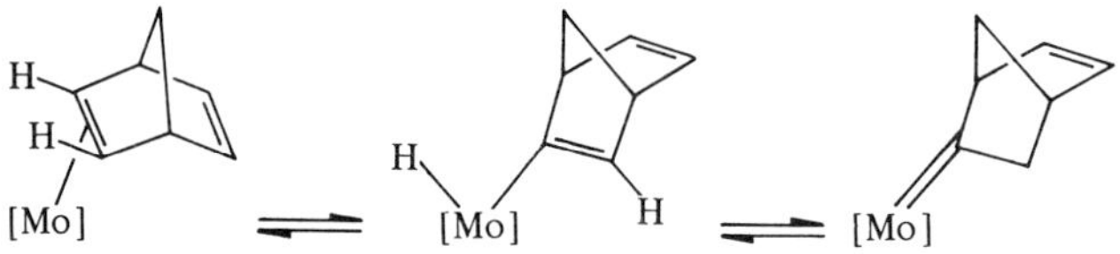


(4.96)

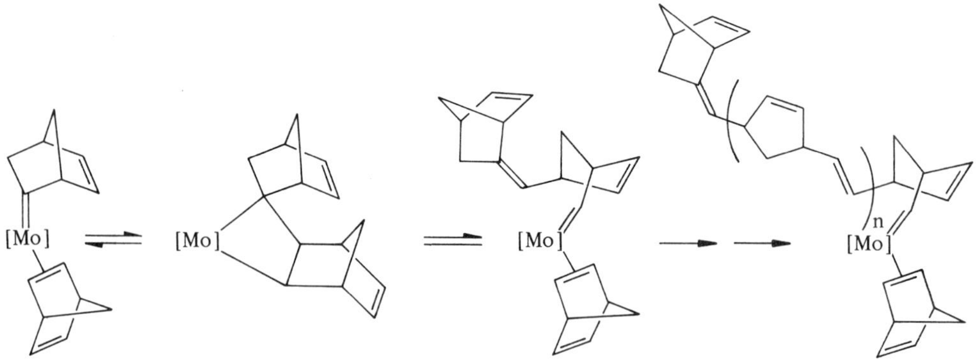


(4.97)

## 4.13 Alkatrienes

Alkatrienes, like alkadienes can be conveniently divided into those without and those with active hydrogens: examples of each are the arenes and cycloheptatriene respectively.

Arenes, on complexing to a metal atom normally maintain their integrity, and bond $\eta^6$ to the metal atom though there is evidence of lower bonding modes. Alkylbenzenes have exocyclic activated hydrogens, but the chemistry of such benzylic hydrogens is usually swamped by the great stability of the aromatic ring, though there is some evidence of benzylmetalhydrides and hydrogen transfer.

Cycloheptatriene (ChtH) also can form sandwich complexes with metal atoms, but like CpH its chemistry is dominated by its activated hydrogen. There are few examples of $\eta^6$-ChtH bonding, rather $\eta^7$-Cht, $\eta^5$-$ChtH_2$ or $\eta^4$-$ChtH_3$ bonding. Furthermore, during a systematic survey of the reactions of ChtH with atoms of the first row transition series[221)], it was observed that the products are very similar if not identical to those made by reducing the metal halide with $Pr^iMgBr$ in the presence of excess ChtH. Therefore it has been suggested that reductive methods should be reexamined for some of those compounds which are at present only obtainable by metal atom technique.

### *4.13.1 Arenes*

Arenes can be expected to readily interact *via* their ring $\pi$-systems with metal atoms, as this merely involves facile electron redistribution between ligand and metal. Thus $(\eta\text{-PhH})_2Cr$ is observed at 20 K after cocondensation of Cr atoms, PhH and Ar[376)]. The previously known[222)] complexes $(\eta\text{-PhH})_2M$ (M = V, Cr, Mo, W) can be readily made by direct reaction of PhH and metal atoms (Table 4.13) and there is evidence that previously unknown $\eta$-arene-metal complexes can also be prepared, but with the exception of $(\eta\text{-arene})_2Ti$[374)], and $(\eta\text{-arene})_2Nb$[223)] they decompose at sub-ambient temperatures. Attempts to synthesise $\eta$-arene-metal complexes from atoms of non-transition metals have met with limited success. Warming the PhH/Zn[391)] and PhH/Al[121)] cocondensates leads only to isolation of PhH and the solid metal, though there is spectral evidence for a weak PhH-Na complex, stable up to *ca.* $-75\ ^\circ C$ in solid PhH[225)]. This suggests that valence d-orbitals are necessary for binding. Certainly, the benzene $\pi$-system and transition metal valence orbitals form favourable bonding combinations[23)] (*vide infra*).

A systematic study of the reaction between PhH or $C_6F_6$ and atoms of the first row transition elements shows that the early transition metals prefer to bond to PhH, while the later transition metals prefer to bond to $C_6F_6$[224)] (Table 4.14).

In the case of the arene-metal complexes of the early transition metals, where the prefered bonding mode is the symmetrical sandwich structure $(\eta^6\text{-arene})_2M$ (M = Ti, (Zr? ), V, Nb, Cr, Mo, W) *4.50* these trends can be explained in terms of a simple M.O. model[244)] used to assign the photoelectron spectra of such species[234,244,245)]. This leads to the generalised M.O. scheme (Fig. 4.5) which is filled as for an 18e complex such as $(\eta^6\text{-PhH})_2Cr$. The orbitals in the box are the most interesting as they correspond to the metal valence orbitals. For a typical early transition metal the valence (i.e. d) orbital energies are quite high and thus form a strong $e_{2g}$ (metal donor) and weaker $e_{1g}$ (metal acceptor) interaction

**Table 4.13.** Synthesis of ($\eta$-arene)$_2$M (M = Ti, V, Nb, Cr, Mo, W, Fe) from metal atoms

| Metal | Arene | Yield (%) | Observations | Ref. |
|---|---|---|---|---|
| Ti | PhH | 30 | Maroon v. air sensitive crystals[a] | 226,374) |
| | PhMe | 30 | Burgundy solutions | 374) |
| | Mesitylene | 20 | | 374) |
| | DC 510[b] | | Bright pink air-sensitive liquid | 372) |
| | Menaphth[d] | | Deep red air-sensitive solution in $Et_2O$ | 227) |
| V | PhH | 2 | Dark red v. air-sensitive solid | 4) |
| | PhF | 13 | Dark red v. air-sensitive solid | 4) |
| | p-$C_6H_4F_2$ | 1 | Red-amber m. air-inert solid | 4,228) |
| | PhCl | 7 | Orange-red air-sensitive solid | 4) |
| | $PhCF_3$ | 22 | Orange m. air-inert, odour of organic sulphides | 4) |
| | DC 510[b] | | Dark brown liquid | 372) |
| | Naphth[d] | | Deep red, air-sensitive | 227) |
| Nb | PhH | 40 | Deep red-purple v. air-sensitive | 223,229) |
| | PhMe | 40 | crystals | 229) |
| | Mesitylene | 40 | | 229) |
| Cr | PhH | 60 | Dark green, brown or black air-sensitive solid[c] | 376,380) 230,231) |
| | PhMe | 50–60 | Green waxy or black air-sensitive solid | 115,230) |
| | PhEt | 30–50 | Black liquid or green waxy solid, air-sensitive | 115,230) |
| | $PhPr^n$ | 16 | Black air-sensitive liquid | 230) |
| | $PhPr^i$ | 20 | Yellow-brown air-sensitive solid | 91,230) |
| | $PhBu^n$ | 14 | Black air-sensitive liquid | 230) |
| | $PhBu^i$ | 7 | Black air-sensitive liquid | 230) |
| | $PhBu^t$ | 27 | Black air-sensitive solid | 230) |
| | PhCy | 14 | Dark yellow air-sensitive solid | 230) |
| | $Ph(CH_2)_4Ph$ | 29 | Air-sensitive | 232) |
| | o-$C_6H_4Me_2$ | 25 | Black air-sensitive solid | 230) |
| | m-$C_6H_4Me_2$ | 25 | Black air-sensitive solid | 230) |
| | p-$C_6H_4Me_2$ | 29 | Black air-sensitive solid | 230) |
| | o-$C_6H_4(CH_2)_3$ | 17 | Olive-green air-sensitive solid | 230) |
| | m-$C_6H_4Pr^i_2$ | 20 | Dark green-yellow air-sensitive solid | 230) |
| | Mesitylene | 13–20 | Black or dark green air-sensitive solid | 230,231) |
| | 1,2,3-$C_6H_3Me_3$ | 5 | Black air-sensitive solid | 230) |
| | 1,2,3,4-$C_6H_2Me_4$ | 7 | Olive-green air-sensitive solid | 230) |
| | 1,2,4,5-$C_6H_2Me_4$ | 21 | Olive-green air-sensitive solid | 230) |
| | o-$C_6H_4MeF$ | 16–18 | Black waxy or olive-green air-sensitive solid | 230,235) |
| | m-$C_6H_4MeF$ | 7–16 | Black waxy air-sensitive solid | 230,235) |
| | p-$C_6H_4MeF$ | 21–46 | Dark green or dark purple air-sensitive solid | 230,235) |
| | o-$C_6H_4MeCl$ | 23 | Black air-sensitive solid | 230) |
| | m-$C_6H_4MeCl$ | 17 | Black air-sensitive solid | 230) |
| | p-$C_6H_4MeCl$ | 17 | Green-black air-sensitive solid | 230) |
| | p-$C_6H_4Me(OMe)$ | 12 | Brown-black air-sensitive solid | 230) |
| | PhF | 15–32 | Yellow, yellow-green or black air-inert solid | 119,230) 235) |
| | o-$C_6H_4F_2$ | 17–19 | Orange or dark violet air-inert solid | 230,235) |
| | m-$C_6H_4F_2$ | 18 | Brown air-inert solid | 235) |

Table 4.13 (continued)

| Metal | Arene | Yield (%) | Observation | Ref. |
|---|---|---|---|---|
| | p-$C_6H_4F_2$ | 12 | Bright yellow, red-brown or dark violet air-inert solid | 119,230, 235) |
| | o-$C_6H_4FCl$ | 23 | Brown air-inert solid | 235) |
| | m-$C_6H_4FCl$ | 17 | Brown air-inert solid | 235) |
| | p-$C_6H_4FCl$ | 17 | Light brown air-inert solid | 235) |
| | m-$C_6H_4FCF_3$ | 20 | Red-brown air-inert solid | 235) |
| | p-$C_6H_4FCF_3$ | 7 | Dark green air-inert solid | 235) |
| | PhCl | 34–47 | Olive-green m. air-inert solid | 119,230) |
| | m-$C_6H_4Cl_2$ | 10 | Olive-green m. air-inert solid | 230) |
| | o-$C_6H_4ClCF_3$ | 33 | Yellow-green air-inert solid | 4) |
| | p-$C_6H_4ClCF_3$ | 12 | Rust brown air-inert solid | 230) |
| | $PhCF_3$ | 17–26 | Yellow or dark brown-black m. air-inert solid | 4,230) |
| | m-$C_6H_4(CF_3)_2$ | 17 | Yellow-green air-inert solid | 230) |
| | p-$C_6H_4(CF_3)_2$ | 38 | Amber air-inert solid | 230) |
| | PhCOOMe | 4 | Red m. air-inert solid | 230) |
| | PhCOOEt | 30 | Air-sensitive | 237) |
| | $PhCH_2COOEt$ | | Olive viscous oil | 236) |
| | $PhCH_2CH_2COOEt$ | | | 236) |
| | PhOMe | 8–11 | Brown-black air-sensitive solid | 230,238) |
| | PhOPh | 3 | Olive-green air-sensitive solid | 230) |
| | 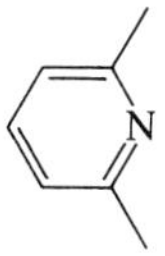 | 2 | Red-brown air-sensitive solid | 239) |
| | DC 510[b] | | Red-brown air-sensitive liquid | 268a) |
| | Naphth[d] | | Dark red-brown, air-sensitive | 227) |
| | Menaphth | | Dark red-brown, air-sensitive | 227) |
| | 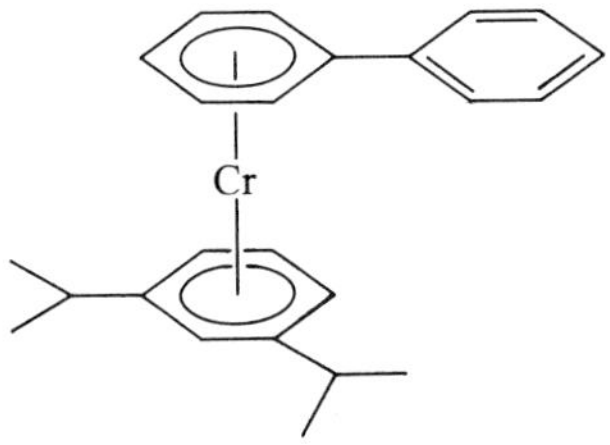 | | | 240) |
| | ($\eta$-PhH)($\eta$-$C_6F_6$)Cr | 8.5 | Air-inert solid | 19,238) |
| | ($\eta$-PhH)($\eta$-$C_6F_5H$)Cr | 22 | Air-inert solid | 241) |
| | 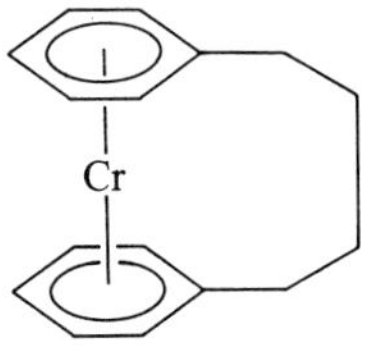 | 6 | Air-sensitive | 232) |

Table 4.13 (continued)

| Metal | Arene | Yield (%) | Observations | Ref. |
|---|---|---|---|---|
| | 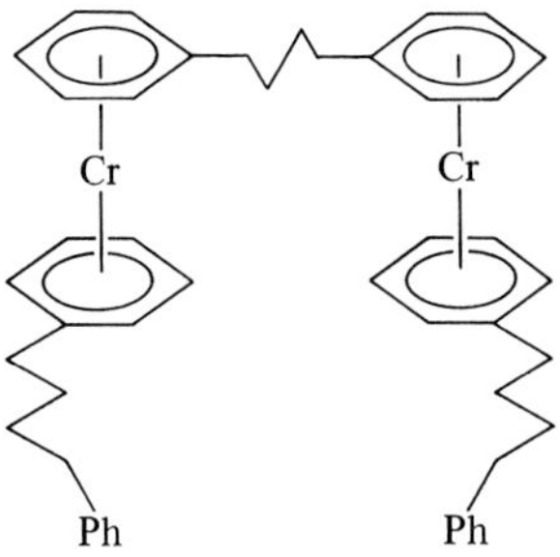 | 15 | Air-sensitive | 232) |
| | ($\eta^6$–py)($\eta$–Mes) Cr | | | 102) |
| Mo | PhH | 20–40 | Green air-sensitive solid | 226, 242) |
| | PhMe | 20–40 | Green air-sensitive solid | 226, 242) |
| | Mesitylene | 20 | Green air-sensitive solid | 226) |
| | PhF | 30–40 | Green air-sensitive solid | 242) |
| | PhOMe | 30–40 | Green air-sensitive solid | 242) |
| | PhCOOMe | 30–40 | Red air-sensitive solid | 242) |
| | $PhNMe_2$ | 30–40 | Red air-sensitive solid | 242) |
| | DC 510[b] | | Golden brown air-sensitive liquid | 372) |
| | Naphth[d] | | | 227) |
| | Menaphth | | | 227) |
| W | PhH | 30–40 | Green air-sensitive solid | 242, 243) |
| | PhMe | 30–40 | Green air-sensitive solid | 242, 243) |
| | o-$C_6H_4Me_2$ | 30–40 | Green air-sensitive solid | 242) |
| | Mesitylene | 30 | Green air-sensitive solid | 243) |
| | PhF | 30–40 | Green air-sensitive solid | 242) |
| | PhOMe | 30–40 | Green air-sensitive solid | 242) |
| Fe | PhH | | $T_{dec}$ −50 °C | 382) |
| | PhMe | | $T_{dec}$ −60 °C | 382) |
| | Mesitylene | | $T_{dec}$ −40 °C | 382) |

[a] v. = very, m. = moderately.

[b] Dow Corning 510 silicone diffusion pump oil (a phenylmethylsiloxane polymer).

[c] Different authors often cite differing colours for ($\eta$-arene)$_2$Cr complexes: this is paralleled in the literature of more conventional syntheses. Different colours are most often due to different purification procedures (recrystallisation versus sublimation) leading to different crystal sizes, habits etc.

[d] Naphth = naphthalene, Menaphth = 1-methylnaphthalene.

**Table 4.14.** Approximate stabilities of arene-metal species formed from metal atoms[224)]

| | Ti | V | Cr | Mn | Fe | Co | Ni | Cu | Zn |
|---|---|---|---|---|---|---|---|---|---|
| $C_6H_6$ | 100[a] | 300 | 300 | (–)[b] | –50 | (–) | (–) | *[c] | (–) |
| $C_6F_6$ | –50 | 100 | 40 | (–) | –40 | 10 | 70 | (–) | (–) |

a From ref. 226.
b (–) indicates no complex detected above –80 °C.
c A red charge-transfer complex formed at low temperatures. Allowing the cocondensate to warm forms a little $C_6F_5–C_6F_5$.
(Klubunde, K., et al.: J. Fluorine Chem. *4*, 114 (1976) with permission of Elsevier S. P. Co.)

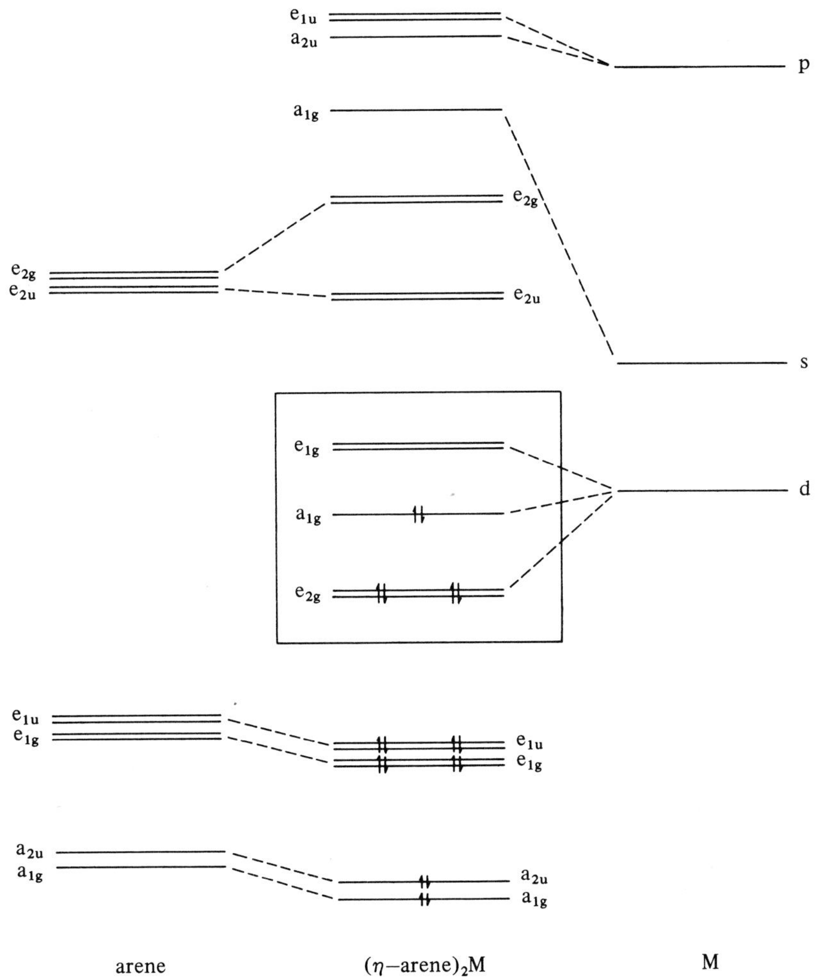


**Fig. 4.5.** Generalised M.O. Scheme of ($\eta$-arene)$_2$M

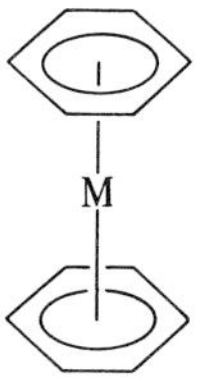

*4.50*

leading to a net drift of electron charge to the rings. (The $a_{1g}$ orbital is essentially non-bonding which explains why it can contain 0, 1 or 2 electrons as in $(PhH)_2Ti$, $(PhH)_2V$ or $(PhH)_2Cr$ respectively with little effect on stability.) More involved calculations[246] and experiments[247] lead to the same conclusion. Replacement of the two PhH ligands by $C_6F_6$ may lead to too high a positive charge building up on the metal. In moving to the later transition elements, the $(\eta^6\text{-arene})_2M$ structure is destabilised owing to the extra electrons being accommodated in antibonding $e_{1g}^*$ orbitals, weakening the metal-ligand bonds. There is thus a tendence towards $(arene)_1M$ bonding, for example there is evidence for the 14e, 15e, and 16e species $(PhH)_1M$ (M = Fe, Co, Ni) in PhH matrices at 77 K[231]. They decompose at *ca.* −50 °C while, in contrast, the unusual material $C_6F_6Ni$ is stable to 70 °C[224]. Its relative stability is most likely due to the relative ease of $C_6F_6$ to distort to form ligands more amenable to bonding to Ni (*vide infra*).

## $(\eta^6\text{-Arene})_2Ti$

Cocondensation of Ti atoms and PhH forms the maroon, air and impurity sensitive new binary arene-metal complex $(\eta\text{-PhH})_2Ti$[226]. $^1H$ nmr, ir and mass spectrometry[374] indicate it is diamagnetic and isostructural with the isotypic $(\eta\text{-PhH})_2Cr$ which has the molecular structure *4.50*[248]. $(\eta\text{-PhMe})_2Ti$ and $(\eta\text{-mesitylene})_2Ti$ may be similarly made[374]. The photoelectron spectrum (pes) shows to the limit of the spectrometer's resolution (0.05 eV) only two bands in the valence region (ionisation energies less than 10 eV). Comparison with the corresponding spectrum of $(\eta\text{-PhMe})_2Cr$ indicates that the two bands are due to ionisations from $(e_{1g} + e_{1u})$ and $e_{2g}$ orbitals and therefore that the ordering and occupancy of the highest energy orbitals is $e_{1g}^4e_{1u}^4e_{2g}^4a_{1g}^0$. In contrast the isoelectronic $(\eta\text{-mesitylene})_2V^+$ cation is paramagnetic with the orbital scheme $e_{1g}^4e_{1u}^4e_{2g}^3a_{1g}^1$ [241]. This is due to the increased d orbital energy of Ti, which will raise by a corresponding amount the essentially non-bonding $a_{1g}$ orbital energy in going from $(\eta\text{-arene})_2V^+$ to $(\eta\text{-arene})_2Ti$ but the $e_{2g}$ orbitals by a lesser amount owing to increasing metal interaction with the ligand $e_{2g}$ orbitals. Therefore the $a_{1g} - e_{2g}$ energy difference ($\Delta$) is larger in $(arene)_2Ti$ and in fact must be sufficiently large to overcome electron-pairing requirements.

$(\eta\text{-PhH})_2Ti$ is stable for at least one year in the solid state or as a dilute solution (*ca.* 5 mmol/litre) in hexane. In addition, a more concentrated (*ca.* 50 mmol/litre) solution in thf is apparently unchanged after 20 days at 100 °C. However, very concentrated solutions (*ca.* 250 mmol/litre in e.g., PhH) decompose readily and apparently autocatalytically. Solutions formed immediately after warming the cocondensate are therefore better speedily filtered at low temperatures, as impurites, perhaps finely divided metal, can decompose the complexes. Furthermore some decomposition always occurs on sublimation, with PhH (ir, ms) as the volatile product. Variation in the relative intensities of Ti-containing ions in the mass spectrum with temperature suggests decomposition in the vapour state and a dissociation equilibrium as shown in Eq. (4.98). Moreover, "second bodies" such as Ar atoms increase the decomposition rate of the vapour[103].

$$(\text{arene})_2\text{Ti(g)} \rightleftharpoons (\text{arene})\text{Ti(g)} + \text{arene(g)} \qquad (4.98)$$

The 16e ($\eta$-arene)$_2$Ti complexes seemingly do not react with electron donor ligands such as CO, $PF_3$, $PPh_3$, $Ph_2PCH_2CH_2PPh_2$, and $CF_3 \cdot C{\equiv}C \cdot CF_3$ except under conditions where decomposition occurs[250]. However, the lowest energy vacant orbital in ($\eta$-arene)$_2$Ti is the $a_{1g}$ orbital which will be mostly metal $d_{z^2}$ in character[244] and is thus oriented along the molecule's z axis completely protected by the arene rings or the filled $e_{1g}$ orbitals.

On the other hand, ($\eta$-arene)$_2$Ti complexes are electron-rich as their filled $e_{2g}$ orbitals are of high energy (e.g. 5.4 eV for ($\eta$-PhMe)$_2$Ti, c.f. 6.19 eV for ($\eta$-PhMe)$_2$Cr[1] and 6.03 eV for ($\eta$-PhMe)$_2$Mo[234] $e_{2g}$ orbitals) and are available for bonding. Therefore they can be expected to react with Lewis acids. ($\eta$-PhMe)$_2$Ti reacts with deoxygenated 1 M · $H_2SO_4$ in water to form $Ti^{III}$ salts. ($\eta$-PhH)$_2$Ti, however, is rapidly precipitated under the same conditions and is difficult to react. Addition of pure $CF_3COOH$ to ($\eta$-PhH)$_2$Ti in PhMe results in precipitation of metal.

Treatment of the burgundy PhMe solutions of ($\eta$-PhH)$_2$Ti with $Et_2AlCl$ gives a violet solution which even at –80 °C rapidly forms a red-brown solution and separation of an extremely air- and water-sensitive red-brown precipitate. Treatment of the solution with butadiene causes rapid cyclotrimerisation to cyclododeca-1,5,9-triene. $EtAlCl_2$ and $AlCl_3$ behave in the same manner, while addition of $Et_3Al$ causes no apparent change. Addition of cycloheptatriene to the red-brown solutions formed from (PhH)$_2$Ti and $Et_2AlCl$ gives a green solution from which fine apple green crystals can separate. However, these are insoluble in hydrocarbon solvents and are decomposed by solvents such as ethers, $CH_2Cl_2$ and MeCN. They dissolve in a solution of $Et_2AlCl$ in PhMe to give a solution very active in the trimerisation of Bd to Cdt and dodecatrienes. The formation of the latter may be indicative of Ti-H formation (see the section on butadiene). Significantly cyclopentadiene forms similar bottle green solutions while cyclooctatetraene does not. The apple-green crystals contain Ti, Al and Cl and hydrolysis forms ethane and a trace of butane[108]. Further reactions of ($\eta$-arene)$_2$Ti systems with Bd are discussed in the section on butadiene.

Alumina, impregnated by a solution of ($\eta$-PhMe)$_2$Ti, gives a supported catalyst, active in forming very high molecular weight polyethylene and polypropene. Bd is also polymerised to Pbd. Supported catalysts derived from ($\eta$-PhMe)$_2$Ti and $MgCl_2$ also form polyolefins but not of such a high molecular weight, and act more efficiently if promoted by an organoaluminium compound, a halogen source and (for propene and higher olefins) a Lewis base[251].

Ti, like most of the early transition metals has a high affinity for oxygen. Thus ($\eta$-arene)$_2$Ti has proved a very effective deoxygenating agent for a variety of organic substrates under relatively mild conditions (see the section on organic compounds containing oxygen).

Cocondensation of a wide variety of arenes with Zr atoms forms deep burgundy cocondensates, strikingly similar to Ti/arene cocondensates. However, the colour is discharged on warming. The presence of ($\eta$-arene)$_2$Zr is infered[252].

## ($\eta$-Arene)$_2$Nb

Cocondensation of Nb atoms and PhH, PhMe or mesitylene (Mes) forms the new deep red-purple highly air-sensitive crystals ($\eta^6$-arene)$_2$Nb (40%)[223]. ($\eta^6$-PhH)$_2$Nb can also be observed in a Nb/PhH/Ar matrix at 30 K[229]. The ms, ir and ers spectra, and pes of these compounds are wholly consistent with a symmetrical sandwich structure *4.50*. The pes of (arene)$_2$Nb shows an energy order of the ion states $^1A > {}^3E > {}^1E$ in the d$^5$ paramagnetic species, which differs from that found for (arene)$_2$V ($^3E > {}^1A > {}^1E$)[234], and may be interpreted as arising from a greater ligand field splitting for the second row metal.

## Arene Complexes of V, Cr, Mo, and W

Cocondensation of arenes with the atoms of V, Cr, Mo and W gives the corresponding ($\eta$-arene)$_2$M compounds in good yields (see Table 4.13). Such types of compounds have been made previously and in greater quantities[222] but MVS offers a very ready synthesis of known complexes in gram quantities and of many complexes not accessible by more conventional means.

One of the most successful conventional methods of ($\eta$-arene)$_2$M synthesis is that devised by Fischer and Hafner[253] involving an aluminium halide, and Al powder if reduction is required, e.g., Eq. (4.99) (M = Cr, V)[254].

$$3\,MCl_3 + 2\,Al + AlCl_3 + 6\ \text{arene} \longrightarrow 3[(\eta^6\text{-arene})_2M]^+[AlCl_4]^- \qquad (4.99)$$

However this method is not applicable when the arene is itself reactive to such (Friedel-Crafts) conditions. Thus "($\eta$-cumene)$_2$Cr" prepared in this way has been reported as a liquid[255] and on thermal decomposition gives PhH (4%), PhEt (6%), PhPr$^i$ (28%), $C_6H_4(Pr^i)_2$ (46%) and $C_6H_3(Pr^i)_3$ (16%)[256]. However ($\eta$-cumene)$_2$Cr prepared from cumene and Cr atoms is crystalline and gives PhPr$^i$ (100%) on decomposition by 2 M · $HNO_3$[91] or thermally[230]. Similarly ($\eta$-arene)$_2$Cr (arene = PhMe, PhBu$^t$, *p*-$C_6H_4Me_2$, 1,2,3,4-$C_6H_2Me_2$[230], *m*-$C_6H_4(Pr^i)_2$[91], and 1,3,5-$C_6H_3Me_3$[233]) formed from Cr atoms have been shown to be pure. Advantage of this high purity synthesis has been taken for detailed nmr[230] and pes[234] studies.

($\eta$-arene)$_2$W complexes can now be prepared in gram quantities from electron-beam evaporated W atoms and arenes[243]. Such complexes have been prepared previously by a classical route[257] but it is lengthy, difficult and very inefficient (< 2%) leading only to small quantities. The larger quantities now available will allow a systematic study of ($\eta$-arene)$_2$W complexes to be made. For example the complexes can be protonated in dilute acid to form the orange diamagnetic cation ($\eta$-arene)$_2$WH$^+$ which is the first example of a bisarenemetalhydride[243].

($\eta$-arene)Cr($PF_3$)$_3$ complexes can be simply made by condensation of Cr atoms with a mixture of the arene and $PF_3$. The arenes include PhH, mesitylene, PhPr$^i$ and $C_6F_6$. Cr($PF_3$)$_6$ is always also produced[91]. Cocondensation of Cr atoms, PhH and NO does not lead to the unknown 18e compound ($\eta$-PhH)Cr(NO)$_2$[91]. Cocondensation of Cr atoms, pyridine and $PF_3$ results in two new complexes other than Cr($PF_3$)$_6$, involving respectively pyridine $\eta^2$- and $\eta^6$-bonded to chromium. The latter appears to be the first $\pi$-pyridine complex (Eq. (4.100))[19].

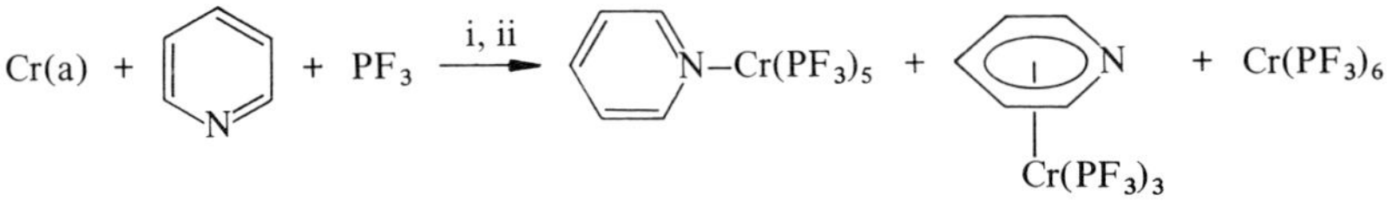


(4.100)

(i) cocondense, 77K; (ii) 25°C

The synthesis of a binary pyridine-Cr complex *4.51* from reaction of Cr atoms and 2,6-dimethylpyridine has since been published[239]. Its crystal structure[258] shows that the two rings can take two attitudes to each other in the crystal. Reaction of Cr atoms with other heterocycles such as pyridine, 2,6-difluoropyridine and borazine gives only involatile complex products while N-methylpyrrole does not react[239].

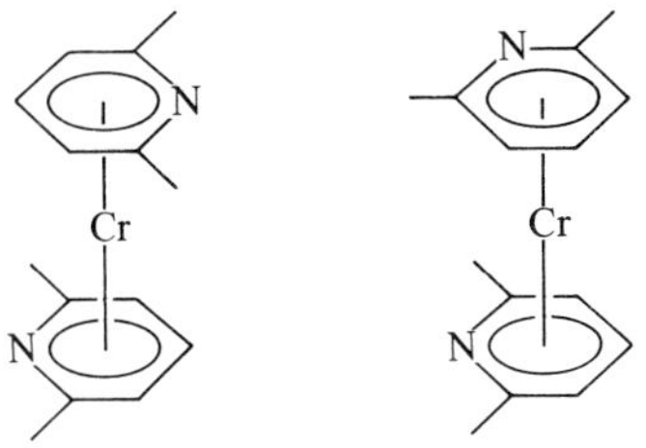


*4.51*

Haloarene Complexes of V, Cr, Mo, and W

The ease of formation by more conventional methods of ($\eta$-arene)$_2$M complexes is drastically reduced for arenes containing electron-withdrawing substituents. This is not so for syntheses involving metal atoms, at least for arenes having only 1 or 2 electron-withdrawing substituents. In fact it can be shown that the yields of ($\eta$-arene)$_2$Cr is higher for some halobenzenes than for PhH itself[115]. To study this effect further, an equimolar mixture of PhH, PhF, *p*-$C_6H_4F_2$, 1,3,5-$C_6H_3F_3$, 1,2,3,5-$C_6H_2F_4$, $C_6HF_5$ and $C_6F_6$ has been condensed with $PF_3$ and Cr atoms. The yield of (arene)Cr$(PF_3)_3$ is a maximum for *p*-$C_6H_4F_2$ and falls to a minimum for $C_6F_6$[19]. Thus there seems to be two opposing trends. Firstly an increasing rate of complexation with increasing F-substitution, and secondly what is probably a competing defluorination reaction. Whatever the mechanisms involved they are kinetically controlled as ($\eta$-*p*-$C_6H_4F_2$)$_2$Cr is less stable than ($\eta$-PhH)$_2$Cr. Similar trends have found for ($\eta$-arene)$_2$V complexes as well as for ($\eta$-arene)$_2$Cr[4]. In addition there is evidence that binary polyhalogenobenzene-metal complexes are unstable especially at the beginning of the transition series[224] (*vide supra*). It has been shown that the 1,500 cm$^{-1}$ band in benzene shifts to 1,428 cm$^{-1}$ upon complexation with Cr ($\Delta\nu_{C=C}$ = 72 cm$^{-1}$)[259]. In ($\eta$-*p*-$C_6H_4(CF_3)_2$)$_2$Cr the analogous band at 1,480 cm$^{-1}$ is found at 1,529 cm$^{-1}$ in the free arene ($\Delta\nu_{C=C}$ = 49 cm$^{-1}$) infering less ring C-C bond weakening and therefore a weaker metal-ring interaction upon bonding[389]. There is also a tendency for such species to explode, presumably via facile exothermic formation of metal-halide bonds[4,91,224]. It seems that when more than two fluorines are attached to an arene ligand only very explosive organochromium or organovanadium complexes are obtained[4]. Luckily the explosive nature of e.g., $C_6F_6$ can be tempered by addition of other perhaps less electron demanding ligands and in this way the more stable ($\eta$-$C_6F_6$)($\eta$-PhH)Cr and ($\eta$-$C_6F_6$)Cr $(PF_3)_3$ complexes have been prepared[91]. Contrarily other cocondensations involving $C_6F_6$, 1,2,3,4-$C_6H_2F_4$ and –$C_6H_2Cl_4$ and Cr atoms are reported to give only the starting substrate and Cr solid, with no evidence of unstable complexes, while PhBr or PhI form only Ph-Ph and what is most likely a chromium halide. Therefore it has been postulated that (in the case of Cr) explosions may in fact be due to the finely divided Cr always formed in such reactions, perhaps on coming into contact with oxygen[230].

Whatever their thermal stability, the ($\eta$-halogenoarene)$_2$M complexes (M = V, Cr) are less air-sensitive than their halogen-free counterparts. Thus ($\eta$-*p*-$C_6H_4F_2$)M (M = Cr, V) are inert to air[4]. Trifluoromethyl substituted arenes are even more effective and have the added advantage of not imparting any explosive qualities on their cocondensates with metal atoms. Other electron-withdrawing substituents have similar effects. This enhanced air-inertness is not difficult to explain. The first step in oxidative decomposition is probably one-electron oxidation of e.g. Cr to give a ($\eta$-arene)$_2$Cr$^+$ complex. Such electron loss is less likely if the metal is bonded to an arene possessing electron withdrawing substituents; a view supported by polarographic studies on these systems[260]. The molecular structure of ($\eta$-*p*-$C_6H_4F_2$)$_2$V shows an interesting distortion of the arene ring[228]. The fluorine-substituted carbons are distorted slightly away from the V atom (deviation from planarity of arene ring, 0.06 Å) *4.52*. The unsubstituted sites therefore form better bonds with the V atom in agreement with prediction.

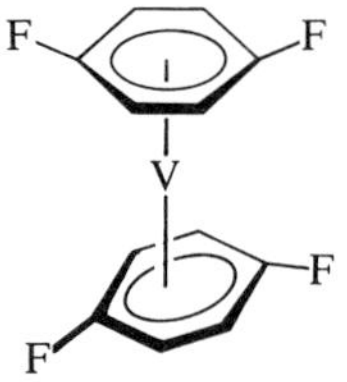


*4.52*

No less than 38 ($\eta$-arene)$_2$Cr complexes have been synthesised by reaction Cr atoms in order to determine by a systematic $^{13}$C and $^1$H nmr study the extent of perturbation on an arene complexing to a metal, and maybe to estimate the $\pi$-electron density of such ligands. The $^{13}$C nmr of complexes of monosubstituted arenes, ($\eta$-PhX)$_2$Cr, show that the chemical shift at "the position" *para* to substitution hardly changes with varying substitution. The correlation between these shifts and the transmission of resonance effects is well established[261], thus this apparent insensitivity seems due to a drastic reduction of transmission of resonance effects and therefore loss of aromatic character. Using the empirical relation (Eq. (4.101))[262] of the $^{13}$C chemical shift para to substitution [$f_H^X C(4)$] of H by X to the field (including inductive) and resonance factors ($F_X$, $R_X$) of X[263], the weighting factors f and r for PhX in ($\eta$-PhX)$_2$Cr, ($\eta$-PhX)Cr(CO)$_3$ and free PhX have been calculated and are shown in Table 4.15.

$$f_H^X C(4) = fF_X + rR_X \qquad (4.101)$$

**Table 4.15.** Field and resonance effects using Eq. (4.101)[230]

| | % | |
|---|---|---|
| | f | r |
| Uncomplexed arene | 11.0 | 89.0 |
| ($\eta$-arene)Cr(CO)$_3$ | 11.0 | 89.0 |
| ($\eta$-arene)$_2$Cr | 57.9 | 42.2 |

(Lagowski, J. J., et al.: J. Amer. Chem. Soc. *15*, 577 (1976) with permission, copyright the American Chemical Society)

They show the $\pi$-system in ($\eta$-PhX)$_2$Cr is strongly perturbed, presumably by interaction with the metal. The corresponding $^1$H chemical shifts give the same interpretation[264]. (There are arguments against deceptively simple explanations of coordination shifts of carbon atoms bound to metal atoms, and care is needed in attempting to treat total screening constants[265]. However it must be pointed out that while this is true for gross coordination shifts, only the *trends* in substituent effects of a closely related series of compounds *relative* to already complexed PhH in ($\eta$-PhH)$_2$Cr have been investigated, and that such deductions are therefore reasonable. Furthermore, complications such as inductive and neighbour anisotropy perturbations are generally considered smallest at the para position.)

This loss of aromaticity may help explain the reluctance of ($\eta$-arene)$_2$Cr complexes to undergo electrophilic substitution characteristic of free arenes, though other effects such as competition for the incoming electrophile by the relatively electron-rich metal centre will be very important. There should be much less competition for nucleophiles and therefore nucleophilic substitution may be possible. However, the net electron drift to the arene rings, especially those with electron-withdrawing substituents in ($\eta$-arene)$_2$M complexes may make initial nucleophilic attack difficult. On the other hand, assuming substitution *via* an associative mechanism, intermediates represented by *4.53* should have the extra negative charge delocalised over what is still an 18e compound, especially if there are electron-withdrawing substituents on both rings. In either case, nucleophilic substitution will be enhanced by a suitable leaving group present on the complexed arene ring. Unfortunately such arenes cannot be complexed to metal centres by conventional means. For example, the very nature of their being good leaving groups facilitates reactions with Lewis acids such as AlCl$_3$ encountered in the Fischer-Hafner syntheses. Luckily, MVS largely circumvents this problem.

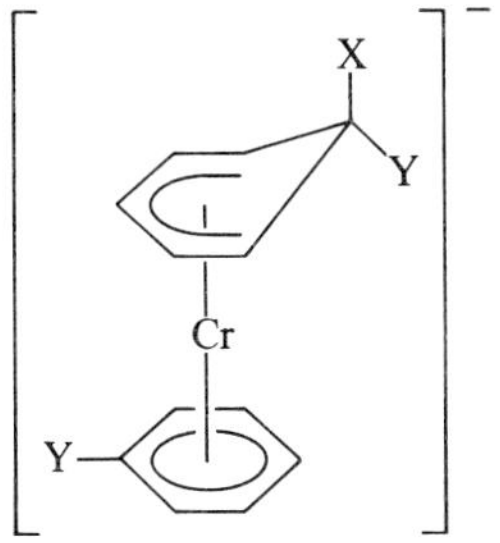


*4.53*

($\eta$-$C_6F_6$)($\eta$-PhH)Cr, synthesised from reaction of $C_6F_6$, PhH and Cr atoms[91] is not affected by hard nucleophiles such as NaOMe or piperidine even after several hours at 60 °C[238] while free $C_6F_6$ is substituted by these reagents at moderate temperatures[266]. However with softer nucleophiles such as $Bu^nLi$ and $Bu^tLi$, substitution of the chromium complex is at least as high as in the free arene[389]. Thus the empirical evidence shows that in some cases complexation to the ($\eta$-PhH)Cr moiety of $C_6F_6$ reduces its susceptibility to nucleophilic attack over the free arene in accord with the fact that complexation of $C_6F_6$ increases its electron density in the ground state. However when substitution at the complexed $C_6F_6$ does occur, it occurs at least as completely as in the free arene supporting stabilisation of the suggested intermediate complexation.

The yields from the more sterically demanding *tert*-butyl carbanion are low presumably because of the competing reaction to form *4.53* most probably *via* the anion [($\eta^6$-Ph)($\eta^6$-$C_6F_6$)Cr]$^-$ formed from proton abstraction by the *tert*-butyl anion. The acidity of the PhH ring in this complex is understandable in terms of delocalisation of change over the whole complex anion, especially the electron drift towards the more electron-demanding $C_6F_6$ ligand. There is clear evidence of the two rings interacting as the appropriate signals in the $^{19}F$ and $^1H$ nmr spectra appear as septets[238]. The net electron drift to the $C_6F_6$ ligand in at least the starting material ($\eta$-$C_6F_6$)($\eta$-PhH)Cr is also shown by ESCA measurements[267] which show that the difference in $C_{1s}$ binding energies is less between $C_6F_6$ and PhH complexed in ($\eta$-$C_6F_6$)($\eta$-PhH)Cr than between the free arenes.

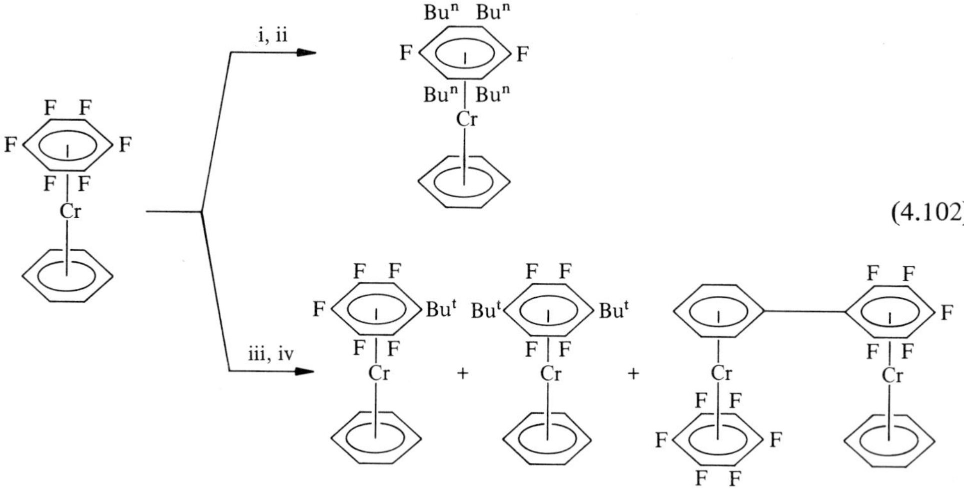


(4.102)

4.54

(i) $Bu^nLi$, –78°C, 90 min; (ii) RT, 10 h; (iii) $Bu^tLi$, –78°C to RT; (iv) RT, 2 h

Reaction of ($\eta^6$-$C_6F_6$)($\eta^6$-PhH)Cr with LiPh forms ($\eta^6$-PhH)($\eta^6$-$C_6F_5$Ph)Cr in good yield. Interestingly, the chromium bonds to the fluorinated ring of the $C_6F_5$-Ph ligand[238].

Another, similarly acidic arene-chromium complex ($\eta$-$C_6F_5H$)($\eta$-PhH)Cr has been prepared from reaction of Cr atoms, PhH and $C_6F_5H$. The complex reacts with $Bu^tLi$ at –78 °C to form the corresponding carbanion which can undergo further reaction[241].

The reactions are summarised in Scheme 4.10.

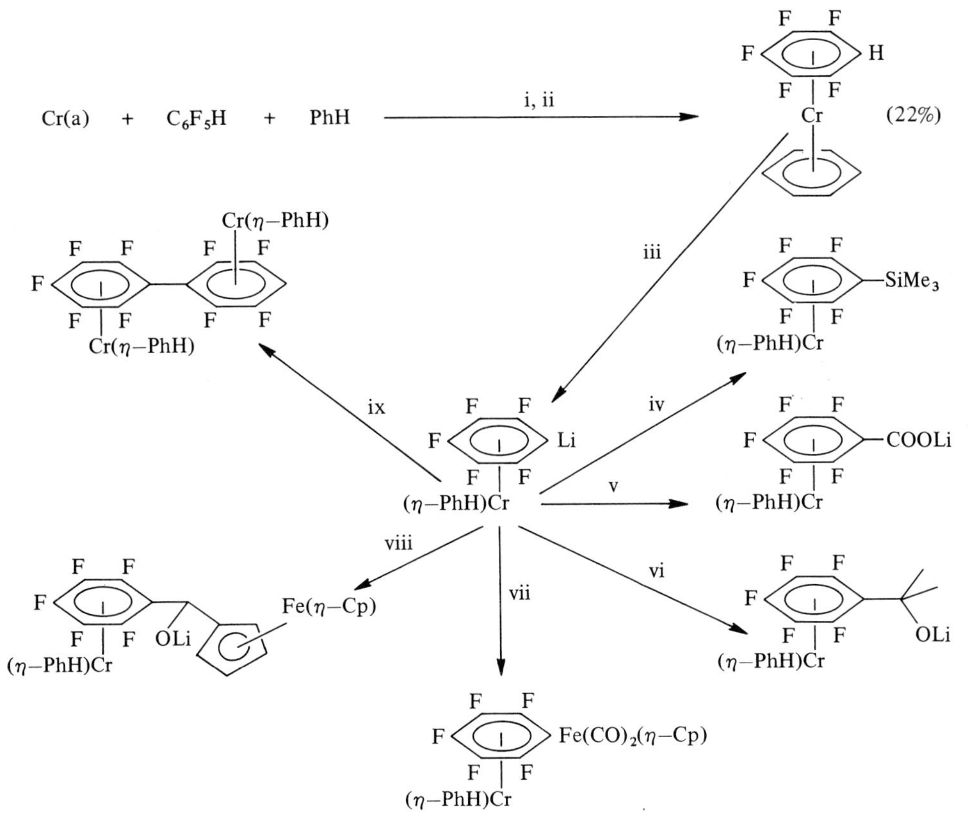


**Scheme 4.10.** (i) Cocondense, 77 K, $C_6F_5H$:PhH = 3:2; (ii) Δ; (iii) $Bu^tLi$, −78 °C; (iv) $Me_3SiCl$; (v) $CO_2$; (vi) AcMe; (vii) (η-Cp)Fe$(CO)_2$I; (viii) (η-Cp)Fe(η-$C_5H_4$ · CHO); (ix) Δ, higher homologues may also be formed

## Complexes with Polymeric Arenes

In a novel experiment to attempt to prepare bridged $(\eta\text{-arene})_2Cr$ derivatives, Cr atoms have been condensed with 1,4-diphenyl butane (Eq. (4.103))[232].

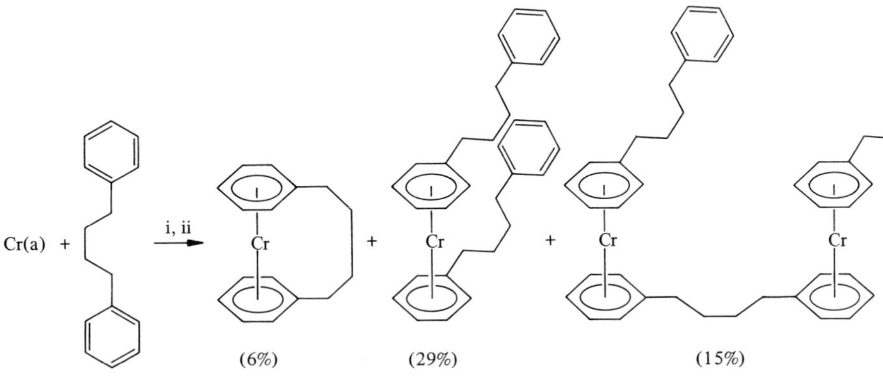


(4.103)

(i) cocondense, 77K; (ii) warm

The use of 1,4-diphenylbutane in conventional syntheses has not been successful as under reducing Friedel-Crafts conditions it dealkylates to form ($\eta$-tetraline)$_2$Cr and ($\eta$-PhH)($\eta$-tetraline)Cr[268].

The MVS method has proved successful in fixing metal atoms to polymer-bound arenes. For example condensation at 0 to $-10$ °C of Cr atoms into rotating Dow Corning 510 silicone diffusion pump oil (a mobile liquid methylphenylsiloxane with a Me : Ph ratio of ca. 17 : 1 and $OSiMe_3$ end groups) gives on filtration a clear red-brown solution[372]. The $^1$H nmr of this liquid shows multiplets at 5.27 $\tau$ and 2.23 $\tau$ and a singlet at 9.64 $\tau$ which correspond to phenyl groups coordinated to Cr, free phenyl groups and methyl groups respectively. Integration gives the ratio of free to coordinated arene of 2.4 : 1. The liquid is thermally stable to at least 300 °C but in air is oxidised to a clear yellow liquid which has an esr spectrum typical for $[(\eta\text{-arene})_2Cr]^+$. Further oxidation forms a fine yellow water-soluble precipitate which on treatment with $NH_4PF_6$ gives ($\eta$-arene)$_2$CrPF$_6$. The residual silicone liquid now shows a strong band at 3,660 cm$^{-1}$ in its ir spectrum indicative of SiOH. Oxidation of the DC 510-Cr product with 7,7,8,8-tetracyanoquinodimethane (TCNQ) forms a bright green paramagnetic complex with spectral properties reminiscent of ($\eta$-arene)$_2$Cr(TCNQ)$_2$[272]. Thus the DC 510-Cr complex is most likely *4.54* and a reasonable scheme for its reactions is shown in Scheme 4.11. The electron properties of *4.54* are under investigation.

Reaction between Cr atoms and DC 510 is efficient till at least 50% of the phenyl groups have been coordinated. Other methylphenylsiloxanes, e.g., Silicone 704, a tetramethyltetraphenyltrisiloxane, react in a similar way to DC 510 with Cr atoms[372].

In addition similarly bound ($\eta$-arene)$_2$M complexes can be made by reaction of DC 510 with Ti atoms (a bright pink liquid), with V atoms (a dark brown paramagnetic liquid) and with Mo atoms (a golden-brown liquid). Reaction of Mn, Fe or Ni atoms with DC 510 result in the formation of fine metal slurries.

Cocondensation of Cr atoms and styrene leads to Cr atoms bonded to the phenyl groups of polystyrene[142]. This reaction will be discussed in the section on styrene.

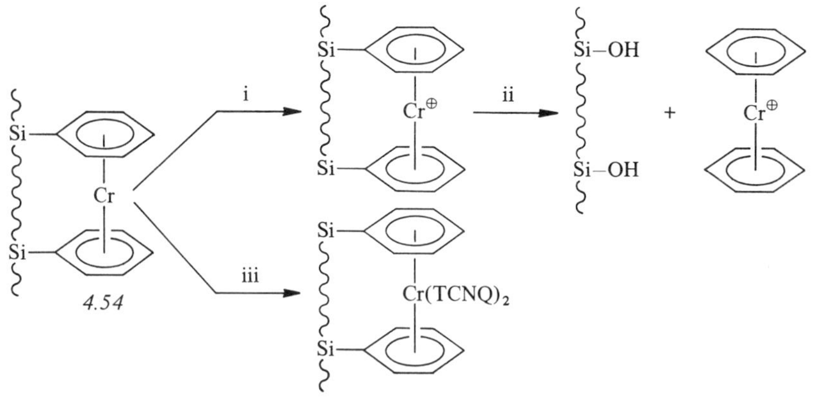


**Scheme 4.11.** (i) $O_2$; (ii) extended $O_2$; (iii) TCNQ

## ($\eta$-Naphthalene)$_2$M

($\eta$-naphthalene)$_2$Cr is not amenable to reducing Friedel-Crafts synthesis owing to hydrogenation, giving a production of approximate composition ($\eta$-tetraline)$_2$Cr[273]. However it is readily synthesised by cocondensation of naphthalene and Cr atoms[227]. ($\eta$-1-methylnaphthalene)$_2$Cr may be similarly prepared in up to 70% yield by condensing Cr atoms into cooled liquid 1-methylnaphthalene. Contrary to all other reported ($\eta$-arene)$_2$Cr complexes, naphthalene ligands in these complexes are easily displaced under mild conditions as shown in Scheme 4.12. This facile displacement can be attributed to the lower loss of aromatic delocalisation energy relative to that of a monocyclic arene, as the $\eta^6$-naphthalene ligand is displaced to a $\eta^4$-mode by an incoming ligand (Eq. (4.104)). There are still some steric constraints; bulky ligands such as $PPh_3$ or $P(OPh)_3$ do not react under mild conditions.

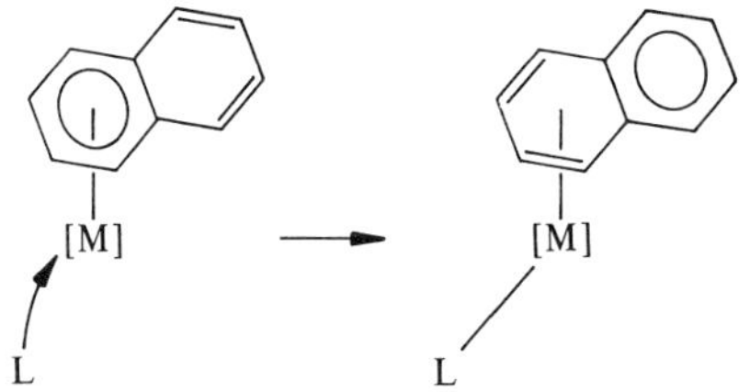


(4.104)

Similar cocondensation reactions form $\eta$-naphthalene derivatives of Mo, V and Ti. These complexes too have labile naphthalene ligands which suggest that these complexes will be excellent starting materials for the synthesis of low-valent compounds of Ti, V, Cr, and Mo[227].

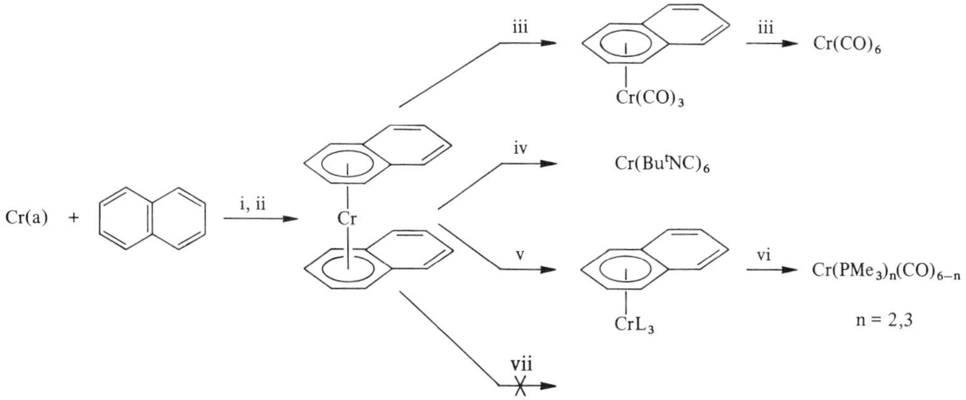


**Scheme 4.12.** (i) Cocondense, 77 K; (ii) Δ; (iii) CO, 1 atm, 20 °C; (iv) $Bu^tNC$, hexane, 20 °C; (v) L (L = $PF_3$, $P(OMe)_3$, $PMe_3$); (vi) CO (L = $PMe_3$); (vii) $PPh_3$ or $P(OPh)_3$, mild conditions

## Arene Complexes of the Later Transition Metals

Cocondensation of Mn atoms and arenes forms no isolable product. Compounds analogous to the Re complex *4.55* may have been expected as it is formed from the very unstable ($\eta$-$C_6Me_6)_2Re$ complex detected on sublimation probes at 77 K above heated mixtures of $[(\eta\text{-}C_6Me_6)_2Re]^+$ salts and molten Na or Li under vacuum (Eq. (4.105))[269].

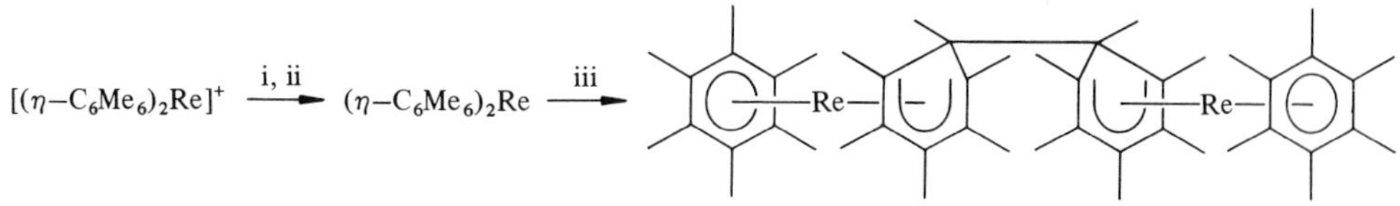


*4.55* (4.105)

(i) Na or Li, heat; (ii) collect on probe, 77K; (iii) warm

Cocondensation of Mn atoms with arenes and cyclopentadienes forms the expected ($\eta^6$-arene)($\eta^5$-cyclopentadienyl)Mn complexes but in disappointingly low yields[17,19]. The reasons for these poor results may be competing side-reactions. There is evidence that Mn aggregation is a particularly facile process in Mn/substrate cocondensates[378].

Cocondensation of Fe atoms and PhH at 77 K forms an unstable material which on warming to $-50\,^\circ C$ explodes. Fe, PhH, $H_2$ and a trace of $Ph_2$ can be detected after the explosion[381]. The Fe(a)/PhMe is also unstable[383]. However, addition of ligand precursors at low temperatures to Fe(a)/arene cocondensates leads to isolation of ($\eta^6$-arene)Fe derivatives (Scheme 3.1). Infra-red studies show that ($\eta$-arene$)_2$Fe (arene = PhH, PhMe, Mesitylene) exist in low temperature Fe(a)/arene cocondensates but decompose below $-40\,^\circ C$[382]. It is presumed that one arene ligand in this apparently 20e-complex is labile to substitution to form the 18e ($\eta^6$-arene)$FeL_2$ complexes. However, there is also evidence that the species ($\eta$-PhH$)_1$M (M = Fe, Co, Ni) exist in M(a)/PhH cocondensates at 77 K. This stoichiometry is based on spectroscopic examination of mixed ligand cocondensates M(a)/PhH/$C_6D_6$ and M(a)/PhH/PhF: no ir absorbance lines due to a mixed ligand-metal complex can be observed in the cases of Fe, Co, Ni but can in the case of Cr[231]. Comparison of the change in $\nu_{C=C}$ of PhH on complexation ($\Delta\nu$) in the species (PhH$)_2$Cr and (PhH)M (M = Fe, Co, Fe) shows that the magnitude of $\Delta\nu$ decreases in the order Cr $>$ Fe $>$ CoNi inferring that the M-arene bond strength decreases in the order Cr $>$ Fe $>$ Co $>$ Ni[231].

Decomposition of the PhH/Fe and PhMe/Fe cocondensates precipitates Fe and forms free arene, but the Fe/mesitylene complex forms *4.56* in modest yields, presumably *via* H-abstraction (Eq. (4.106))[385]. It is not clear whence the hydrogen came but observations on the paramagnetic complex ($\eta$-$C_6Me_6)_2Fe$ are illuminating[274]. An equilibrium involving a benzylironhydride species has been proposed on the evidence of deuterium exchange and the detection of coupled product *4.57* on decomposition in alkanes (Eq. (4.110)). It is unlikely that hydrogen is abstracted from the excess arene solvent as decomposition of $(C_6Me_6)_2Fe$ in $C_6D_6$ forms some ($\eta^6$-$C_6Me_6$)Fe($\eta^4$-$C_6D_6H_2$) analogous to *4.56*[274].

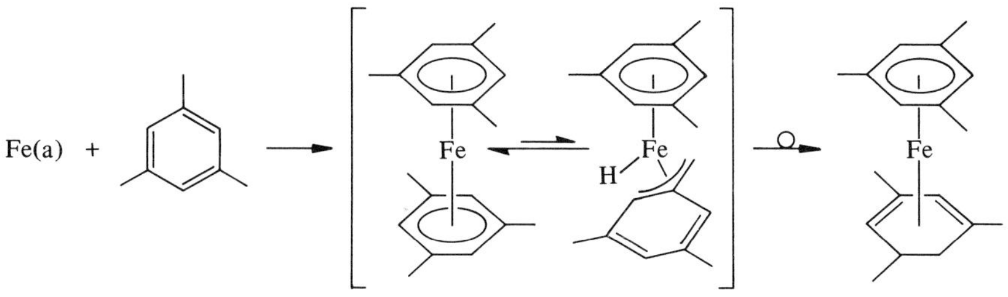


(4.106)

4.56

$$(C_6Me_6)_2Fe \longrightarrow Fe + C_6Me_6 + C_6Me_5CH_2CH_2C_6Me_5 \qquad (4.107)$$

4.57

Addition of ligand precursors to Arene/Co, Ni (or Pd) cocondensates can lead to isolation of organometallic species though none of them has retained the arene ligand: for example addition of $P(OEt)_3$ to the PhMe/Ni cocondensates leads to the isolation of $Ni[P(OEt)_3]_4$ (40%) on warming[390]. If such matrices are warmed in the absence of ligand precursors, finely divided metal is precipitated. In contrast, cocondensation of Co or Ni atoms with $C_6F_6$ leads to arene-metal complexes which survive to ambient temperatures: There is also evidence of an unstable Cu-$C_6F_6$ complex[224]. The Ni/$C_6F_6$ cocondensate is initially green-black but turns black on warming. After excess $C_6F_6$ has been pumped off, the residue is extremely susceptible to explosive decomposition (percussion, static charge, uneven warming) though by careful warming a temperature of 70 °C can be reached before decomposition. At room temperature the material reacts explosively with most compounds including $H_2$, CO, allyl bromide and $CF_3I$ though not with Ar. However, smooth reaction can be affected by recooling the cocondensate to 77 K, after removal of excess $C_6F_6$, then adding a substrate and slowly warming. Some of its reactions are shown in Scheme 4.13. The main type of reaction is displacement suggesting that $C_6F_6$ maintains

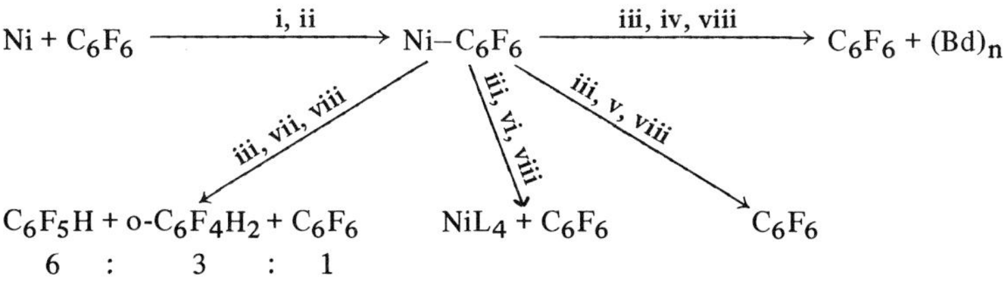


**Scheme 4.13.** (i) Cocondense, 77 K; (ii) remove excess $C_6F_6$, *ca.* −40 °C; (iii) recool to 77 K; (iv) Bd; (v) HBr, $H_2O$, or warm to ambient temperatures; (vi) L = CO or $P(OEt)_3$; (vii) $H_2$; (viii) Δ

its integrity and no defluorination has occurred. The one exception is reaction with $H_2$ which forms $C_6F_5H$, $C_6F_4H_2$ and $C_6F_6$ (6:3:1)[224,389]. The Ni-$C_6F_6$ compound has not been isolated in a pure state owing to its insolubility or reactivity with organic solvents. However *ca.* 1 mole of $C_6F_6$ is released in forming 1 mole $Ni(CO)_4$ suggesting a stoichiometry of Ni · $C_6F_6$. Structures such as *4.58*[275], *4.59*[276], *4.60*[276], and *4.61*[277] have been proved or suggested for various complexes of $C_6F_6$ or similar arenes with the later transition metals. A polymeric structure involving distorted $C_6F_6$ can thus be envisaged for Ni · $C_6F_6$.

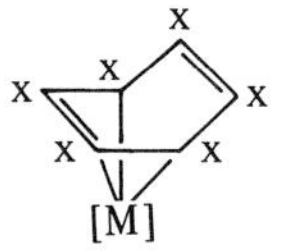

*4.58*

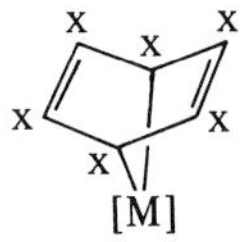

*4.59*

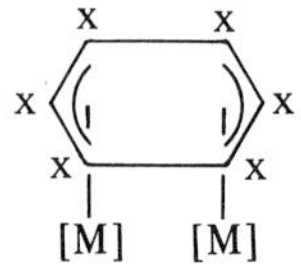

*4.60*

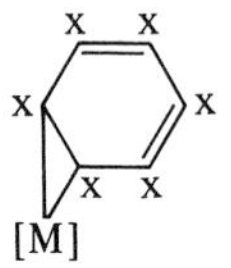

*4.61*

### *4.13.2 Cycloheptatriene*

Like that of cyclopentadiene (CpH), the metal atom chemistry of cycloheptatriene (ChtH) is dominated by its active hydrogen. The driving force in both cases is much the same; the increase in coordination and electron number, and the formation of the compact, robust, fully unsaturated and delocalised ring ligands cyclopentadienyl (Cp) and cycloheptatrienyl (Cht).

Cocondensation of Ti atoms and ChtH forms the known[278] compound ($\eta^7$-Cht)-($\eta^5$-ChtH$_2$)Ti (46%) *4.62a*[221]. The prefered formation of the 16e-($\eta$-Cht)($\eta$-ChtH$_2$)Ti complex from a presumed ($\eta$-Cht)($\eta$-ChtH)TiH intermediate rather than the 18e- but unknown complex ($\eta$-Cht)$_2$Ti may be because hydrogen transfer is a more facile process than hydrogen elimination and once the relatively inert ($\eta$-Cht)($\eta$-ChtH$_2$)Ti complex has been formed there is no further reaction. Alternatively ($\eta$-Cht)$_2$Ti is not detected because of some inherent instability or lability although one ($\eta$-Cht)$_2$M complex is known, the 17e-($\eta$-Cht)$_2V^{2+}$ dication[279], formed by hydride abstraction from ($\eta$-ChtH)$_2$V. In many ways ($\eta$-Cht)$_2$Ti should not be expected. Pes experiments on early transition metal complexes show an increasing proportion of ligand character in the $e_2$ and $e_1$ ring-metal interactions on increasing ligand ring size and on increasing the metal valence orbital energy[234]. Thus the Cht ligand $e_2$ and $e_1$ orbitals are respectively stronger metal electron acceptors and weaker ligand electron donors than those of PhH and Cp ligands. This is especially so for Ti which has a high valence orbital energy, thus a strong $e_2$ interaction and large metal to ligand electron transfer can be expected for the postulated ($\eta$-Cht)$_2$Ti. A negative charge of 0.7 to 0.8 electrons has been estimated from ESCA experiments, to reside on Cht in ($\eta$-Cht)($\eta$-Cp)Ti[280] while the Ti-C (Cht) distance in that compound is found to be "anomalously" low at 2.19 Å[281] indicating a strong bond. Therefore, by analogy, ($\eta$-Cht)$_2$Ti can be expected to have two Cht rings each with a negative charge of up to 0.7 to 0.8 electrons and a C-C distance between rings of *ca.* 3 Å, even in the staggered conformation. Assuming a carbon van der Waals radius of 1.7 Å considerable inter-ring repulsion can be expected leading to a weakening of the metal-ring bonds. In comparison the ($\eta$-Cht)$_2V^{2+}$ cation will have a decreased $e_2$ Cht-V interaction owing to the lower valence orbital energy of V leading to an increased V-C (Cht) bond length (2.25 Å in ($\eta$-Cht)($\eta$-Cp)V[282]) and reduced ligand charge estimated at 0.3 to 0.4 eV for Cht in ($\eta$-Cht)($\eta$-Cp)V[280]

Similar arguments have been used to explain the diamagnetism of ($\eta$-mesitylene)$_2$Ti and the paramagnetism of ($\eta$-mesitylene)$_2$V$^+$ (*vide supra*).

Cocondensation at 77 K of Ti atoms with an equimolar mixture of ChtH and PhH leads only to formation of approximately equal amounts of ($\eta$-Cht)($\eta$-ChtH$_2$)Ti and ($\eta$-PhH)$_2$Ti. No mixed ligand systems such as ($\eta$-ChtH)($\eta$-PhH)Ti, ($\eta$-Cht)($\eta$-PhH)Ti or ($\eta$-Cht)($\eta$-PhH$_2$)Ti are observed[221].

V atoms condensed at 77 K with ChtH form a yellow brown material of composition $C_{14}H_{16}V$ (ms and analysis) which has very similar spectral properties to the purported ($\eta$-ChtH)$_2$V *4.63a* synthesised from reduction of $VCl_4$ by Pr$^i$MgBr in the presence of ChtH[279], but appears much more thermally stable (dec. > 300 °C). It has therefore been postulated that there may well be two isomers of $C_{14}H_{16}V$, ($\eta$-ChtH)$_2$V and ($\eta$-Cht)($\eta$-ChtH$_2$)V *4.62b* which may be impossible to distinguish spectroscopically. An X-ray structure determination is therefore to be undertaken[221].

Condensation of Cr atoms into rotating solutions of ChtH (10%) in CyMe at −130 °C[221] or with ChtH at 77 K[18,283] forms the known[284] 17e-complex ($\eta^7$-Cht)($\eta^4$-ChtH$_3$)Cr *4.64*. No 18e compounds such as ($\eta^7$-Cht)($\eta^5$-ChtH$_2$)Cr or ($\eta^6$-ChtH)$_2$Cr are apparently formed, though at lower temperatures (< −20 °C) there are indications of the species ($\eta$-ChtH)$_2$Cr[18,115] but these are tentative. Cocondensation of Cr atoms, ChtH and $PF_3$[221] or addition of $PF_3$ to the Cr/ChtH cocondensate[118] traps the ChtH ligand as ($\eta^6$-ChtH)Cr($PF_3$)$_3$ *4.65*. In comparison condensation at 77 K of Mo or W atoms with ChtH forms the 18e-complex ($\eta^7$-Cht)($\eta^5$-ChtH$_2$)M [M = Mo *4.62c*, W *4.62d*][11]. In addition ($\eta$-Cht)($\eta$-ChtH$_2$)Mo can be converted to ($\eta^7$-Cht)($\eta^5$-Cp)Mo by reaction of NaCp in refluxing thf over 18 h.

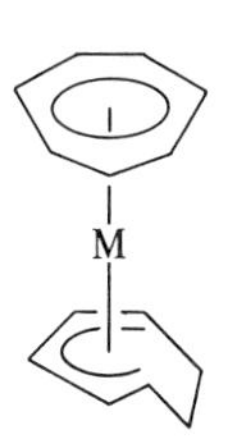

(*4.62* a = Ti, b = V, c = Mo, d = W)

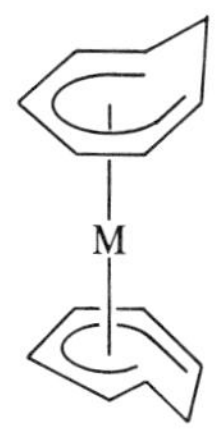

(*4.63* a = V, b = Cr)

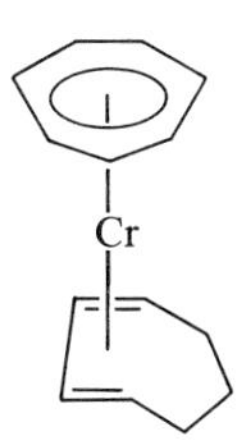

*4.64*

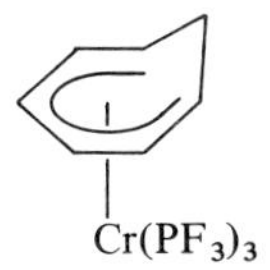

*4.65*

Condensation of Mn vapour onto a rotating solution of ChtH (10%) in CyMe at −130 °C forms golden-yellow solutions containing no free bulk metal[221]. However, allowing the solution to warm results in precipitation of bulk metal. Thus it seems that Mn atoms (or small Mn clusters) are weakly bonded to ChtH at low temperatures but no complex apparently survives to room temperature. No complex is also apparently formed on cocondensation at 77 K of Mn vapour, ChtH and $PF_3$, even though species such as ($\eta^5$-Cht)$Mn(CO)_3$[285] are known.

Condensation of Fe atoms onto rotating solutions of ChtH in CyMe[221] or with ChtH at 77 K[286] forms the dark-red complex ($\eta^5$-Cht)($\eta^5$-$ChtH_2$)Fe whose structure *4.66* is derived from X-ray analysis[286], which shows clearly the ($\eta^5$-Cht)($\eta^5$-$ChtH_2$) sandwich structure and the open faces of the two rings skew (60 °C) to each other.

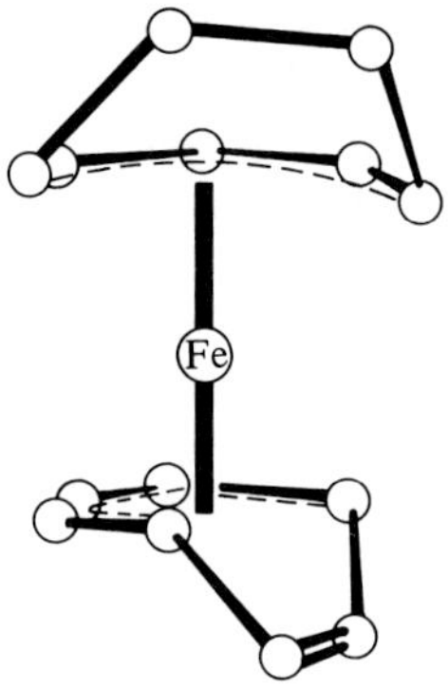


*4.66*

At low temperatures (–80 to –30 °C) the $^1$H and $^{13}$C nmr spectra show 16 and 14 resonances respectively, consistent with a skew structure similar to that found in the solid state. At more ambient temperatures (–30 to +50 °C) the $^{13}$C proton noise decoupled nmr spectrum shows pairwise coalescence of 12 of the 14 resonances has occurred. The $^1$H nmr spectrum shows a similar pairwise coalescence. The spectra are consistent with the ($\eta^5$-Cht) and ($\eta^5$-$ChtH_2$) rings each developing a mirror plane through C6 and C8 respectively. Therefore the rings are either no longer skew or more likely they rock or rotate with respect to each other at a rate sufficient to induce pairwise equivalence in each ring on the nmr time scale (Eq. (4.108))[285,287]. At higher temperatures (50 °C and higher) further fluxional behaviour is observed. Resonances attributable to the Cht ring coalesce to form one signal while those attributable to $ChtH_2$ remain unchanged indicating rapid exchange of the unbonded double bond in the $\eta^5$-Cht ligand while the $ChtH_2$ ring is unchanged. Other fluxional $\eta^5$-Cht-containing complexes are known[285,288].

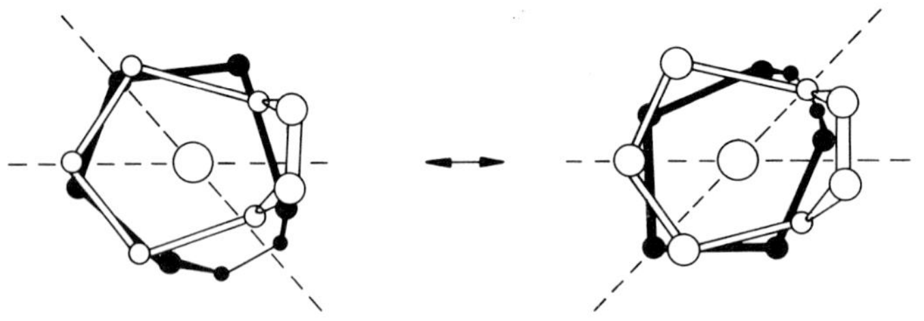

(4.108)

A second product of the reaction of Fe atoms and ChtH, $C_{14}H_{18}Fe$, is observed in trace amounts in the mass spectrum of ($\eta$-Cht)($\eta$-$ChtH_2$)Fe[221] and is probably ($\eta^5$-$ChtH_2$)$_2$Fe. In comparison the reduction of $FeCl_3$ by Pr$^i$MgBr in the presence of Cht forms $(\eta^5\text{-ChtH}_2)_2\text{Fe}$ and $(\eta^5\text{-Cht})(\eta^5\text{-ChtH}_2)\text{Fe}$, their relative yields depending on experimenter[289,290], which suggests the free double bond in $(\eta^5\text{-Cht})(\eta^5\text{-ChtH}_2)\text{Fe}$ is labile. No $(\eta^4\text{-ChtH})\text{Fe}(\text{PF}_3)_3$ analogous to the known complex $(\eta^4\text{-ChtH})\text{Fe}(\text{CO})_3$[288] can be detected on cocondensation of Fe atoms and a mixture of $PF_3$ and ChtH[221].

Condensation of Co atoms onto a solution of ChtH (10%) in CyMe gives an air-sensitive orange oil containing an inseparable mixture of products in poor yields. The mass spectrum of this oil shows it contains complexes of composition $C_{14}H_nCo$ (n = 15, 17 or 19). A mixture of three complexes of the same compositions in thought to be formed on reduction of $CoCl_2$ by $Pr^iMgBr$ in the presence of ChtH[289], and one of them has been shown to be $(\eta^5\text{-ChtH}_2)(\eta^4\text{-ChtH}_3)\text{Co}$ *4.67*. Cocondensation of Co atoms, ChtH and $PF_3$ forms the air-sensitive orange oil $(\eta^3\text{-Cht})\text{Co}(\text{PF}_3)_3$ *4.68* and $HCo(PF_3)_4$[221].

Ni and Pd atoms react similarly to Mn atoms with ChtH in CyMe at −130 °C as again no tractable complex seems to survive to ambient temperatures, the major product being Ni or Pd slurries. Cocondensation at 77 K of Ni atoms, ChtH, and $PF_3$ only forms $Ni(PF_3)_4$[221].

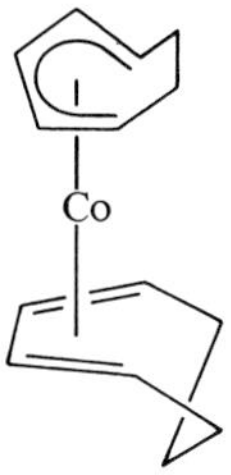


*4.67*

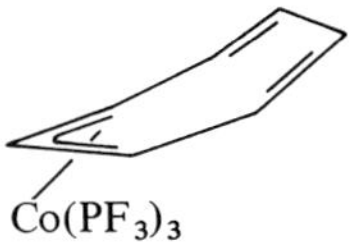


*4.68*

## 4.14 Styrene

Styrene is treated separately owing to its unique ability to act as a monoene, conjugated diene or arene. In addition, its polymerisation is catalysed by some metal atoms which can lead to polymer-bound metal derivatives.

Cr atoms (supported on silica) polymerise ethylene, and are known to form $(\eta\text{-arene})_2\text{Cr}$ complexes with a wide variety of arenes (see Arenes section). Thus reaction of Cr atoms and styrene may lead to Cr bound to polystyrene. Cocondensation of Cr atoms and styrene at 77 K onto a pentane matrix does indeed produce polystyrene and an intractable material in which Cr is bound to polystyrene (mass spectrum $(C_8H_8)_nH_mCr$; n = 2, m = 0; n = 3 to 5, m = 4). Warming the cocondensate in the presence of CO or $PF_3$ leads to isolation of $CrL_3$ fragments (L = CO, $PF_3$) bonded to the arene portion of styrene or to polystyrene (e.g. mass spectrum $(C_8H_8)_nH_2Cr(CO)_3$; n = 3–7). In addition the $^1H$ nmr spectrum of polystyrene-$Cr(CO)_3$ shows signals at 2.4 ~ 2.7, 5.7, 8.5 ~ 9.0 ppm in the intensity ratio 4:1.2:3.2 assigned to phenyl, $\eta^6$-phenyl bonded to $Cr(Co)_3$ and methylene hydrogens respectively. Thus polymer-bound species of the type *4.69* and *4.70* are being formed[142].

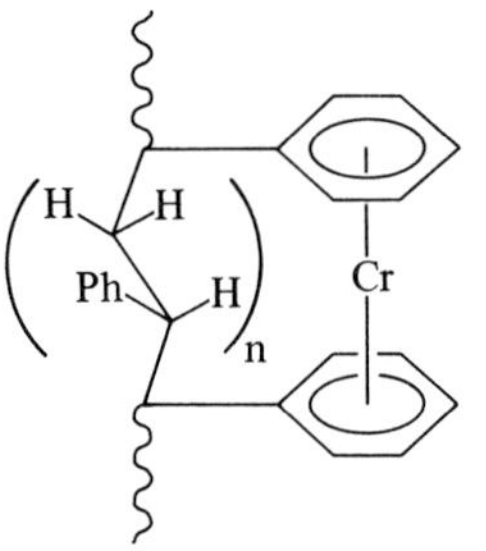

4.69

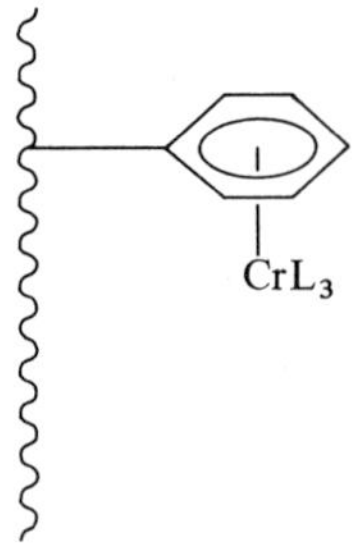

4.70

L = CO, $PF_3$

Cocondensation of styrene and Fe atoms leads to very low conversion (*ca.* 1%) to polystyrene and no identifiable organometallic compounds. However, if the cocondensate is warmed in an atmosphere of CO till –20 °C, ($\eta^4$-styrene)$Fe(CO)_3$ *4.71*, ($\eta^2$-styrene)-$Fe(CO)_4$ *4.72* and $Fe(CO)_5$ can be isolated[117].

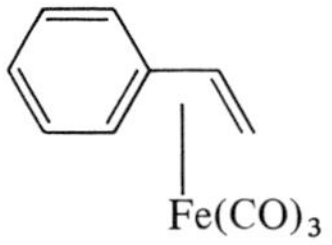

4.71

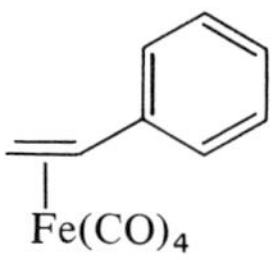

4.72

Reaction of styrene and Ni atoms gives ($\eta^2$-styrene)$_3$Ni which decomposes at –20 °C. The more stable derivative bipy Ni($\eta^2$-styrene) can be trapped by addition of bipyridyl to the cocondensate at temperatures below –30 °C[142].

## 4.15 Alkatetraenes

The only reactions of metal atoms with alkatetraenes presently published are those with cyclooctatetraene (Cot), the next highest in the family of ring ligand precursors CpH, PhH, ChtH. Unfortunately its metal atom chemistry is proving not as extensive as that of the others in the series owing to Cot's tendency to form polymeric products with metal atoms (e.g. those of Ti, Fe, Co)[19,221]. However Cot can be used as a secondary reagent to form Cot-metal complexes when added to unstable cocondensates (e.g. $(Cod)_2Fe$[385]).

Cocondensation of Cr atoms and Cot at 77 K forms the moderately air-stable black complex $(Cot)_3Cr_2$ (43%)[221]. A unit cell determination by X-ray diffraction gives identical results to those obtained for *4.71*[291] made originally by reduction of $CrCl_3$ by either $CotNa_2$ or PriMgBr in the presence of Cot[292]. The $^1H$ and $^{13}C$ nmr of this complex show fast intermolecular exchange leading to all CH positions being equivalent. It does not react with CO or $PF_3$ but reacts rapidly with $Bu^tNC$ in refluxing n-hexane to give $Cr(CNBu^t)_6$[221].

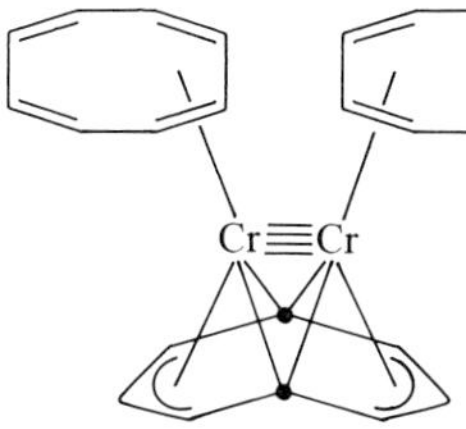


*4.71*

Cocondensation at 77 K of Mn atoms and Cot gives trace amounts of an air-stable purple complex whose nature is yet to be determined[221].

Cocondensation of Fe atoms and Cot forms only polymeric material, but cocondensation at 77 K of Fe atoms, Cot and $PF_3$ forms ($\eta^4$-Cot)Fe$(PF_3)_3$, analogous to ($\eta^4$-Cot)-Fe$(CO)_3$[293] but in low yield[19,221].

Cocondensation of U atoms and Cot leads to uranocene (($\eta^8$-Cot)$_2$U)[294]: this may have been formed *via* highly activated bulk metal which is known to afford Cot$_2$U.

Cocondensation at 77 K of Lanthanon atoms and Cot leads to the new class of organolanthanides Ln$_2$(Cot)$_3$[295] which show chemical and physical properties similar to those of ($\eta$-Cot)$_2$U and KLn($\eta$-Cot)$_2$[296]. The Nd member can be purified by Soxhlet extraction into thf to form bright green crystals of the ion pair [($\eta$-Cot)Nd(thf)$_2$]$^+$[($\eta$-Cot)$_2$Nd]$^-$ which has the strucutre *4.72*[295]. The anion [($\eta$-Cot)$_2$Nd]$^-$ is a sandwich with two planar Cot rings but uniquely, Nd$^I$ is assymetrically positioned. The average Nd$^I$-C distances are respectively 2.660 (24) and 2.787 (19) Å for rings 1 and 2, which are also tilted towards each other (8.25° angle of intersection). Furthermore Nd$^{II}$ completes its coordination sphere by bonding to ring 2 [Nd$^{II}$-C$_a$, C$_b$, C$_c$ bond lengths 3.059 (19), 2.700 (18) and 2.896 (20) Å respectively; Nd-$^{II}$C (ring 3) = 2.673 (16) Å (average)] and as such is similar to the ligand bridged species ($\eta$-Cot)$_3$Ti$_2$[297] and ($\eta$-Cp)$_3$Sc[298]. Such structures are to be expected in species with electron-defficient centres and high electron transfer to the ligands.

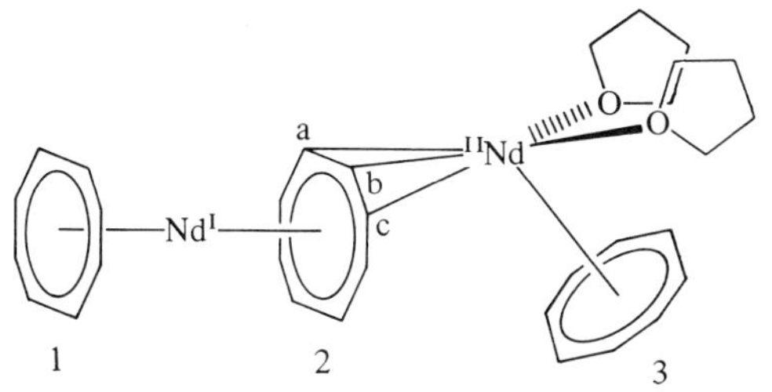


*4.72*

The Na or K/Cot cocondensates after removal of excess Cot at low temperatures show four ir-active absorbance lines identified as those expected for $(M^+)_2Cot^{2-}$[299]. Previous ir spectra of salts, for example K$_2$Cot(thf)$_n$, have assigned the expected C-C stretch to a line at 1,430 cm$^{-1}$[300] rather than at 1,295 (K$_2$Cot) or 1,290 cm$^{-1}$ (Na$_2$Cot) as found in the low temperature matrices. Judicious oxidation of the matrices leads to formation of a line at 1,430 cm$^{-1}$ suggesting it is due to an oxidised product. Esr spectra of the original cocondensates show no evidence for paramagnetic species such as $M^+Cot^-$[299].

## 4.16 Heterocycles

Unsaturated heterocycles such a furan, thiophene, pyrrole, and pyridine possess ring $\pi$-systems similar to those of PhH and Cp; and may thus form similar $\pi$-complexes with transition metals. Unfortunately these species undergo many side reactions during attempts to make such $\pi$-complexes using conventional synthetic methods. The gentle conditions of the MVS experiment in principle ought to allow complexation without too many competing reactions, but, to a large extent this has not proved the case as metal atoms – even at low temperatures – induce unwanted reactions such as fragmentation and defunctionalisation. For example, cocondensation at 77 K of Fe atoms and thiophene, or 2-methyl- or 2,5-dimethyl-thiophene leads to plum coloured solutions at 20 °C from which no tractable organometallic products can be isolated. Warming these cocondensates in the presence of CO yields small amounts (*ca.* 5%) of the known complex *4.73* while addition of Bd, 1,5-Cod or $PMe_3$ to such cocondensates forms intractable unstable red solutions. Addition of CpH to the deposit at 77 K forms no $Cp_2Fe$[301].

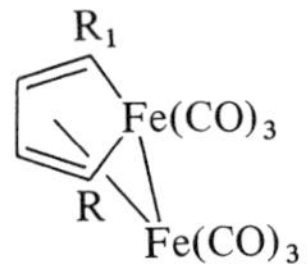


*4.73* ($R = H, R^1 = H$; $R = Me, R^1 = H$; $R = Me, R^1 = Me$)

The Cr/thiophene cocondensate similarly shows extensive signs of desulphurisation while warming the deposit in the presence of CO forms no ($\eta^6$-Thiophene)$Cr(CO)_3$ or ($\eta^4$-Thiophene)$Cr(CO)_4$[301].

Some $\pi$-heterocycle-metal complexes, however, have been synthesised. While cocondensation at 77 K of Cr atoms and pyridine (py) forms an involatile, intractable solid, reactions of Cr atoms, py, and $PF_3$ or mesitylene lead to the new complexes, ($\eta^2$-py)Cr-$(PF_3)_5$, ($\eta^6$-py)$Cr(PF_3)_3$ and ($\eta^6$-py)($\eta^6$-Mes)Cr respectively. ($\eta^6$-$B_3N_3Me_6$)$Cr(PF_3)_3$ may be similarly prepared[102]. 2,5-dimethylpyridine, however, reacts with Cr atoms to form the first bis-$\pi$-pyridine-metal sandwich complex, ($\eta^6$-2,5-dimethylpyridine)$_2$Cr[239]. Reaction of Cr atoms with 2,5-difluoropyridine or borazine only forms involatile species, while N-methylpyrrole does not react[239].

## 4.17 Boranes and Carboranes

Cocondensation of Ni atoms and *closo*-$C_2B_9H_{11}$ forms the known complex[302] $(C_2B_9H_{11})_2Ni$[19]. This complex is a derivative of the *nido*-$C_2B_9H_{11}^{2-}$ dianion, thus Ni atoms have forced the *closo*-carborane to distort giving an open-faced *nido*-structure to which the metal can bond: similar syntheses are known involving transition metal complex fragments[303]: for example, Eq. (4.109)[304].

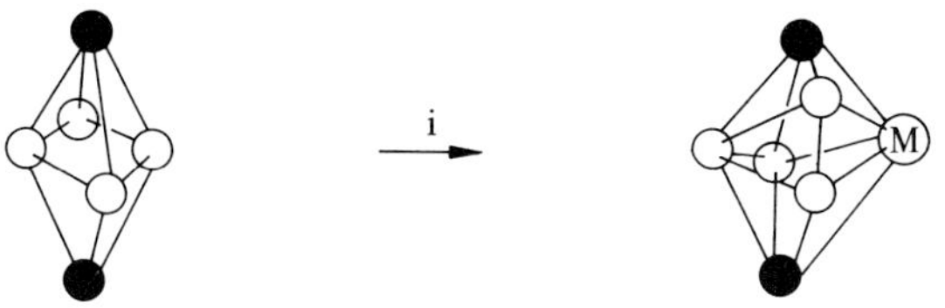

$C_2B_4H_6$ → $C_2B_4H_6$ Ⓜ

● = CH
○ = BH (4.109)

(i) a) $Fe(Co)_5$, Δ — Ⓜ = $Fe(Co)_3$

b) $CpCo(Co)_2$, Δ — Ⓜ = CoCp

c) $(C_2H_4)Ni(PEt_3)_2$, Δ — Ⓜ = $Ni(PEt_3)_2$

Such syntheses are likely because experiment and theory suggest that addition of electrons (and hence metal atoms or complex fragments) to (for example) *closo* structures forces distortion to (for example) a *nido* fragment[305]. Indeed the reduction of the *closo*-$C_2B_9H_{11}$ octadecahedron to the *nido*-$C_2B_9H_{11}^{2-}$ icosahedral fragment is relatively easy[306]. Metal atom reactions may therefore prove a useful general synthesis of heteroboranes. $Ni(B_nC_2H_nMe_2)_2$ (n = 9, 10) has been similarly prepared[105].

A further, useful extension of metal atoms' synthetic usefulness is that more than one metal atom may add to a (car)borane structure and that any remaining coordination requirement may be met by second ligands. For example, reaction of $NaB_5H_8$, $CoCl_2$ and NaCp in cold thf forms a number of metalloborane (and carbametalloborane) clusters which contain 3, 4 and 9 boron units[307]. No metalloborane containing an intact $B_5$-unit was detected. However, Co atoms react with CpH and $B_5H_9$ to form not only the previously discovered clusters but also a series of new complexes, many of which contain $B_5$-units, two of which are the dark green air stable $B_5H_5(CoCp)_3$ and the yellow-green cyclopentyl-$B_5H_4(CoCp)_3$[308]. Two possible structures have been put forward on the basis of spectroscopic evidence and electron-counting rules *4.74* and *4.75* (Eq. (4.110)). A crystallographic study is underway.

Co(a) + CpH +

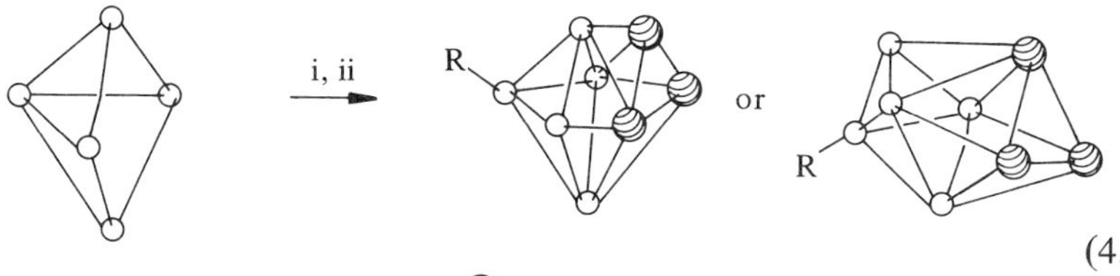

$B_5H_9$ — *4.74* ◉ = CoCp; R = H, $C_5H_9$ — *4.75* (4.110)

(i) cocondense, 77K; (ii) warm

## 4.18 Organic Halides

Many metal atoms react with halogenated hydrocarbons by breaking the polar R-Hal bonds to form oxidative addition products containing R-M and M-Hal bonds as shown in Eqs. (4.111) to (4.114).

$$Mg(a) + Pr^nI \xrightarrow{i,ii} Pr^nMgI\ (76\%)^{309)} \qquad (4.111)$$

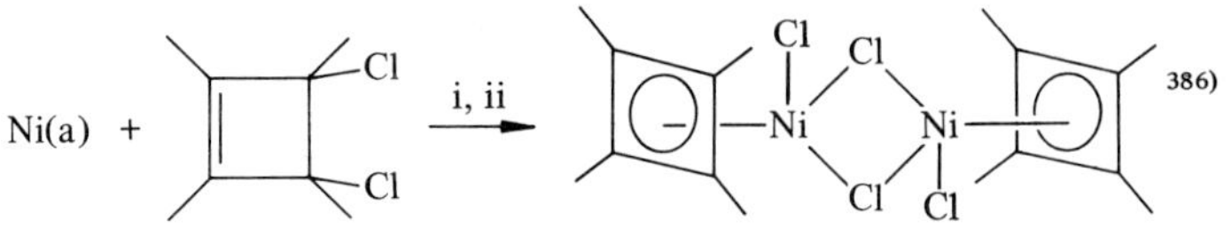

(4.112)

$$Ag(a) + CF_3CF_2I \xrightarrow{i,\ iii} CF_3CF_2Ag + AgI^{310)} \qquad (4.113)$$

$$Ni(a) + EtBr \xrightarrow{i,ii} NiBr_2 + [Et^{\cdot}]\ C_2H_4 + C_2H_6 + n\text{-}C_4H_{10}{}^{8)} \qquad (4.114)$$

(i) cocondense, 77 K; (ii) 25 °C; (iii) –20 °C

Some species are unstable but can be trapped by the addition of ligands at low temperatures (Eqs. (4.115) to 4.116).

$$Co(a) + C_6F_5Br \xrightarrow{i,ii} (C_6F_5)_2Co + CoBr_2 \xrightarrow{iii,v} (\eta^6\text{-PhMe})Co(C_6F_5)_2 + CoBr_2{}^{18)} \qquad (4.115)$$

$$Pd(a) + CF_3Br \xrightarrow{i} [CF_3PdBr] \xrightarrow{iv,v} \textit{trans}\text{-}CF_3Pd(PEt_3)_2Br \qquad (4.116)$$

(i) cocondense, 77 K; (ii) *ca.* –100 °C; (iii) PhMe, *ca.* –100 °C; (iv) $PEt_3$, 77 K; (v) 25 °C

The reactivity of R-Hal increases as F > Cl > Br > I corresponding to the decrease in R-Hal bond strength. Furthermore, only atoms of the most electropositive metals, with high M-F bond strengths can break the strong R-F bonds as shown in Eqs. (4.117) and (4.118).

$$Ca(a) + CF_3(F)C{=}C(CF_3)F \xrightarrow{i} CF_3(F)C{=}C(CF_3)CaF^{9)} \qquad (4.117)$$

$$Pd(a) + CF_3(F)C{=}C(F)CF_3 \xrightarrow{i} Pd\left(\begin{matrix}CFCF_3\\ \|\\ CFCF_3\end{matrix}\right)_3^{45,141)} \qquad (4.118)$$

(i) cocondense, 77 K

Reactivity can often be increased by the presence of unsaturation such as a C=C double bond, which may act as a site of initial bonding for an incoming metal atom, decrease the R-Hal bond strength, and increase the stability of any resulting organometallic compound by being involved in bonding. As an example of the relative importance of unsaturation in bonding with metal atoms, it has been observed that 6-bromo-1-hexene is catalytically converted to 6-bromo-2-hexene on condensation with Pd atoms without formation of any $PdBr_2$ or other products attributable to RPdBr formation[311].

The mechanism of the "insertion" or "oxidative addition" reactions common to metal atom/halogenated hydrocarbon reactions is yet to be clarified, but one point is certain. It is unlikely that there is only one mechanism for such a wide variety of halides and metal atoms. Nonetheless, metal atom reactions, without the further complications of for example ligand effects or surface effects, should be simpler to understand thus contributing towards understanding of more complicated systems as for example the insertion reactions of complexes and solid metals. Preliminary observations are now being reported[16,311,312].

### *4.18.1 Reactions with Main Group Metal Atoms*

The first ever deliberate attempt at a matrix reaction between metal atoms and organic substrates was condensing Li atoms and MeI in an Ar matrix to produce the Me$^\bullet$ radical in the same matrix cage as LiI[313] presumably by dissociative electron transfer. Later experiments show that the Me$^\bullet$ radical is strongly perturbed by the LiI ion pair[314], a conclusion supported by similar reactions of Na and K atoms with MeI and MeBr.

Li atoms have since been condensed in noble gas matrices with $CCl_4$[315,316], $CBr_4$[316], $CHX_3$ (X = Cl, Br)[317], $CHBrF_2$[318], and $CH_2X^1X^2$ ($X^1$ = Cl, Br; $X^2$ = F, Cl, Br, I)[319]. In these cases the substituted methyl radicals are thought essentially free, thought other products are formed. For example Eqs. (4.119) to (4.121)[315,316].

$$Li(a) + CCl_4 \xrightarrow{i} LiCl + {}^\bullet CCl_3 \quad (4.119)$$

$$2\,Li(a) + CCl_4 \xrightarrow{i} 2\,LiCl + :CCl_2 \quad (4.120)$$

$$2\,Li(a) + CCl_4 \xrightarrow{i} LiCl + LiCCl_3 \quad (4.121)$$

(i) cocondense in Ar matrix, *ca.* 15 K

However, more recently[320], the reactions of Li, Na, K or Cs atoms with $CCl_4$, $^{13}CCl_4$, $CCl_3Br$ and $CCl_2Br_2$ at 15 K in Ar have been reexamined and evidence for $^\bullet CCl_3$ perturbed by the metal halide found. In addition there is evidence for MHal and $:CCl_2$ in which the relative orientation of MHal and $:CCl_2$ is non-random. The structure most consistent with the data and which makes the most chemical sense is the carbenoid structure *4.76*.

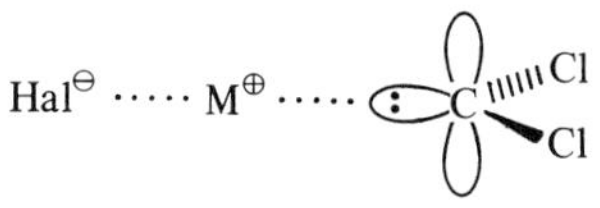


*4.76*

A rotating cryostat has been used to masterfully study a wide range of radicals formed by dehalogenation of organic halides by Na atoms at 77 K[74)]. The radicals, identified by ir and esr spectroscopy at low temperatures and glc analysis of warm-up products, include the cyclohexyl *4.77*, phenyl *4.78*, acetyl *4.79*, and allyl *4.80* radicals.

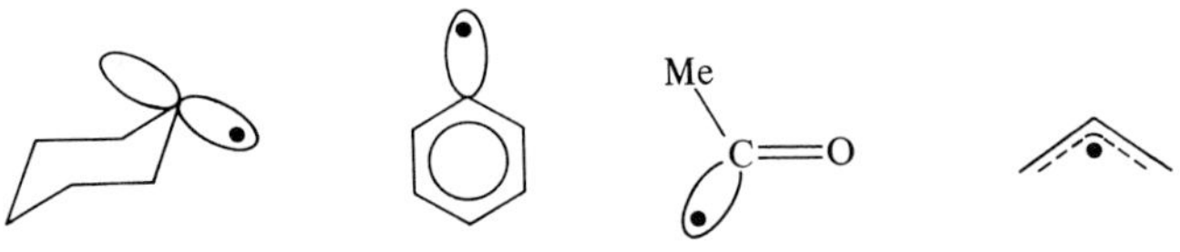


*4.77* *4.78* *4.79* *4.80*

Rotation allows subsequent reaction of the radicals to be studied. For example Eqs. (4.122) to (4.124).

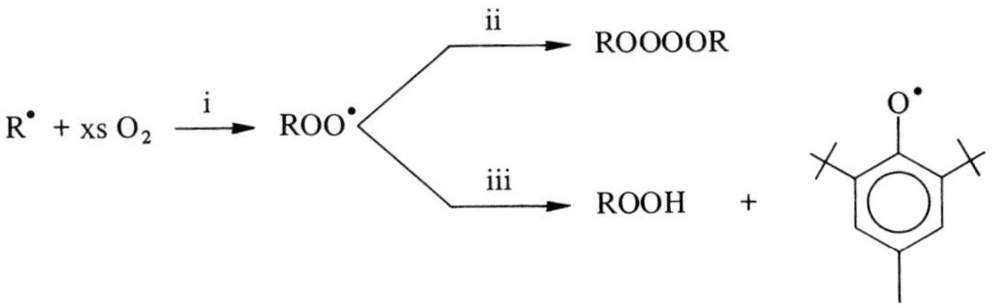


(4.122)

$$R^\bullet + \text{xs } CH_2{=}CH_2 \xrightarrow{\text{iv}} RCH_2CH_2^\bullet \qquad (4.123)$$

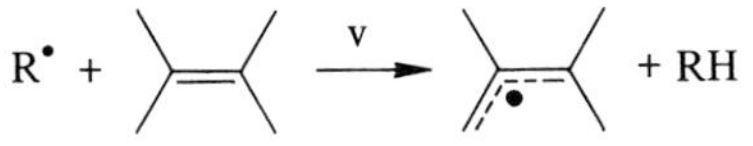


(4.124)

(i) 77C, $R^\bullet$ can be a wide variety of organic radicals; (ii) 77K, R = Me, Et, $Bu^t$, $Cl_3C$, $F_3C$, but *not* other alkyls or Ph; (iii) 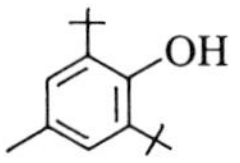  , 77K; (iv) 77K; (v) 77K, R = Me, Et, n–$C_7H_{15}$, Ph, $F_3C$

In contrast, Li atoms (in excess) react with perchlorocarbons such as $CCl_4$ and $C_2Cl_6$ in the vapour phase at elevated temperatures to form heterogeneous, amorphous, multicoloured solids whose deuteriolysis products suggest the stoichiometry $CLi_4$ and $C_2Li_6$[321)]. Similarly, halohydrocarbons can give perlithio-derivatives even though saturated hydrocarbons are unreactive under these conditions, and it is presumed that though Li atoms will not react with ground state saturated hydrocarbons, the exothermic Cl-Li reaction excites the halohydrocarbon such that hydrogen is also substituted[322)]. Comparison of the deuteriolysis products of a series of substituted propanes (1,3-dichloro-, 1,2,3-trichloro-, 1-bromo- and 2-bromo-propane) suggests that both fragmentation to give $C_2Li_6$, $C_2Li_4$, and $CLi_4$, and eliminations to give $C_3Li_6$ and $C_3Li_4$ compete against the reaction to form $C_3Li_8$. The highest proportion of $C_3D_8$ is found for monohalopropanes, therefore it has been contended that key positioning of halogen substituents in higher hydrocarbons may lead to the appropriate perlithiocarbon. However $Bu^sBr$ for example forms mainly $C_4H_9D$,

$C_3D_4$ and $C_2D_4$. $Bu^tCl$ and 1-chlorohexene behave similarly while 1,4-dichlorobutane gives only fragmentation products[322)]. Extensive fragmentation is found in all cases and is to be expected for the forcing conditions (excess Li atoms, 850 °C).

Similarly, K atoms in the vapour phase (200–300 °C) can act as effect dehalogenating agents. Coupling of the radicals then occurs (Eqs. (4.125) to (4.127))[323)].

$$K(a) + Me_3SiCl \xrightarrow{i,vi} Me_3Si^{\cdot} \longrightarrow Me_3SiSiMe_3\ (73\%) \qquad (4.125)$$

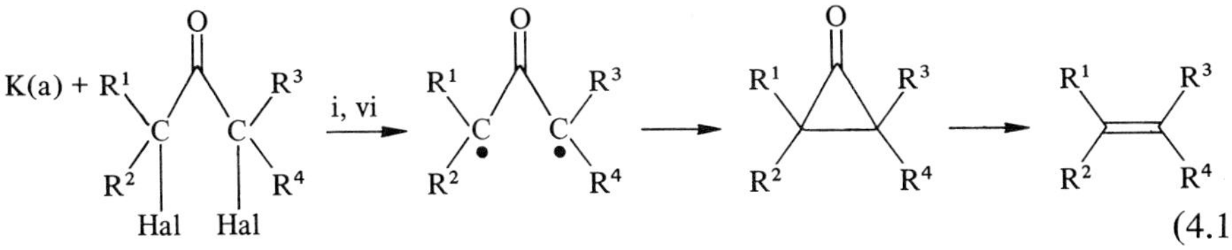


(4.126)

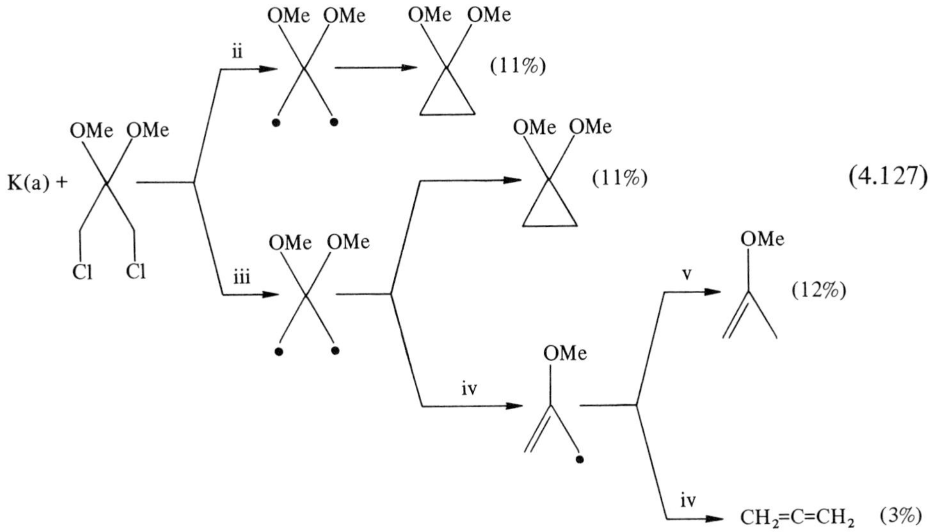


(4.127)

(i) vapour phase, 200–300°C; (ii) vapour phase, *ca.* 275°C; (iii) vapour phase, *ca.* 250°C; (iv) vapour phase, *ca.* 250°C; (v) hydrogen abstraction; (vi) condense, 77K

In all reactions some $CH_4$, $C_2H_6$, $C_2H_4$, $C_3H_8$ and $C_3H_6$ is also formed.

Mg vapour produced by resistance heating of a crucible contains essentially $^1S$ ground state Mg atoms. However, atoms of Mg have a low-lying long-lived excited $^3P$ state and it has been estimated that 60–70% of the atoms in the vapour produced by arcing are in the excited state[324)]. The two states have different reactions with organic halides, which might be expected owing to their differences in multiplicity.

Cocondensation of $^1S$ Mg atoms and halohydrocarbons (e.g. $Pr^nCl$) forms a black or deeply coloured matrix at 77 K. Warming to *ca.* −120 °C discharges the colour to form the corresponding species $Pr^nMgCl$. In contrast, cocondensation of $^3P$ Mg atoms and $Pr^nCl$

forms on warming, $MgCl_2$, propene and propane with a trace of hexane. The intermediacy of the propyl radical is assumed[324].

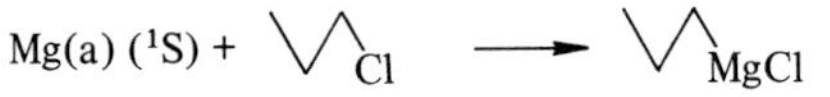


(4.128)

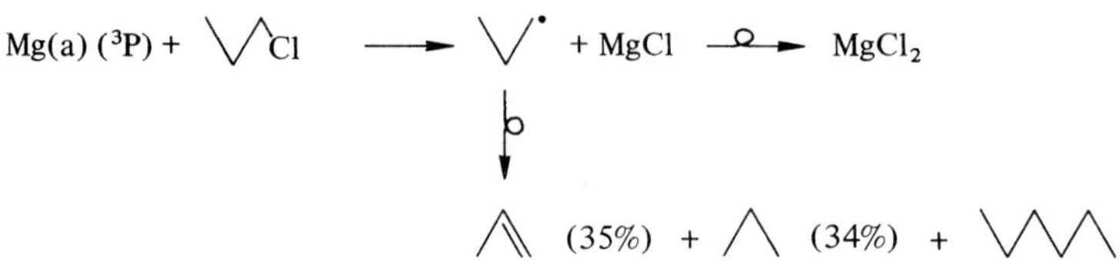


(4.129)

(i) cocondense, 77K; (ii) warm

The RMgBr species are apparently not produced at 77 K but on warm-up, because if MeBr is condensed onto a Mg/$Pr^nBr$ cocondensate at 77 K, followed by warm-up, it is MeMgBr that is formed. However, if MeBr is added after $Pr^nMgBr$ has been made by warming, no exchange occurs. Thus only a weak Mg-RHal interaction must occur at low temperatures and reaction can only proceed on warming, in this case with MeBr which is more reactive than $Pr^nBr$. Whether the reactive Mg species are a solution of solvated atoms or a slurry of finely divided metal solid[390] (or something in between) is unknown, but it is known that Mg slurries formed by cocondensation of Mg atoms and thf forms Grignard reagents at temperatures of $-78\ ^\circ C$ or less[325].

Yields of RMgHal from metal atoms are high – $Pr^nMgI$ (76%), $Pr^iMgBr$ (55%), $Bu^tMgBr$ (5%), PhMgCl (58%) and vinylMgBr (78%)[326] – though not as high as the yields from the most efficient conventional method which uses "KI-activated Mg" formed by reduction of $MgCl_2$ by K in refluxing thf in the presence of KI. However, these solvent-free RMgHal compounds are of interest in that they have different properties to the conventionally formed solvated Grignard reagents. Thus, whereas a solvated Grignard reagent may react with acetone to form a tertiary alcohol on hydrolysis, reaction of its unsolvated counterpart with perdeuterioacetone forms monodeuteriopropane (90%) and diacetone alcohol on hydrolysis presumably *via* enolisation to the magnesioenolate (Eq. (4.130))[326,328].

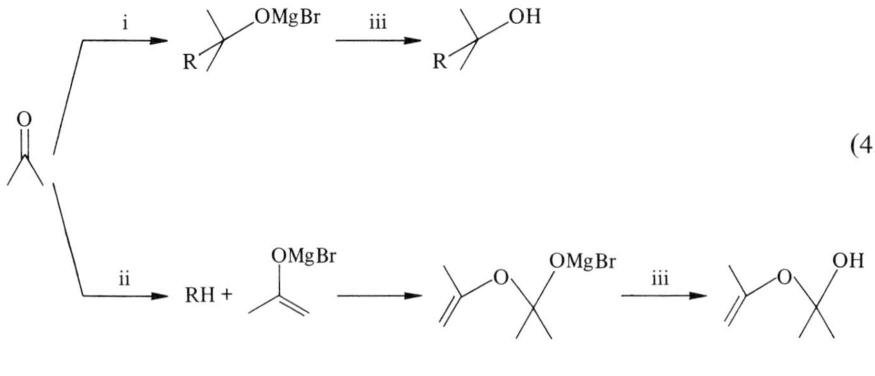


(4.130)

(i) solvated RMgBr; (ii) unsolvated RMgBr; (iii) hydrolysis

Also, solvated Grignards add 1,4 to crotonaldehyde while unsaturated MeMgBr adds 1,2.

Reactions of Ca or Ba atoms with alkyl halides have been briefly reported and are similar to those of Mg atoms[16].

Cocondensation of Ca atoms and perfluorocarbons leads to defluorination and formation of $CaF_2$ only if vinyl or aryl fluorines are present[9].

$$Ca(a) + CF_3 \cdot CF{=}CF \cdot CF_3 \xrightarrow{i,ii} CF_3 \cdot CF{=}CF \cdot CF_3 + CaF_2 + \text{polymers} \quad (4.131)$$

$$Ca(a) + CF_3 \cdot CF_3 \xrightarrow{i,ii} \text{no reaction} \quad (4.132)$$

(i) cocondense, 77 K; (ii) warm

Reactions of Ca atoms and perfluoropropene, perfluoro-cyclobutane, $C_6F_6$ and PhF all lead to polymeric products and $CaF_2$. If these vapours are saturated with $H_2O$ before condensation with Ca atoms, the same products are formed except in the cases of $C_6F_6$ and PhF when $C_6F_5H$ and PhH are detected. Zn and Fe atoms do not defluorinate $CF_3 \cdot C{=}C \cdot CF_3$[9]. On the basis of this evidence it can be proposed that for defluorination to occur an unsaturated centre must be present on the perfluorocarbon for some stabilising complexation with a metal atom which in turn must have readily available electrons.

Zinc atoms are unreactive towards most alkyl halides[116], though Zn (or Cd) slurries formed by cocondensing Zn (or Cd) atoms with a solvent, react efficiently with RHal at reflux temperatures (Eq. (4.133)). The efficacy of the solvents is in the order diglyme > dioxane > thf > hexane > toluene for Zn, though surprisingly hexane is more effective for Cd than thf[391].

$$M(a) + \text{solvent} \xrightarrow{i,ii} M(\text{slurry}) \xrightarrow{iii} R_2M \; (M = Zn, Cd) \quad (4.133)$$

(i) cocondense, 77 K; (ii) room temperature; (iii) RHal, reflux.

Reaction of Zn atoms with perfluoroalkyliodides ($R_fI$) appears to form the unsolvated species $R_fZnI$, which are much less stable, decomposing at *ca.* 150 °C lower than their solvated counterparts. Decomposition is presumed *via* radicals as good sources of $H^{\bullet}$

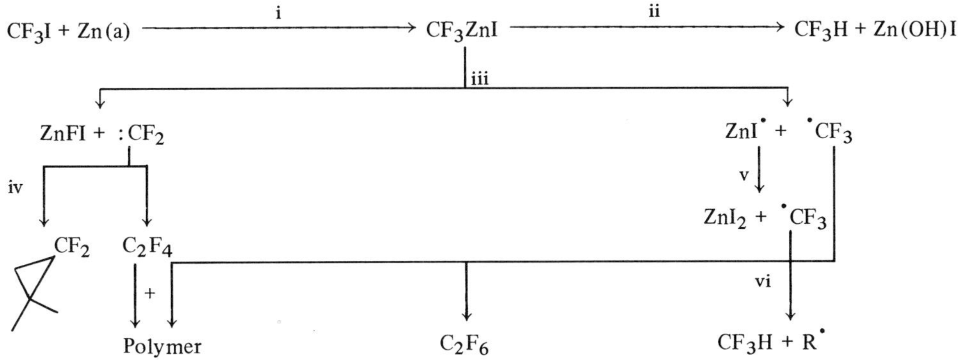


**Scheme 4.14.** (i) Cocondense, 77 K; (ii) $H_2O$, −100 °C; (iii) −80 °C; (iv) isobutene; (v) excess $CF_3I$; (vi) AcMe

(e.g. AcMe) increase $R_fH$ formation. A likely decomposition route of $CF_3ZnI$ is shown in Scheme 4.14[329].

Germanium vapour fails to react with EtCl but reacts with $CCl_4$ and $CHCl_3$ to form $CCl_3GeCl_3$ and $CHCl_2GeCl_3$ respectively in moderate yield[330]. An insertion-abstraction mechanism involving triplet Ge atoms and a triplet germylene (Eq. (4.134)) has been suggested[330].

$$Cl_3C{-}\overset{\uparrow}{Cl} + :\overset{\uparrow}{\underset{\cdot\cdot}{Ge}}\uparrow\ ({}^3P) \longrightarrow Cl_3C{-}\underset{\uparrow}{\overset{\uparrow}{Ge}}{-}Cl \xrightarrow{CCl_4} Cl_3C{-}\overset{\uparrow}{GeCl_2} + \overset{\uparrow}{CCl_3} \qquad (4.134)$$

$$\downarrow CCl_4$$

$$Cl_3C{-}GeCl_3 + \overset{\uparrow}{CCl_3}$$

However there is evidence that the intermediate $Me_3SiGeH$ formed by reaction of $Me_3SiH$ and Ge atoms behaves as a singlet (Eq. (4.135))[12]. Thus relaxation of triplet $Cl_3CGeCl$ to a singlet followed by abstraction cannot be ruled out.

$$Me_3\overset{\uparrow}{SiH} + :\overset{\uparrow}{\underset{\cdot\cdot}{Ge}}\uparrow \longrightarrow Me_3Si\underset{\uparrow}{\overset{\uparrow}{Ge}}H \longrightarrow Me_3Si\underset{\downarrow}{\overset{\uparrow}{Ge}}H \xrightarrow{Me_3SiH} (Me_3Si)_2GeH_2 \qquad (4.135)$$

Generally tin atoms are only slightly more reactive than finely divided tin when reacting with RHal. Alkyl chlorides and bromides do not react at all with Sn atoms though alkyliodides will. MeI forms mainly polymeric $(MeSnI)_n$ and some $Me_nSnI_{4-n}$ ($n = 1-3$). EtI and 1-iodooctane also form some $R_nSnI_{4-n}$ ($n = 1-3$) while allyl chloride gives $(C_3H_5)_3SnCl$ in small yield. The reaction of Sn atoms, $C_2H_2$ and HCl suggests the intermediacy of HSnCl species (Eq. (4.136)) though the corresponding reaction between Sn(a), propene and HCl forms no isolable organometallic products[102].

$$Sn(a) + C_2H_2 + HCl \xrightarrow{i,ii} (CH_2{:}CH)_3SnCl\ (1\%) \qquad (4.136)$$

(i) cocondense, 77 K; (ii) warm

### *4.18.2 Reactions with Transition Metal Atoms*

A well-studied and profitable area of research has been the reactions of atoms of the nickel triad with RHal to form the oxidative addition products RMHal. Often these products are unstable and decompose on warming. However, in many cases, addition of a ligand such as $PEt_3$ can lead to isolation of ligand-stabilised intermediates which in turn give an indication of the decomposition mechanisms as summarised for Pd in Scheme 4.15[389]. The reactions are given in Table 4.16. It is interesting to note that RPdHal and RCOPdHal have been argued as catalytic intermediates[331].

The products involving Pd are generally much more stable than those of Ni. Those of Pt have been less studied but seem similar to those of Pd. One difference is that addition

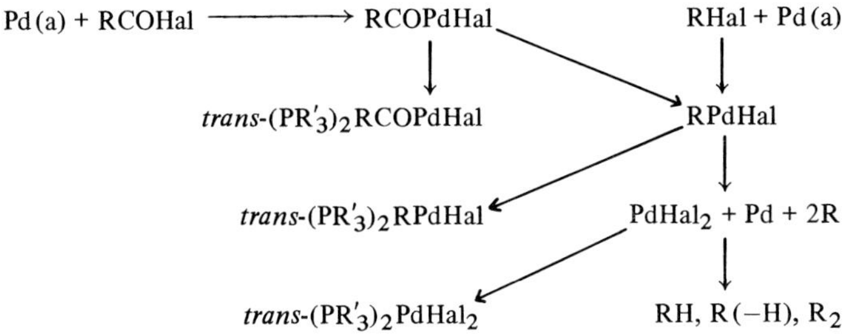


**Scheme 4.15.**

**Table 4.16.** Oxidative addition reactions of organic halides to Ni, Pd, and Pt atoms

| RHal | Initial product (% yield) | $T_{dec}$ (° C) | Typical ligand-stabilised products (% yield)[a] | Ref. |
|---|---|---|---|---|
| PhBr | PhNiBr? | | | 25) |
| $CF_3Br$ | $CF_3NiBr$? | −78 | $CF_3NiBr(PEt_3)_2$ | 95) |
| $C_6F_5Cl$ | $C_6F_5NiCl$? | −80 | $C_6F_5NiCl(PEt_3)_2$ (10) | 95, 111) |
| $C_6F_5Br$ | $C_6F_5NiBr$? | −80 | $C_6F_5NiBr(PEt_3)_2$ (10) | 25, 95) |
| $C_3H_5Cl$ | $[\eta^3\text{-}C_3H_5NiCl]_2$ (75) | stable[b] | | 382) |
| $C_3H_5Br$ | $[\eta^3\text{-}C_3H_5NiBr]_2$ (60) | stable | | 382) |
| 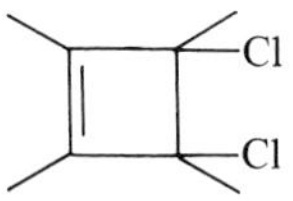 | 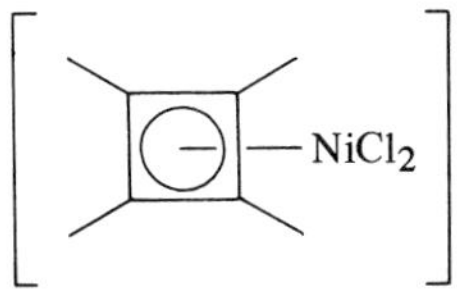 | stable | | 394) |
| AcCl | AcNiCl? | −100 | $NiCl_2(PEt_3)_2$ (10) | 95) |
| $CF_3COCl$ | $CF_3CONiCl$? | −78 | $NiCl_2(PEt_3)_2$ (20) + $CF_3NiCl(PEt_3)_2$ | 95) |
| MeBr | MePdBr? | −100 | $MePdBr(PEt_3)_2$(trace) | 95) |
| MeI | MePdI? | −100 | $PdI_2(PEt_3)_2$ (15) | 393) |
| EtI | EtPdI? | −100 | $PdI_2(PEt_3)_2$ (28) | 393) |
| $Bu^tCH_2Br$ | $Bu^tCH_2PdBr$? | −100 | $PdI_2(PEt_3)_2$ (6) | 393) |
| $CF_3Br$ | $CF_3PdBr$ | stable | $CF_3PdBr(PEt_3)_2$ (3) | 112, 393) |
| $CF_3I$ | $CF_3PdI$ (20) | stable | $CF_3PdI(PEt_3)_2$ (10) + $PdI_2(PEt_3)_2$ | 112, 393) |
| $C_2F_5I$ | $C_2F_5PdI$ (49) | stable | $C_2F_5PdI(PEt_3)_2$ (10) + $PdI_2(PEt_3)_2$ (25) | 112, 393) |
| n-$C_3F_7I$ | n-$C_3F_7PdI$ | stable | n-$C_3F_7PdI(PEt_3)_2$ (9) | 112) |
| $CCl_3Br$ | $CCl_3PdBr$? | 0 | $PdBr_2(PEt_3)_2$ (40) | 112) |
| $CF_2Br_2$ | $CF_2BrPdBr$? | −78 | $PdBr_2(PEt_3)_2$ (13) | 112) |
| PhCl | PhPdCl? | −116 | $PhPdCl(PEt_3)_2$ (2) + $PdCl_2(PEt_3)_2$ (3) | 95, 112) |
| PhBr | PhPdBr? | −116 | $PhPdBr(PEt_3)_2$ (11) + $PdBr_2(PEt_3)_2$ (14) | 25, 112) |
| $C_6F_5Cl$ | $C_6F_5PdCl$ | stable | $C_6F_5PdCl(PEt_3)_2$ (15) | 112) |

Table 4.16 (continued

| Rhal | Initial product (% yield) | $T_{dec}$ (°C) | Typical ligand-stabilised products (% yield) | Ref. |
|---|---|---|---|---|
| $C_6F_5Br$ | $C_6F_5PdBr$ | stable | $C_6F_5PdBr(PEt_3)_2$ (8) + $PdBr_2(PEt_3)_2$ (5) | 25, 112) |
| $C_6F_5I$ | $C_6F_5PdI$ | stable | $C_6F_5PdI(PEt_3)_2$ (23) | 112) |
| $C_3H_5Br$ | $[\eta^3\text{-}C_3H_5PdBr]_2$ | stable | | 142, 395) |
| $PhCH_2Cl$ | $[\eta^3\text{-}PhCH_2PdCl]_2$ (50) | stable | $(\eta^1\text{-}PhCH_2)PdCl(PEt_3)_2$ | 113, 232) |
| 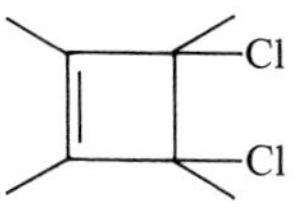 | 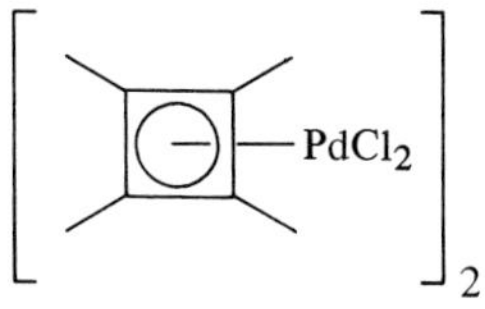 | | | 394) |
| AcCl | AcPdCl? | −100 | $PdCl_2(PEt_3)_2$ (17) | 112) |
| $Pr^nCOCl$ | $Pr^nCOPdCl$? | −100 | $PdCl_2(PEt_3)_2$ (19) | 112) |
| $CF_3COCl$ | $CF_3COPdCl$? | −78 | $CF_3PdCl(PEt_3)_2$ (4) + $PdCl_2(PEt_3)_2$ (5) $CF_3COPdCl(PEt_3)_2$ (7) | 112) |
| $n\text{-}C_3F_7COCl$ | $n\text{-}C_3F_7COPdCl$? | −78 | $n\text{-}C_3F_7COPdCl(PEt_3)_2$ (9) | 111, 112) |
| PhCOCl | PhCOPdCl? | −50 | $PdCl_2(PEt_3)_2$ (8) | 112) |
| $C_6F_5COCl$ | $C_6F_5COPdCl$? | −50 | $C_6F_5PdCl(PEt_3)_2$ (23) | 112) |
| $C_6F_5Br$ | $C_6F_5PtBr$ | stable | *c&t*-$C_6F_5PtBr(PEt_3)_2$ | 95) |
| $C_3H_5Cl$ | $[\eta^3\text{-}C_3H_5PtCl]_4$ (48) | stable | $(\eta^3\text{-}C_3H_5)PtCl(PPh_3)$ | 373, 384) |
| $PhCH_2Cl1$ | $[\eta^3\text{-}PhCHPtCl]_2$ | stable | | 142) |

[a] All products *trans*-except $C_6F_5PtBr(PEt_3)_2$ which is a mixture of *cis*- and *trans*-isomers. $PEt_3$ added at various temperatures.
[b] Stable at 25 °C.

of $PR'_3$ to RPtHal forms *cis*- and *trans*-$(PR'_3)_2$-RPtHal while only *trans*-$(PR'_3)_2$RPdHal is formed[389]. The decomposition temperatures of RMHal have been estimated by determining the highest temperature at which addition to $PR'_3$ forms some $(PR'_3)_2$RMHal. This leads to the following stability sequence[389].

$$C_6F_5 > CF_3 > C_2F_5 > n\text{-}C_3F_7 > C_6H_5 > CH_3 \sim n\text{-}C_3H_7 \sim CH_2Br \sim CF(CF_3)_2.$$

Indeed the compounds $R_fPdHal$, $R_f = C_6F_5, CF_3, C_2F_5, n\text{-}C_3F_7$ and Hal = Cl, Br, I (and $C_6F_5PtBr$) are stable at room temperature. The yield of $R_fMHal$ decreases in the order I > Br > Cl while perfluorocarbons are inert because of thermodynamic considerations already mentioned.

In addition to $R_fPdHal$, other complexes stable at ambient temperatures have been synthesised (Eqs. (4.137) to (4.139)).

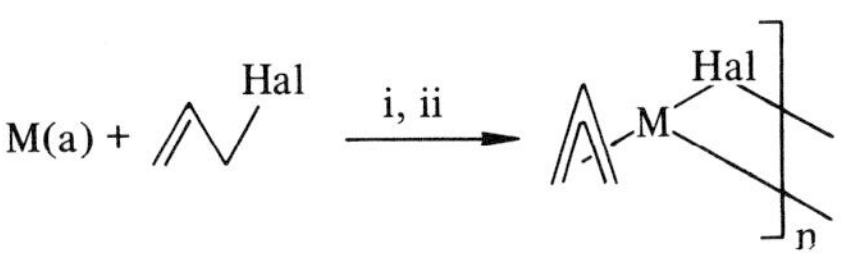


M = Ni, Hal = Cl(75%), Br(60%), n = 2[8)]
M = Pd, Hal = Cl, n = 2[72,141)]
M = Pt, Hal = Cl, n = 4(48%)[126)] (4.137)

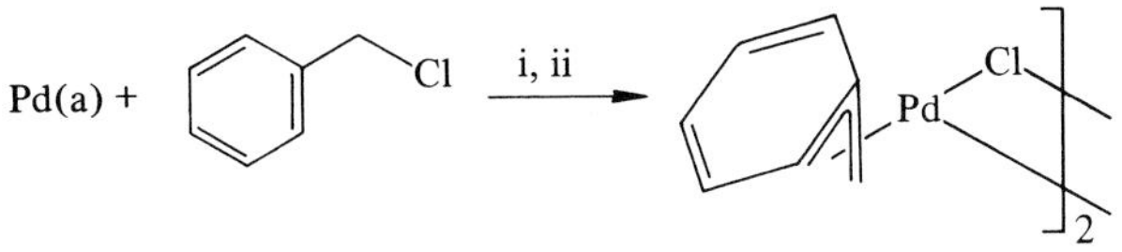


(30–40%)[332,333)] (4.138)

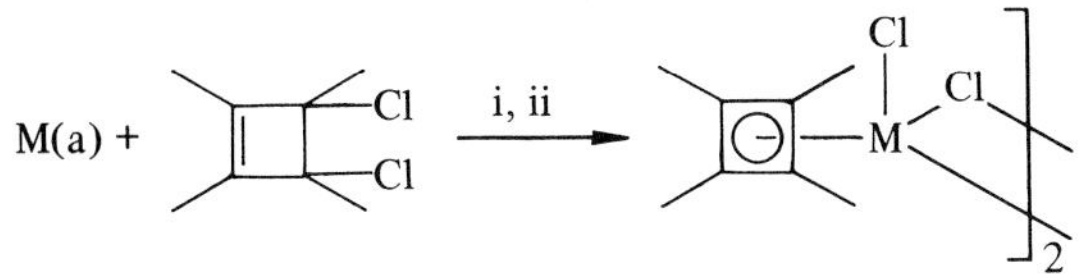


M = Ni, Pd[386)] (4.139)

(i) cocondense, 77 K; (ii) warm

In all cases, the complexes owe some of their stability to their four or higher coordination which inhibits decomposition *via* deinsertion and elimination. [This seems a common decomposition mode for transition metal complexes of the type RMHal having lower coordination numbers (*vide infra*).] However, these complexes are not electronically or coordinately saturated and are thus not inert to ligand addition (Eqs. (4.140) to (4.142)).

$[(\eta^3\text{-}C_3H_5)PtCl]_4 \xrightarrow{\text{i}} (\eta^3\text{-}C_3H_5)Pt(PPh_3)Cl$ (4.140)[126)]

$[(\eta^3\text{-}PhCH_2)PdCl]_2 \xrightarrow{\text{ii}} \textit{trans}\text{-}(\eta\text{-}PhCH_2)Pd(PEt_3)_2Cl$ (4.141)[332)]

$[C_6F_5PdBr]_n \xrightarrow{\text{iii}} \textit{trans}\text{-}C_6F_5Pd(L)_2Br$ $L = NR_3, SR_2, PR_3, AsR_3$, $NH_3$ etc. (4.142)[334)]

(i) $PPh_3$; (ii) excess $PEt_3$; (iii) L

A number of substituted benzylpalladiumhalides have been synthesised[333)].

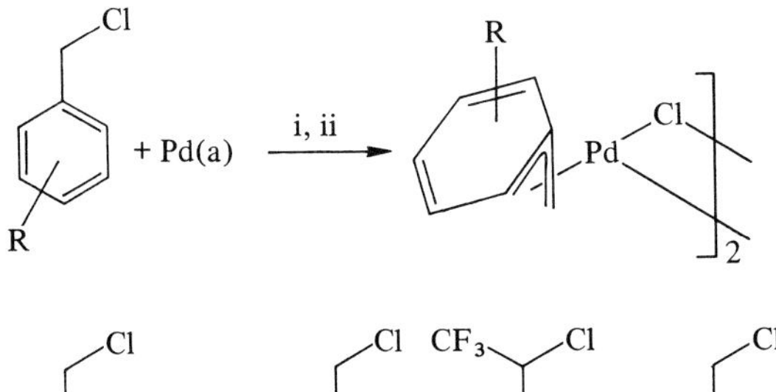


(4.143)

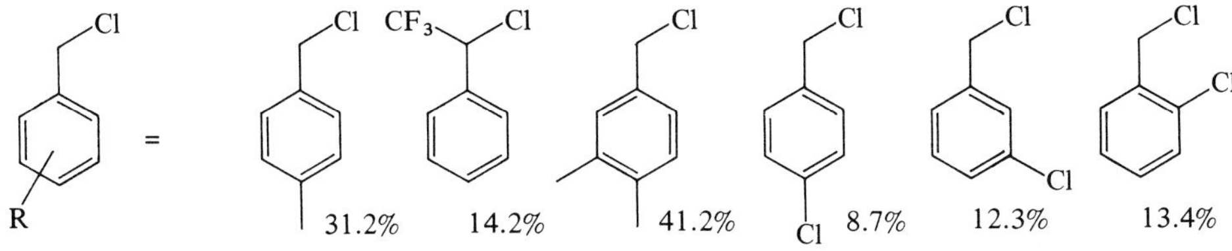


*4.81* *4.82*

(i) cocondense, 77 K; (ii) warm

The last three form thermally unstable complexes. Surprisingly no isolable products were found when *4.83*, *4.84* and *4.85* are employed.

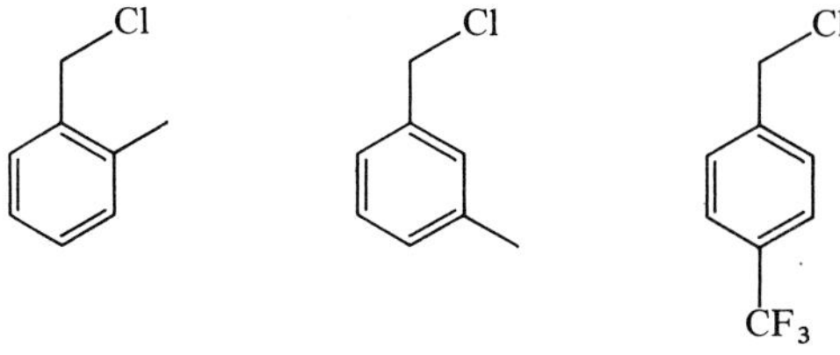


*4.83* *4.84* *4.85*

In the cases of *4.81* and *4.82* isomerisation is observed (Eqs. (4.144) and (4.145)).

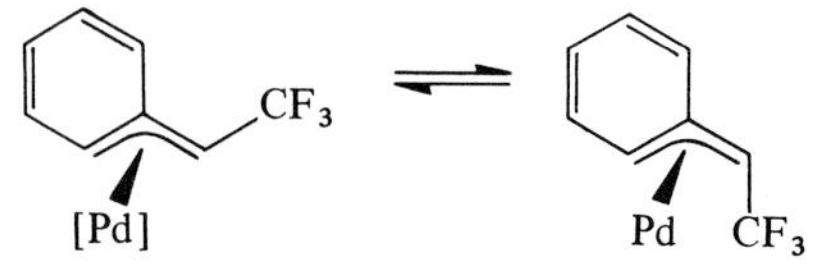


*4.81*

(4.144)

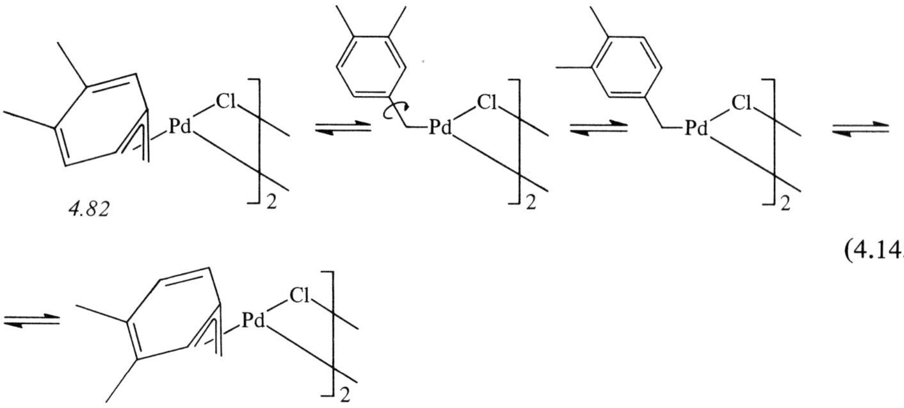


*4.82*

(4.145)

*4.86* reacts with Pd atoms to form an uncertain product.

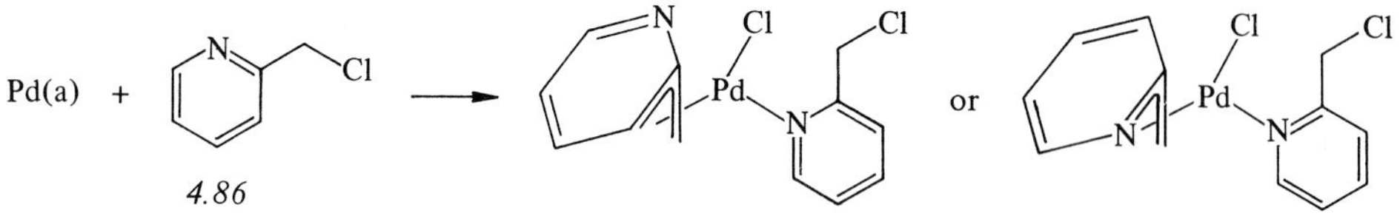


*4.86*

(4.146)

The reactions of $[(\eta^3\text{-PhCH}_2)PdCl]_2$ has been further explored, notably its use as a catalyst (Scheme 4.16)[333].

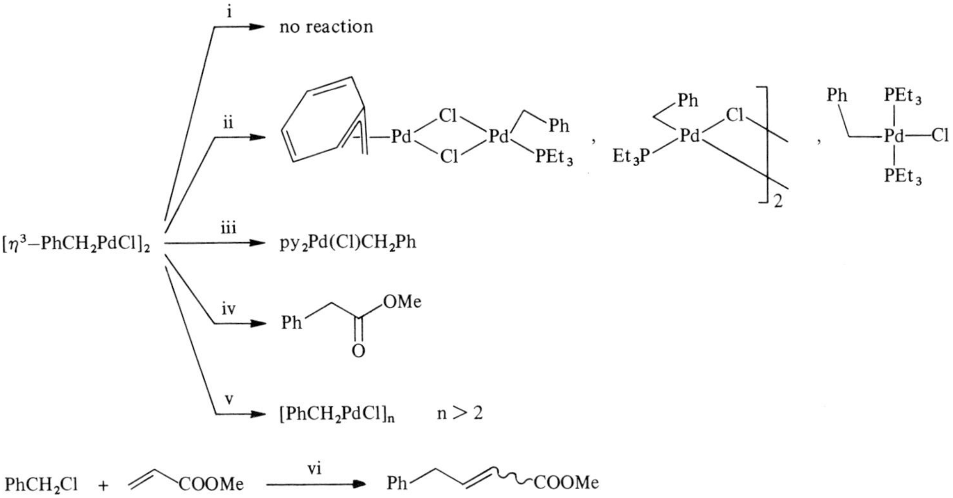

**Scheme 4.16.** (i) Nbd; (ii) $PEt_3$; (iii) py; (iv) CO, MeOH; (v) stand; (vi) $[\eta^3\text{-}PhCH_2PdCl]_2$

An initial investigation of the oxidative addition of organic halides to Pd atoms has been made and the following mechanism proposed (Scheme 4.17)[311].

No evidence for $S_N2$ attack by Pd atoms on RHal (Eq. (4.147)) has been found but no experiments with optically active RHal have yet been carried out.

(4.147)

Attempts to find stable insertion products outside the nickel triad has met with little success. A few reactions between Ti, V and Cr atoms and RHal have been reported, but as yet the highly sensitive products have not been characterised. Some of the products are extremely powerful polymerisation catalysts: for example Bd is very rapidly polymerised at –78 °C by the reaction products of Ti atoms and $C_6F_5Br$[116,389]. Aryl halides form $(\eta\text{-arene})_2M$ complexes with V, Cr, Mo, and W atoms though there is also evidence for dehalogenation[19,230;235,238,241,242,389]. However, reaction of $C_6F_5Br$ and Co atoms followed by removal of excess $C_6F_5Br$ then addition of PhMe forms the orange complex $(\eta^6\text{-PhMe})Co(C_6F_5)_2$ (Eq. (4.148)), stable to 138 °C and whose structure *4.89* has been determined by X-ray crystallography[18].

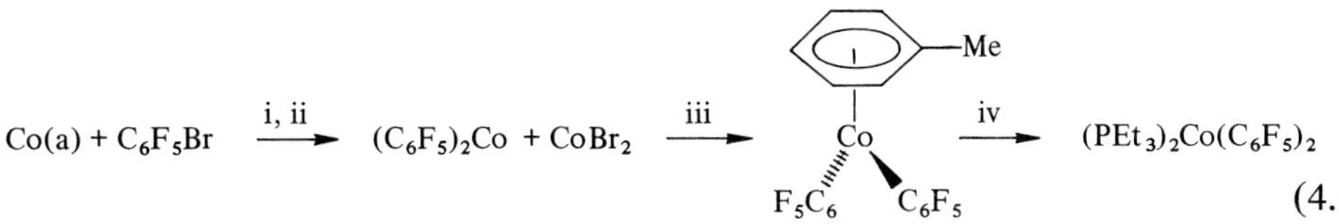

(4.148)

*4.89*

(i) cocondense, 77K; (ii) warm to remove excess $C_6F_5Br$; (iii) PhMe; (iv) $PEt_3$

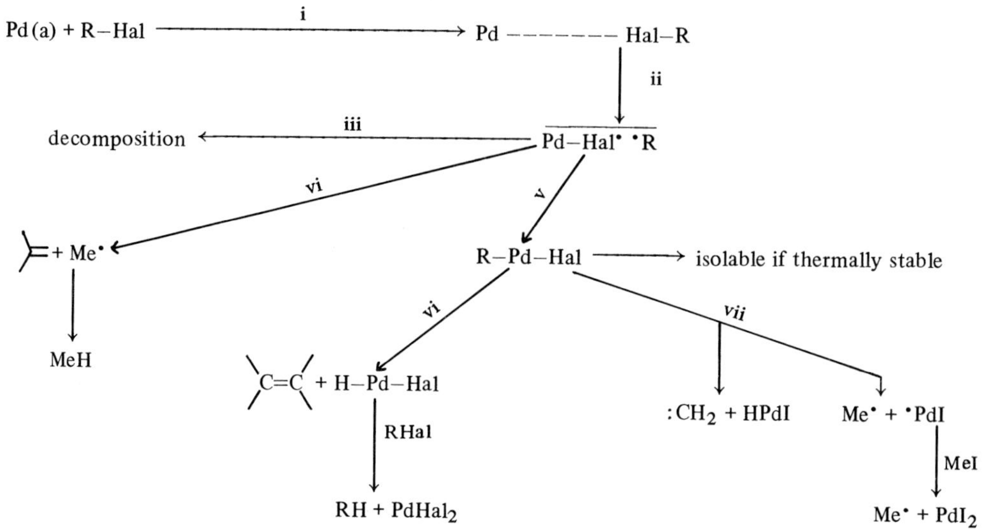


**Scheme 4.17.** (i) Cocondense, 77 K. This type of complex has been proposed for the initial products of reaction of Ag or Mg atoms with RHal. (ii) Warming. Perhaps via dissociative electron transfer: viz M–Hal–R → $M^+$ ––Hal–R → $M^+$ $Hal^-R^\bullet$. (iii) At the moment of formation of the caged radical pair, the $R^\bullet$ portion may possess enough excess vibrational energy to decompose. (iv) R = $Bu^tCH_2$. (v) The radicals do not separate but recombine to form RPdHal, which if stable may be isolated. (vi) Dedomposition depends on RPdHal, but where possible HPdHal and an alkene are formed (presumably via β-elimination). (vii) RHal = MeI

There is evidence that $CF_3PdI$ + PhMe may act similarly but with reduced Pd–PhMe bonding.

Reaction of atoms of the copper triad with alkyl halides leads to dehalogenation and radical products though the exact nature of such products depends on the metal. EtBr forms mainly n-$C_4H_{10}$ with Cu atoms, mainly $C_2H_4$ and $C_2H_6$ with Ag atoms, and approximately equal amounts of all three with Au atoms. This suggests that the metal and radical moieties stay in close contact during decomposition[14]. (The reaction of Ni atoms and EtBr also forms mainly n-$C_4H_{10}$[8].) In addition, reaction of Cu or Ag atoms with R(–)*sec*-butylchloride forms mainly S,S(–)3,4-dimethylhexane (70% optical purity) while the corresponding product from reaction with Na atoms is optically inactive[14]. Au atoms react with MeI and $Me_2S$ to form *4.90* (Eq. (4.149)). The role of $Me_2S$ is not clear but if it is not present, only ethane is formed[335].

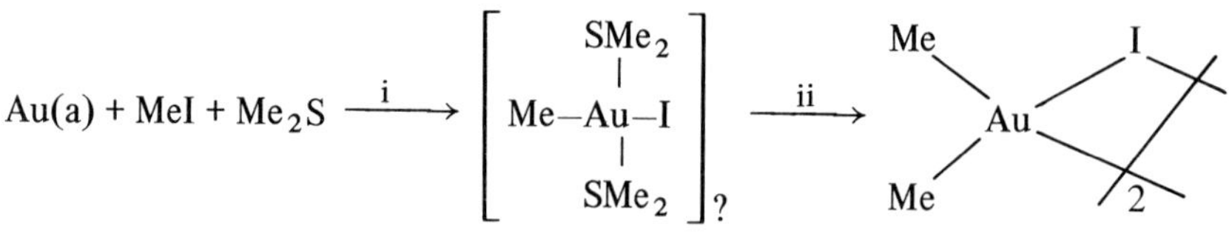


(4.149)

*4.90*

(i) cocondense, 77 K; (ii) 25 °C

As with the nickel triad, perfluororadicals form more stable complexes (Eq. (4.150)). $(CF_3)_2CFAg$ can be prepared in low yield, though it is normally isolated as the more stable MeCN or PhMe solvate. The corresponding complexes n-$C_3F_7Ag$ and $CF_3Ag$ are less stable[310]. High $R_fHal/Ag$ ratios favour $R_fAg$ formation which suggests $R_fHal...Ag$ complexation at low temperatures followed by slow reaction to form $R_f^{\cdot}$ + AgHal on warming. $R_f^{\cdot}$ may then quickly trap Ag from still existing RfHal...Ag.

$$Ag(a) + R_fI \xrightarrow{i} Ag...HalR_f \xrightarrow{ii} AgHal + R_f^{\cdot} \xrightarrow{iii} R_fAg + R_fHal \quad (4.150)$$

(i) cocondense, 77 K; (ii) warming, slow; (iii) $Ag...HalR_f$, fast

Preliminary experiments indicate $C_6F_5Cu$ and $C_6F_5Ag$ can be made, without need of a coordinating solvent[310,392].

## 4.19 Grignard Reagents

It is known that transition metal compounds activate Grignard reagents in a number of organic reactions[336], typically (Eqs. (4.151) and (4.152)):

$$CH_2 = CH_2 + PhMgBr \xrightarrow{i} PhCH_2CH_2MgBr \quad (4.151)^{337)}$$

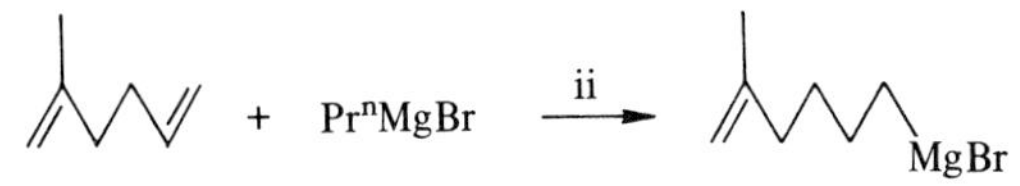


(4.152)[338]

(i) $NiCl_2$; (ii) $TiCl_4$

It is presumed that the transition metal compounds are reduced by the Grignard reagents to a zero-or low-valent species to which further Grignard reagent can oxidatively add to initiate a catalytic cycle (Scheme 4.18).

Metal atoms are another source of zerovalent metal, and many examples are known of oxidative addition of for example organic halides to metal atoms (*vide supra*). Thus it would seem likely that metal atoms should be capable of activating Grignard reagents. Cocondensation of Ni atoms, 1-chloropropene, EtMgBr and thf leads to formation of pent-2-ene[339], directly analogously to reactions catalysed by nickel salts or complexes[340]. Similar reactions with atoms of the first transition series and Pd (Table 4.17) show activity is centred on the latter half of the series. A major competing product is the hydrogenolysis product, propene, which can be explained in terms of $\beta$-elimination in a coordinately unsaturated species EtMMgBr (Eqs. (4.153) to (4.155)).

$$M(a) + EtMgBr \longrightarrow EtMMgBr \rightleftharpoons \underset{H\;\;\;\;MgBr}{\overset{CH_2{=}CH_2}{M}} \rightleftharpoons HMMgBr + CH_2{=}CH_2 \quad (4.153)$$

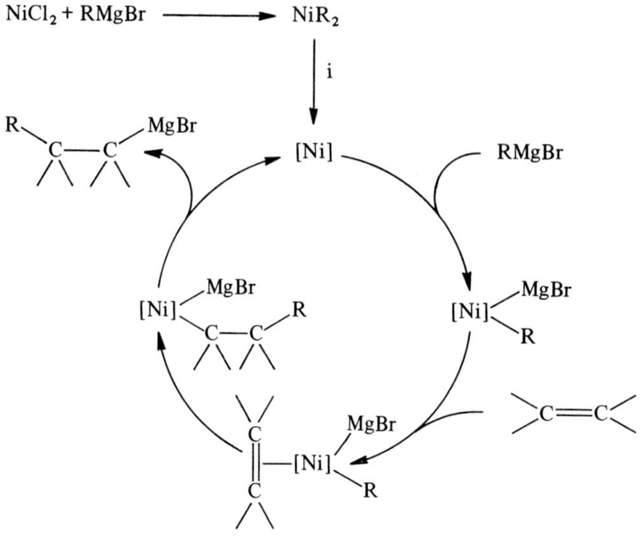


**Scheme 4.18.** (i) decomposition to form RH, R(–H), $R_2$

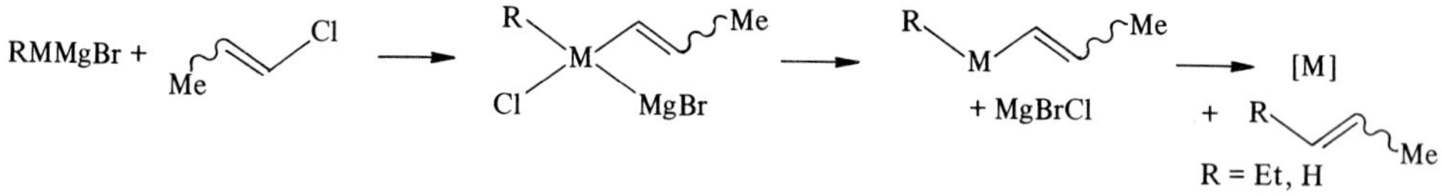


(4.154)

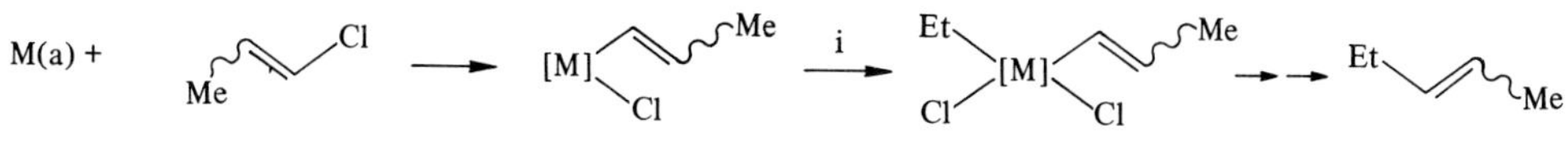


(i) EtMgBr

(4.155)

When MeMgBr is employed, forming MeMMgBr which cannot β-eliminate, the amount of coupled product (2-butene) increases (Table 4.18). By the same token use of $Pr^iMgBr$ leads only to the hydrogenolysis product, owing to ease of β-elimination in species such as $Pr^iMMgBr$. A nickel slurry, formed by cocondensation at 77 K of Ni atoms and thf, followed by warming to ambient temperatures also catalyses the coupling of MeMgBr and 1-chloropropene. Some other coupling reactions are shown in Table 4.19.

2-Pentene can also be formed by *in situ* formation of RMHal by cocondensation of Ni atoms, Mg atoms, 1-chloropropene, EtCl, and thf. This has synthetic application as only the cross-coupled product could be detected. In the corresponding reaction involving Ni atoms, Mg atoms, 1-chloropropene only, and thf, no octadienes are observed.

**Table 4.17.** 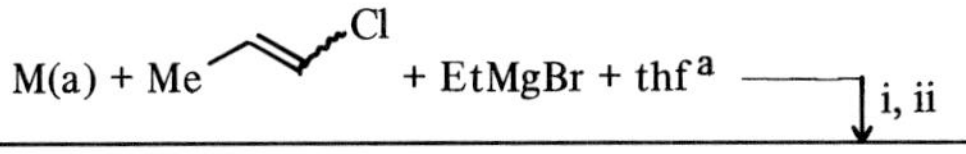

| M | Distribution of products | | | | | | | Conversion[c] |
|---|---|---|---|---|---|---|---|---|
| | $C_2H_6$ | $C_2H_4$ | $C_3H_8$ | $C_3H_6$ | $C_4H_{10}$ | 1-$C_4H_8$ | 2-$C_5H_{10}$[b] | |
| Ti | 50 | | 2 | 46 | 2 | | | 0 |
| V | 81 | 6 | | 13 | | | | 0 |
| Cr | 50 | | 1 | 37 | | 2 | 10 | 18 |
| Mn | 24 | 3 | 1 | 44 | | | 28 | 42 |
| Fe | 16 | 15 | | 15 | | | 54 | 77 |
| Co | 27 | 5 | 3 | 27 | 5 | | 33 | 66 |
| Ni | 21 | 5 | | 13 | 4 | | 57 | 78 |
| Pd | 23 | | | 67 | | | | 0 |

(i) cocondense, 77 K; (ii) 20 °C

[a] Excess halide over Grignard present; *cis*, *trans* mixture of halide.
[b] *cis* and *trans*.
[c] of Grignard to 2-pentenes (%).

**Table 4.18.** 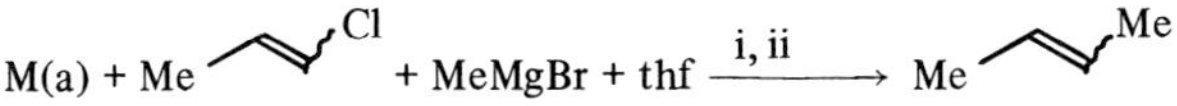

| M | Ti | V | Cr | Mn | Fe | Co | Ni |
|---|---|---|---|---|---|---|---|
| % conversion of MeMgBr to 2-butene | 20 | 27 | 38 | 67 | 59 | 78 | 83 |

(i) cocondense, 77 K; (ii) 20 °C

**Table 4.19.** Reactions of metal atoms, Grignard reagents and various substrates in thf

| M | Substrate[a] | RMgHal | Conversion[b] | Products |
|---|---|---|---|---|
| Ni | MeCH=CHCl | PhMgBr[c] | 0 | |
| Ni | $Bu^tCl$ | MeMgBr | 90 | $Me_2C{=}CH_2$, $CH_4$ |
| Ni | PhBr | MeMgBr | 1 | PhMe |
| Ni | PhBr | $Pr^iMgBr$ | 95 | PhH + $MeCH{=}CH_2$ |
| Ni | t-$MeCH{=}CHCH_2Cl$[d] | MeMgBr | 85 | MeCH=CHEt (c:t = 5:1) |
| Ni | t-$MeCH{=}CHCH_2Cl$ | EtMgBr | 40 | mainly $MeCH{=}CHCH_2Et$ |
| Ni | EtOH | EtMgBr | 1 | n-$C_4H_{10}$ |
| Ni | $EtOCH{=}CH_2$ | EtMgBr | 0 | |
| Ni | MeOPh | MeMgBr | 0 | |
| Ti | MeOPh | MeMgBr | 0 | |

[a] Excess substrate over RMgHal.
[b] of RMgHal (%).
[c] In the case of PhMgBr, this comound was dissolved in $Bu^n_2O$.
[d] t = *trans*, c = *cis*.

## 4.20 Inorganic Halides

Like their organic counterparts, inorganic halides react with metal atoms *via* an oxidative addition process.

Atoms of the Cu triad dehalogenate $BCl_3$ and $PCl_3$ to form $B_2Cl_4$[104,341] and $P_2Cl_4$[104] The reaction between Cu atoms and $BCl_3$ is particularly effective; $B_2Cl_4$ can be formed on a competitive preparative scale (10 g/h)[342]. $B_2Cl_2Me_2$ can also be formed from Cu atoms and $BCl_2Me$[341]. Reaction of Ni atoms and $BCl_3$ forms no $B_2Cl_4$, forming instead an involatile solid containing combined Ni, B, and Cl[104]. Formation of $P_2Cl_4$ is restricted because of complex formation to MCl (Eq. (4.156))[104]. Li atoms and $PCl_3$ form $PCl_2^{\cdot}$ in rare gas matrices[343].

$$M(a) + PCl_3 \longrightarrow MCl(PCl)_x\ (10\text{–}15\%) + P_2Cl_4\ (ca.\ 15\%)\ (M = Cu, Ag) \quad (4.156)$$

Cu atoms if generated by resistance heating does not react with chlorosilanes, but Cu atoms generated by electron beam evaporation reacts with $SiCl_4$ forming perchlorosilanes: an excited species has been suggested[104].

Ge atoms react with $SiCl_4$ to form $SiCl_3GeCl_3$ presumably *via* halide abstraction by an initial $SiCl_3GeCl$ species (see organic halides)[330].

## 4.21 Acids

Metal atoms react with (protonic) acids in a similar manner to solid metal except that atoms have enhanced activity and react with very weak acids and substrates not normally considered as an acid. Hydrocarbons containing activated hydrogens such as CpH and Cod can also be included in this group, but have been treated separately.

Cocondensation of metal atoms with acetylacetone (acacH) is an effective method of forming small batches of anhydrous metal acetylacetonates ($M(acac)_n$) which have use in preparative chemistry and as nmr shift reagents. Details are shown in Table 4.20[18].

Cr atoms react with acids to form divalent salts if air is excluded and trivalent salts if air is present[17,127]. The moment when hydrogen is released can vary. $H_2$ is released on cocondensation of Cr atoms and HCl but when Cr atoms and $H_2O$ or MeOH are cocondensed, half the $H_2$ is released on cocondensation and half on warming the cocondensate. In contrast $Mo/H_2O$ and $W/H_2O$ cocondensates only release $H_2$ on warming[127]. Reaction of Cr atoms with acetic acid (AcOH) in the absence of $O_2$ forms $Cr_2(OAc)_4$[17] which is notable for its quadruple bond. This method may prove of use in the synthesis of similar complexes.

Mg atoms react with $H_2O$ and HBr with release of $H_2$[326]. Triplet but not singlet Mg atoms react with $NH_3$, releasing one equivalent of $H_2$ on cocondensation and a further equivalent on warming[125].

Cocondensation of Sn atoms, $C_2H_2$ and HCl forms $(CH_2 = CH)_3$ SnCl in very small yield and a HSnCl species is suspected[102].

**Table 4.20.** Metal atom synthesis of $M(acac)_n$

| M | n | Yield (%) |
|---|---|---|
| Mn | 2 | 30 |
| Cr | 3[a] | 32 |
| Fe | 2 | 25 |
| Fe | 3[a] | 24 |
| Ni | 2 | 36 |
| Cu | 2 | 10 |
| Zn | 2 | 28 |
| Al | 3 | 14 |
| Sn | 2 | 11 |
| Pb | 2 | 10 |

[a] Worked up in the presence of oxygen.
(Angew. Chem., Int. Ed. *14*, 278 (1975))

## 4.22 Dioxygen

Cocondensation at very low temperatures (typically 15 K) of metal atoms and $O_2$ or $O_2$/rare gas mixtures forms in many cases discrete binary complexes $M(O_2)_{1,2}M_2(O_2)$ and others (see Table 4.21). Isotopic labelling studies show that the bonding is side-on analogous to $\eta^2$-$C_2H_4$ bonding rather than end-on analogous to $\eta^1$-$N_2$ bonding. However, owing to the high electronegativity of $O_2$ there is considerable electron transfer in the $M(O_2)$ complexes which are best described as closely interacting metal-superoxide pairs *4.91*. This is especially true of the electropositive alkali metal atoms. In the case of the transition metal atoms there is less electron transfer and in extreme cases the M–$O_2$ bonding is best described using the Chatt-Dewar-Duncanson model for M–$C_2H_4$ bonds[1]. Thus the complex between $O_2$ and Cu (first IP 7.72 eV) is ionic *4.91* while that of $O_2$ and Au (first IP 9.22 eV) is very similar to $(\eta^2$-$C_2H_4)$Au *4.92*[344]. The bonding in $M(O_2)_2$ complexes tends to be less ionic as two $O_2$ ligands compete for the available electron charge on the metal moiety, and leads to an apparent distinction between the complexes of transition metal and alkali metal atoms. Those incorporating transition metal atoms have a tetrahedral disuperoxide structure *4.93* while alkali metal atoms form a metal-tetraoxide ion pair *4.94*. Ag alone of the transition metals so far explored can form this structure[344]. Alkali metal atoms can also form peroxides $M^+O_2^{2-}M^+$ (see Table 4.22).

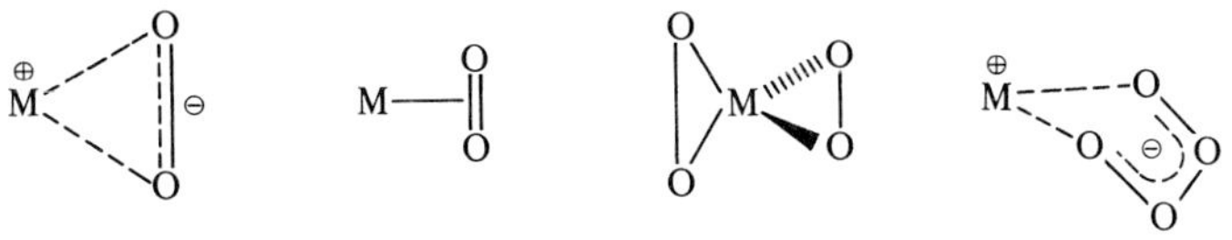


*4.91* *4.92* *4.93* *4.94*

**Table 4.21.** Binary dioxygen and related complexes

| M | Observations | Ref. |
|---|---|---|
| Rh | $Rh(O_2)_{1,2}$, $Rh_2(O_2)_{1-4}$, $Rh_3(O_2)_{2 or 6}$, $Rh(N_2)_m(O_2)_n$, $Rh(CO_2)_n(O_2)_m$ | 342, 345) |
| Ni | $Ni(O_2)_{1,2}$, $Ni(O_2)(N_2)_{1,2}$ | 86, 347, 348, 349) |
| Pd | $Pd(O_2)_{1,2}$, $Pd(O_2)(N_2)_{1,2}$ | 86, 348, 349) |
| Pt | $Pt(O_2)_{1,2}$, $Pt(O_2)(N_2)_{1,2}$ | 86, 348, 349) |
| Cu | $Cu^{\oplus}(O_2)^{\ominus}$, $Cu(O_2)_2$ | 344, 350, 370) |
| Ag | $Ag^{\oplus}(O_2)^{\ominus}$, $Ag^{\oplus}(O_4)^{\ominus}$, $C_2H_4Ag^{\oplus}(O_2)^{\ominus}$, $(CO)Ag^{\oplus}(O_2)^{\ominus}$, $(CO)Ag^{\oplus}(O_4)^{\ominus}$ | 29, 65, 344) |
| Au | $Au(O_2)$, $(CO)\overline{Au\text{-}CO\cdot O\cdot O}$ | 7, 76) |
| Li | $Li^{\oplus}(O_2)^{\ominus}$, $Li^{\oplus}(O_2)^{2\ominus}Li^{\oplus}$ | 351) |
| Na, K, Rb, Cs | $M^{\oplus}(O_2)$, $M^{\oplus}(O_2)^{2\ominus}M^{\oplus}$, $M^{\oplus}(O_2)^{2\ominus}M'^{\oplus}$, $M^{\oplus}(O_4)^{\ominus}$ | 352) |
| Ca | $CaO + O_3$ | 354) |
| Ba | $Ba^{2\oplus}-O_2^{2\ominus}$ | 355) |
| Ge, Sn | O=M=O | 356, 357) |

Rh atoms at high concentrations can form dinuclear and trinuclear complexes with $O_2$. Their structures have been postulated as those shown.

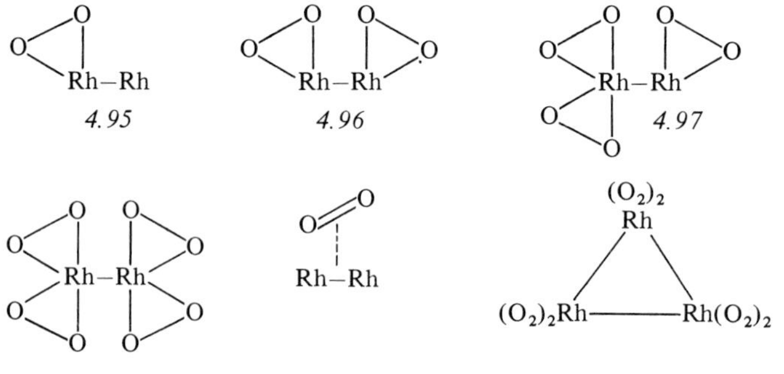


*4.95* *4.96* *4.97* *4.98* *4.99* *4.100*

The $O_2$ ligand has an estimated negative charge δ- of $1 < \delta < 2$ in the complexes *4.95*, *4.96*, *4.97*, and *4.98* which have therefore been proposed as localised models for $O_2$ chemisorption on solid Rh surfaces, while δ- in *4.99* is $0 < \delta < 1$ making it a candidate for a physisorption model[345]. Furthermore, species such as *4.101* have been put forward as localised models for the oxidation of various substrates on metal surfaces (Eq. (4.157))[76].

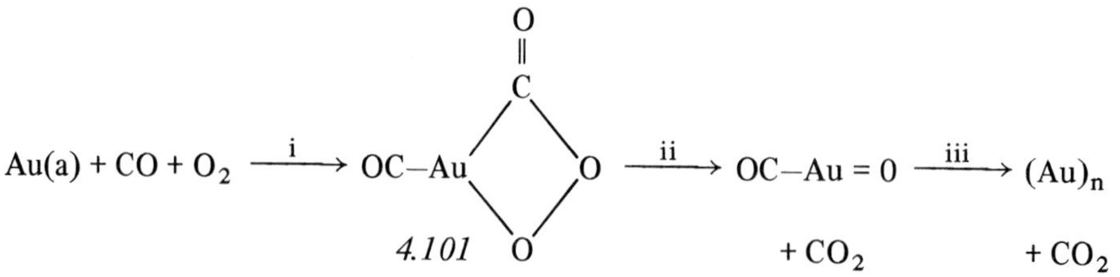


(4.157)

(i) cocondense, 10 K; (ii) 30 K; (iii) 40 K

In a limited number of cases, the reaction between metal atoms and $O_2$ is sufficiently energetic to break the O=O bonds. Reaction of Ge or Sn atoms with $O_2$ forms O=Ge=O[356] and O=Sn=O[357] respectively. Ca atoms react with $O_2$ to form CaO and $O_3$[354] while Ba atoms only form the peroxide $Ba^{2+}(O_2)^{2-}$. Such energy can also be supplied externally as shown in Eq. (4.158)[358].

$$M(a) + O_2 + Xe \xrightarrow{i,ii} MO \; (M = Cu, Ag, Au) \qquad (4.158)$$

(i) cocondense, 14 K; (ii) h$\nu$, 14 K

## 4.23 Dihalogens

Dihalogens react with alkali metals by electron transfer to form dihalogen radical-metal ion pairs in rare gas matrices at low temperatures (Eq. (4.159)). The perturbed centro-symetric radical anions can be detected by Raman spectroscopy[13,369].

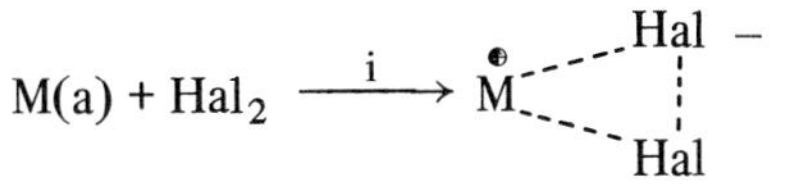

(M = Li, Na, K, Rb, Cs) (4.159)

(Hal = F, Cl)

(i) cocondense with 100 x Ar, 15 K

## 4.24 Oxygen-Containing Organic Compounds

Metal atoms are oxidised by oxygen-containing organic compounds such as ethers, epoxides, alcohols, ketones and aldehydes. In nearly all cases it is clear that complexation occurs through oxygen moieties and it is unlikely that any organometallic species are formed but nonetheless such reactions may prove a useful means of synthesising new organic species.

Cocondensation of excess epoxides with a variety of first transition series metal atoms forms the corresponding olefin *via* a none catalytic process[360]. For example (Eq. (4.160)):

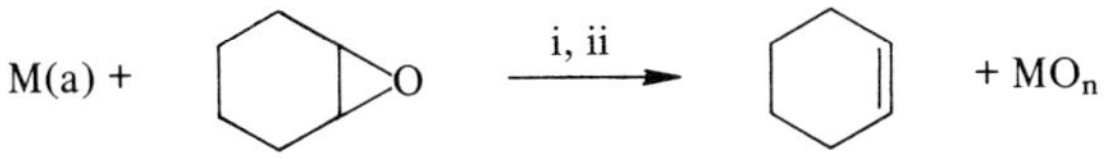

(4.160)

(i) cocondense, 77K; (ii) ambient temperatures

| M | Ti | V | Cr | Co | Ni |
|---|---|---|---|---|---|
| n | 0.9 | 2.8 | 2.7 | 1.2 | 0.6 |

**Table 4.22.** Reactions of Cr atoms with oxygen-containing organic substrates

| Substrate | Product | Ref. |
|---|---|---|
| 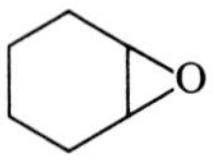 a |  (86 : 11 : 3)[b] | 360) |
| 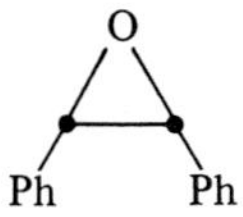 a | 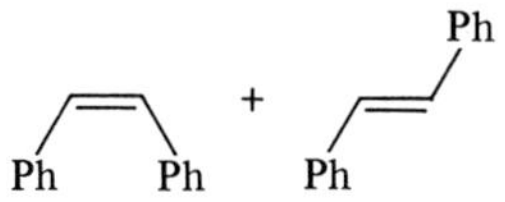 (2 : 3)[b] | 360) |
| 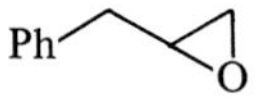 a |  b | 360) |
| 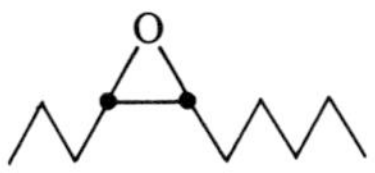 a |  (9 : 16)[b] | 360) |
| 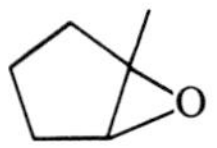 a | 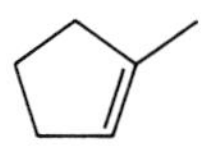 b | 360) |
| 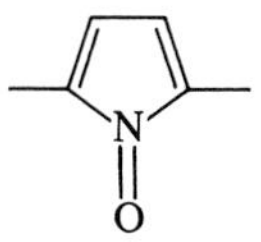 b | 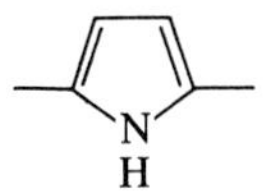 b, c | 360) |
| $Me_2S{=}O$[a] | $Me_2S$[b, d] | 360) |
| $Et_3P{=}O$[a] | $Et_3P$[b, e] | 360) |
| $PhNO_2$[f] | PhNO + 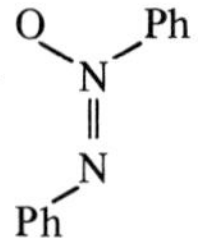 + 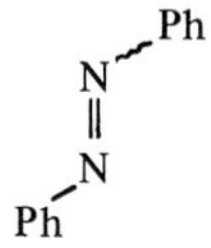 + $PhNH_2$ (3:12:18:10) | 361) |
| PhNO[a] | 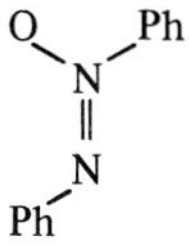 + 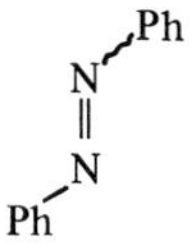 + $PhNH_2$ (10:2:1) | 361) |
| 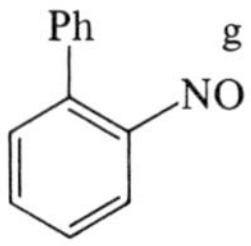 g | 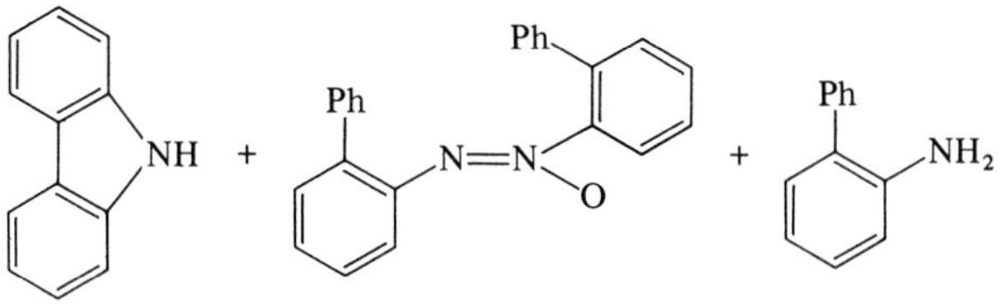 | 361) |

[a] M: substrate = ca. 1 : 10.
[b] total yield 3–10%.
[c] no complex isolated.
[d] initially red matrix.
[e] initially green matrix.
[f] M: substrate = ca. 1 : 15.
[g] M: substrate = ca. 1 : 80.

Some cyclohexadiene and benzene is also formed. The reactions of Cr atoms are particularly effective and some reactions are shown in Table 4.22. No double-bond migration is observed but there is *cis-trans*-isomerisation. In separate experiments *cis*-4-decene is isomerised by Cr atoms thus the initial deoxygenation may have been stereospecific. The requirement of Cr atoms rather than a reactive Cr surface is infered from additional experiments where epoxides are condensed on fresh vapour-deposited Cr surfaces with reduced deoxygenation.

As Table 4.22 shows, Cr atoms deoxygenate nitro- and nitrosoarenes[361]. Interpretation of the product distributions leads to a mechanism of consecutive deoxygenation to an arylnitrene *4.102* which attacks any remaining nitroso-compounds, abstracts hydrogen, or, in the case of 2-nitrosobiphenyl, inserts into a C–H bond to form carbazole *4.103* (Eq. (4.161)).

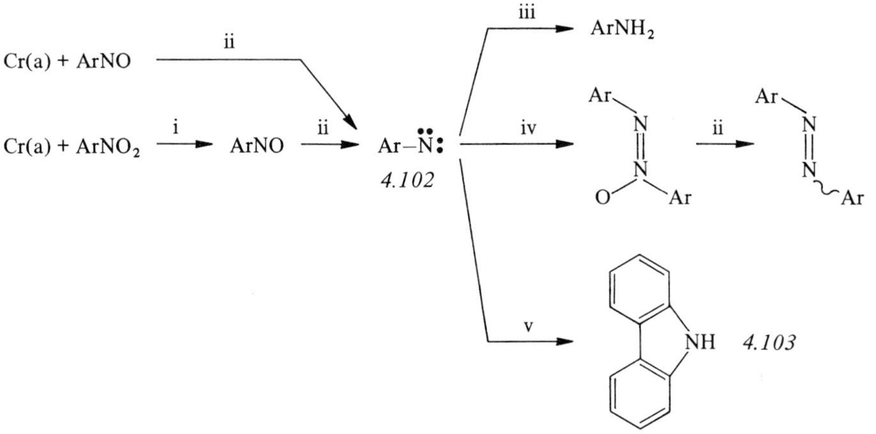


(4.161)

(i) cocondense, 77K; (ii) further oxygen abstraction; (iii) hydrogen abstraction; (iv) ArNO; (v) Ar = o–$PhC_6H_4$

There is a report that condensation of metal atoms with aldehydes leads to alcohols, coupled diols or hydrocarbons depending on the metal used[362]. Coupling of dimethyl maleic anhydride occurs on condensation with Fe atoms (Eq. (4.162))[17].

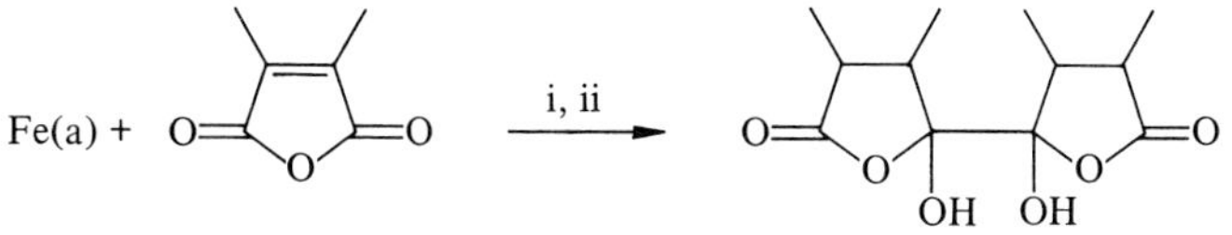


(4.162)

(i) cocondense, 77K; (ii) HCl

Mg atoms are instrumental in the coupling of cyclic ketones (Eq. (4.163))

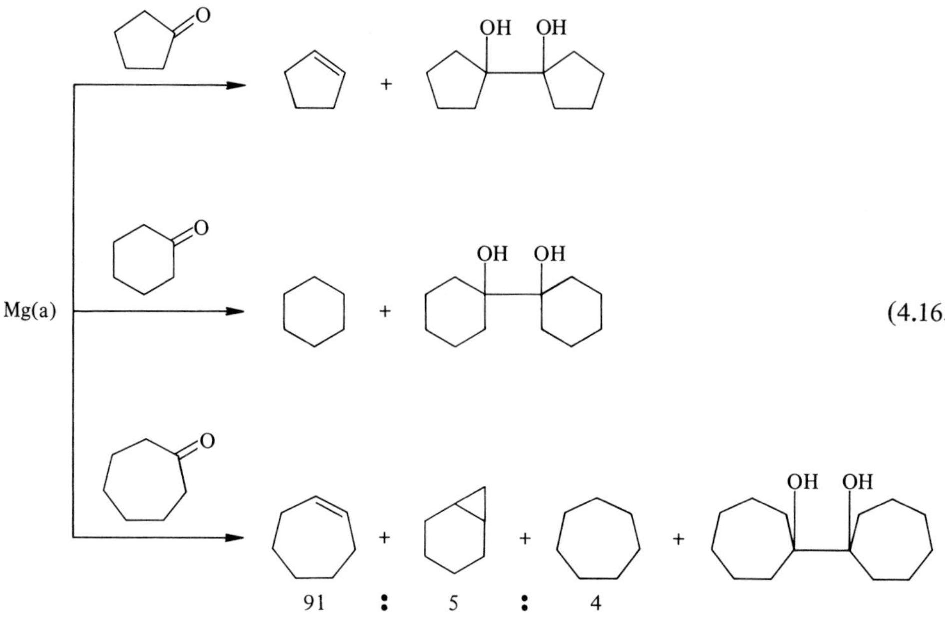


(4.163)

It has previously been reported that carbenacyclo-pentanes and -hexanes form only cyclopentene and -hexene, while carbenacycloheptane forms cycloheptene and norcarane (82:18)[364] which suggests carbene, or more likely, a carbenoid complex involvement. A likely mechanism has been proposed (Scheme 4.19).

Various forms of zerovalent titanium solid have been formed by reduction of $TiCl_3$, which in excess and under forcing conditions reductively couple ketones to olefins[365,366], and deoxygenate epoxides[367]. However it has been found that this reaction can be performed at much less strenuous temperatures by using ($\eta$-arene)$_2$Ti as the source of $Ti^0$ (Eq. (4.164))[368].

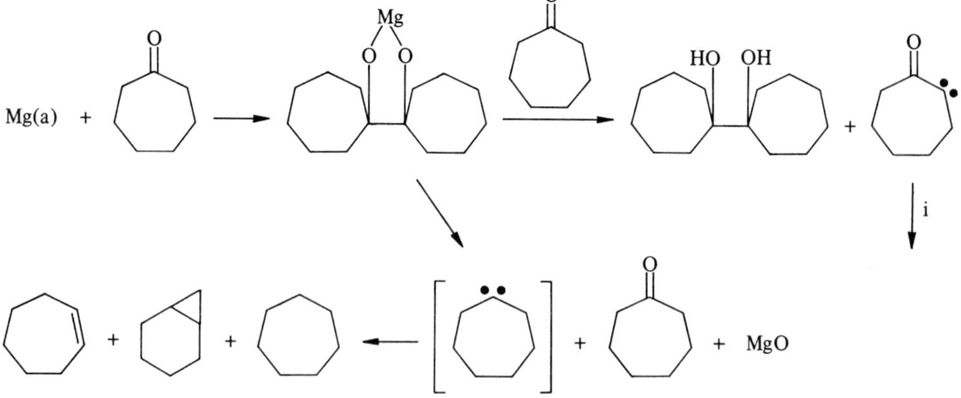


**Scheme 4.19.** (i) Aldol condensation

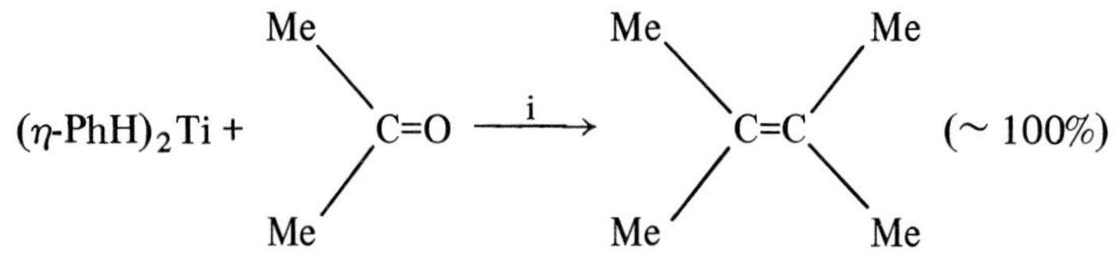 (~ 100%) (4.164)

(i) thf, 70 °C, 2 h

Some related reactions are shown in Table 4.23. Reactions 1–3 are analogous to reactions involving heterogeneous $Ti^0$ and can be explained by a similar mechanism (Eq. (4.165))[365].

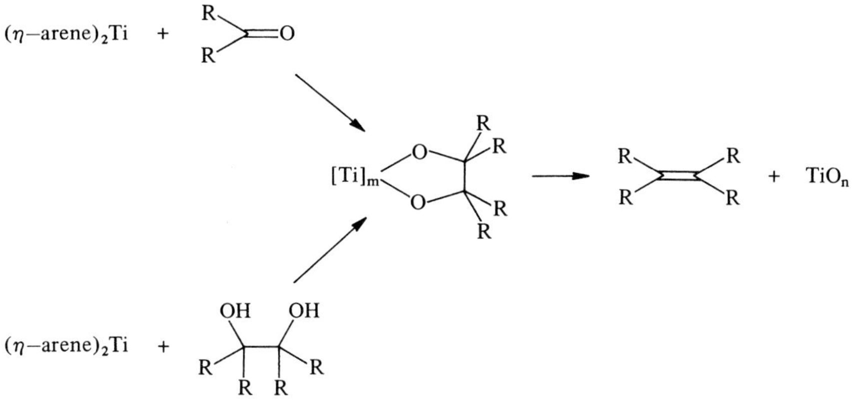 (4.165)

**Table 4.23.** Reaction of ($\eta$-PhH)$_2$Ti with oxygen-containing substrates in thf at 70 °C, 2 h

| | Substrate[a] | Product[b] | |
|---|---|---|---|
| 1. | 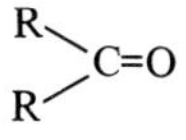 | 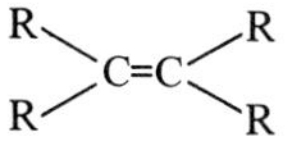 | R = Ph, Me |
| 2. | 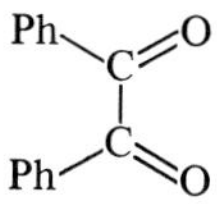 | Ph · C≡C · Ph | |
| 3. | 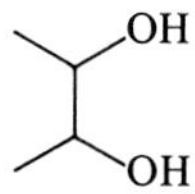 | 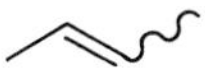 | |
| 4. | 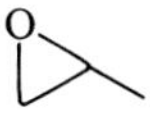 | 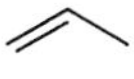 | |
| 5. |  | 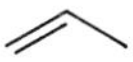 | |
| 6. |  | $PhCH_3$ | |

[a] substrate : ($\eta$-PhH)$_2$Ti = 1 : 2. [b] yield *ca.* 100%.

($\eta$-arene)$_2$Ti, like the Li/TiCl$_3$ couple deoxygenate allyl alcohols (Table 4.23) to the corresponding alkene, contrary to other systems where coupling (of for example allyl alcohol to hexa-1,5-diene) is to be expected[369]. However stereochemical studies show that the isomer distribution of alkene products is different depending on whether ($\eta$-arene)$_2$Ti or Li/TiCl$_3$ is employed and warrants further attention[368].

Ti slurries formed as bye-products of ($\eta$-arene)$_2$Ti also reductively couple ketones in thf but take longer than ($\eta$-arene)$_2$Ti[368].

Cocondensation of Ti atoms and decanal only leads to the coupled diol with no oxygen removal: no reaction occurs with Cr atoms[361].

## 4.25 Sulfur-Containing Organic Compounds

Thiophenes are desulphurised by Cr and Fe atoms. In the case of Cr, no organometallic compound is isolable, even when CO is added at low temperatures. Warming the plum-coloured Fe/thiophene cocondensate to about −20 °C precipitates metallic iron but addition of CO at low temperatures forms ferracyclopentadiene derivatives (Eq. (4.166)). Addition of Cod, Bd, PMe$_3$ or CpH at low temperatures does not lead to any isolable products[301,370].

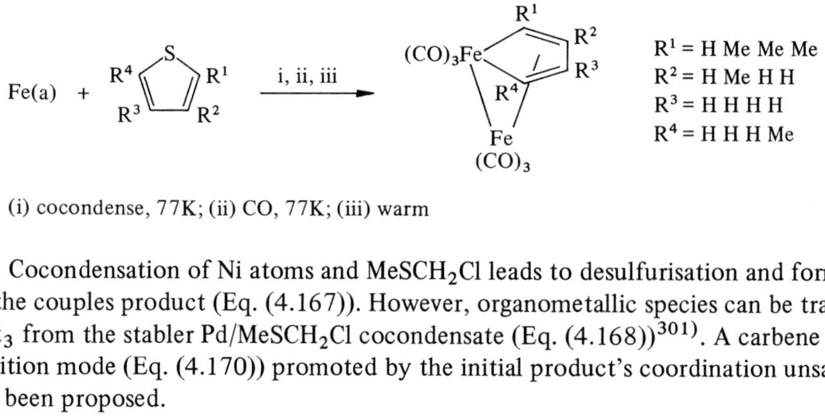


(4.166)

(i) cocondense, 77K; (ii) CO, 77K; (iii) warm

Cocondensation of Ni atoms and MeSCH$_2$Cl leads to desulfurisation and formation of the couples product (Eq. (4.167)). However, organometallic species can be trapped by PEt$_3$ from the stabler Pd/MeSCH$_2$Cl cocondensate (Eq. (4.168))[301]. A carbene decomposition mode (Eq. (4.170)) promoted by the initial product's coordination unsaturation has been proposed.

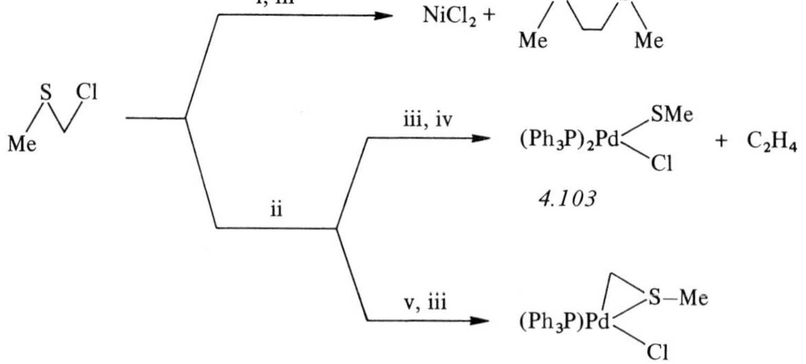


(4.167)

(4.168)

*4.103*

(4.169)

*4.104*

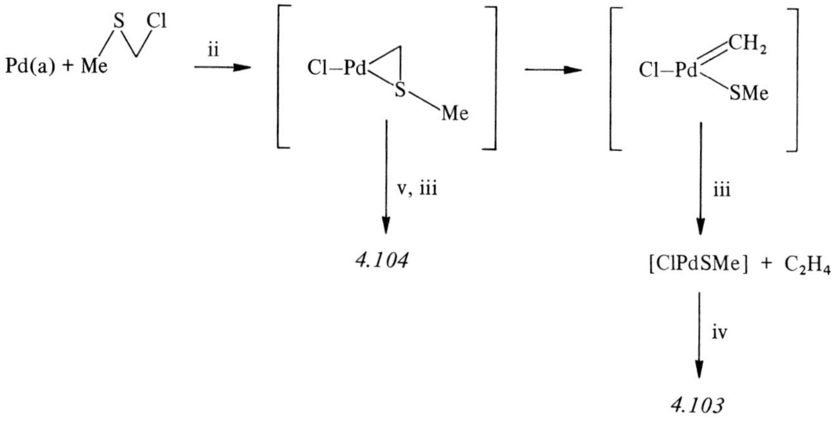


(4.170)

(i) cocondense, Ni(a), 77 K; (ii) cocondense, Pd(a), 77 K; (iii) 20 °C; (iv) $PPh_3$, 20 °C; (v) $PPh_3$, 77 K

## References


1. Dewar, M. J. S.: Bull. Soc. Chim. France *18*, C71 (1951); Chatt. J., Duncanson, L. A.: J. Chem. Soc. *1953*, 2939
2. Huber, H., Kundig, E. P., Ozin, G. A.: J. Amer. Chem. Soc. *96*, 5585 (1974)
3. Skell, P. S., Van Dam, E. M., Silvon, M. P.: J. Amer. Chem. Soc. *96*, 626 (1974)
4. Klabunde, K. J., Efner, H. F.: Inorg. Chem. *14*, 789 (1975)
5. Klabunde, K. J., Low, J. Y. E., Efner, H. F.: J. Amer. Chem. Soc. *96*, 1984 (1974)
6. Ozin, G. A., Klotzbucher, W. E.: J. Amer. Chem. Soc. *97*, 3965 (1975)
7. McIntosh, D., Ozin, G. A.: Inorg. Chem. *55*, 2869 (1976)
8. Piper, M. J., Timms, P. L.: Chem. Comm. *1972*, 50
9. Klabunde, K. J., Low, J. Y. F., Key, M. S.: J. Fluor. Chem. *2*, 207 (1972)
10. D'Aniello, jr., M. J., Barefield, E. K.: J. Organometal. Chem. *76*, C50 (1974)
11. Van Dam, E. M., Brent, W. N., Silvon, M. P., Skell, P. S.: J. Amer. Chem. Soc. *97*, 467 (1975)
12. Conlin, R. T., Lockhart, S. H., Gasper, P. P.: J. C. S. Chem. Comm. *1975*, 825
13. Howard, W. F., Andrews, L.: J. Amer. Chem. Soc. *95*, 3046 (1973)
14. Anthony, M. T.: B. Sc. Thesis, Univ. Bristol 1973
15. Anderson, B. B., Behrens, C. L., Radonovich, L. J., Klabunde, K. J.: J. Amer. Chem. Soc. *98*, 5390 (1976)
16. Klabunde, K. J.: Amer. Lab. *7*, 35 (1975)
17. Koerner v. Gustorf, E. A., Jaenicke, O., Wolfbeis, O., Eady, C. R.: Angew. Chem., Int. Ed. *14*, 278 (1975)
18. Blackborow, J. R., Eady, C. R., Koerner v. Gustorf, E. A., Scrivanti, A., Wolfbeis, O.: J. Organometal. Chem. *108*, C32 (1976)
19. Timms, P. L.: Angew. Chem., Int. Ed. *14*, 273 (1975)
20. Huber, H., Kundig, E. P., Moskovits, M., Ozin, G. A.: J. Amer. Chem. Soc. *97*, 2097 (1975)
21. Hanlan, L. A., Ozin, G. A.: J. Amer. Chem. Soc. *96*, 6324 (1974)
22. Huber, H., Kundig, E. P., Ozin, G. A., Poë, A. J.: J. Amer. Chem. Soc. *97*, 308 (1975)
23. Green, M. L. H.: Organometallic Compounds, Vol. II, The Transition Elements. London: Methuen 1968

24. Tolman, C. A.: Chem. Soc. Rev. *1*, 337 (1972); Mason, R.: Proc. 13th Internat. Congr. Pure Appl. Chem. *6*, 31 (1971)
25. DeVore, T. C.: Diss. Abstr. Int. B. *36*, 3982 (1975)
26. Hanlan, L., Huber, H., Ozin, G. A.: Inorg. Chem. *15*, 2592 (1976)
27. Ford, T. A., Huber, H., Klotzbücher, W., Moskovits, M., Ozin, G. A.: Inorg. Chem. *15*, 1666 (1976)
28. DeVore, T. C., Franzen, H. F.: Inorg. Chem. *15*, 1318 (1976)
29. Ozin, G. A.: Acc. Chem. Res. *10*, 21 (1977)
30. Hanlan, L. A., Huber, H., Kundig, E. P., McGarvey, B. R., Ozin, G. A.: J. Amer. Chem. Soc. *97*, 7054 (1975)
31. Ozin, G. A., Van der Voet, A.: Acc. Chem. Res. *6*, 313 (1973)
32. DeKock, R. L.: Inorg. Chem. *10*, 1205 (1971)
33. Hulse, J. E., Moskovits, M.: Surface Science *57*, 125 (1976); Moskovits, M., Hulse, J. E.: J. Phys. Chem. *81*, 2004 (1977)
34. Kundig, E. P., McIntosh, D., Moskovits, M., Ozin, G. A.: J. Amer. Chem. Soc. *95*, 7234 (1973)
35. Kundig, E. P., Moskovits, M., Ozin, G. A.: Can. J. Chem. *51*, 637 (1973)
36. Darling, J. H., Ogden, J. S.: Inorg. Chem. *11*, 666 (1972)
37. Huber, H., Kundig, P., Moskovits, M., Ozin, G. A.: Nature *235*, 98 (1972)
38. Darling, J. H., Ogden, J. S.: J. C. S. Dalton *1973*, 1079
39. Kundig, E. P., Moskovits, M., Ozin, G. A.: Can. J. Chem. *50*, 3587 (1972)
40. Kundig, E. P., Moskovits, M., Ozin, G. A.: J. Mol. Struc. *14*, 137 (1972)
41. Ogden, J. S.: Chem. Comm. *1971*, 978
42. Moskovits, M., Hulse, J. E.: Surface Sci. *61*, 302 (1976); J. Chem. Phys. *67*, 4271 (1977)
43. McIntosh, D., Moskovits, M., Ozin, G. A.: Inorg. Chem. *15*, 1669 (1976)
44. Brevis, D. R.: personal communication, reported in Ref. 45
45. Darling, J. H., Ogden, J. S.: J. C. S. Dalton *1972*, 2496
46. McIntosh, D., Ozin, G. A.: Inorg. Chem. *16*, 51 (1977)
47. Slater, J. L., DeVore, T. C., Calder, V.: Inorg. Chem. *13*, 1808 (1974)
48. Slater, J. L., DeVore, T. C., Calder, V.: Inorg. Chem. *12*, 1918 (1973)
49. Sheline, R. K., Slater, J. L.: Angew. Chem., Int. Ed. *14*, 309 (1975)
50. Slater, J. L., Sheline, R. K., Lin, K. C., Weltner, W. jr.: J. Chem. Phys. *55*, 5129 (1971)
51. Hague, R. H., Krishan, C. N., Margrave, J. L.: Proc. Amer. Chem. Soc., Meeting, 170th, Chicago, INOR 95 (1975)
52. Hinchcliffe, A. J., Ogden, J. S., Oswald, D. D.: J. C. S. Chem. Comm. *1972*, 338
53. Ogden, J. S.: Private communication (1972); Kundig, E. P., Ozin, G. A.: unpublished, reported in Ref. 54
54. Ozin, G. A., Van der Voet, A.: Prog. Inorg. Chem. *19*, 105 (1975)
55. Lembke, R. R., Ferrante, R. F., Weltner, W. jr.: J. Amer. Chem. Soc. *98*, 416 (1977)
56. Bos, A.: J. C. S. Chem. Comm. *1972*, 26
57. Burdett, J. K.: J. C. S. Faraday Trans. 2, *70*, 1599 (1974)
58. Gilan, M., Hoffmann, R.: Inorg. Chem. *14*, 1058 (1975)
59. Gischens, R. P., Pliskin, W. A., Frances, S. A.: J. Phys. Chem. *60*, 194 (1956)
60. Hair, M. L.: Infra-Red Spectroscopy in Surface Chemistry. New York: Dekker 1967; Little, L. H.: Infra-Red Spectra of Adsorbed Species. New York: Academic 1966
61. Blyholder, G.: J. Phys. Chem. *68*, 2772 (1964)
62. Blyholder, G., Allen, M. C.: J. Amer. Chem. Soc. *91*, 3158 (1969)
63. Eady, C. R., Johnson, B. F. G., Lewis, J.: J. C. S. Dalton *1975*, 2606 and references therein; Johnson, B. F. G.: Ninth Sheffield-Leeds Internat. Symposium on Organometallic and Inorganic Chemistry (1977)
64. Cotton, F. A., Kraihanzel, C. S.: J. Amer. Chem. Soc. *84*, 4432 (1962)
65. Huber, H., Ozin, G. A.: Inorg. Chem. *16*, 64 (1977)
66. Huber, H., McIntosh, D., Ozin, G. A.: Inorg. Chem. *16*, 975 (1977)
67. Huber, H., Moskovits, M., Ozin, G. A.: unpublished, reported in Ref. 54; Yarbrough, II, L. W., Calder, G. V., Verkade, J. G.: J. C. S. Chem. Comm. *1972*, 705

68. Falbe, J.: Carbon Monoxide in Organic Synthesis. Translated by Adams, C. R. Berlin: Springer 1970; Wender, I., Pino, P. (eds.): Organic Synthesis via Metal Carbonyls, Vol. I and II. New York: Interscience 1968, 1977; Volpin, M. E.: Pure Appl. Chem. *30*, 607 (1972); Volpin, M. E., Kolomnikov, I. S.: Pure Appl. Chem. *33*, 567 (1973); Organometallic Reactions *15*, 313 (1975)
69. Huber, H., Ozin, G. A., Power, W. J.: Inorg. Chem. *16*, 2234 (1977)
70. Huber, H., Moskovits, M., Ozin, G. A.: unpublished, reported in Ref. 54
71. Huber, H., Moskovits, M., Ozin, G. A.: Nature *236,* 127 (1972)
72. Timms, P. L.: unpublished reported in Ref. 386
73. Timms, P. L.: Chem. Comm. *1968*, 258
74. Bennett, J. E., Mile, B., Thomas, A., Ward, B.: Adv. Phys. Org. Chem. *8*, 1 (1970)
75. Huber, H., Ford, T. A., Ozin, G. A.: J. Amer. Chem. Soc. *98*, 3176 (1976)
76. Green, D. W., Hodges, R. V., Gruen, D. M.: Inorg. Chem. *15*, 970 (1976)
77. Burdett, J. K., Grahame, M. A., Turner, J. J.: J. C. S. Dalton *1972*, 1620
78. DeVore, T. C.: Inorg. Chem. *15*, 1315 (1976)
79. Burdett, J. K., Turner, J. J.: Chem. Comm. *1971*, 885
80. Barrett, P. H., Montano, P. A.: J. C. S. Faraday II, *73,* 378 (1977)
81. Micklitz, H., Barrett, P. H.: Apl. Phys. Lett. *20*, 387 (1972)
82. Ozin, G. A., Van der Voet, A.: Can. J. Chem. *51*, 637 (1973)
83. Ozin, G. A., Van der Voet, A.: Can. J. Chem. *51*, 3332 (1973)
84. Huber, H., Kundig, E. P., Moskovits, M., Ozin G. A.: J. Amer. Chem. Soc. *95*, 332 (1973)
85. Klotzbücher, W., Ozin, G. A.: J. Amer. Chem. Soc. *97*, 2672 (1975)
86. Ozin, G. A., Klotzbücher, W. E.: J. Amer. Chem. Soc. *97*, 3965 (1975)
87. Moskovits, M., Ozin, G. A.: J. Chem. Phys. *58*, 1251 (1973)
88. Kundig, E. P., Moskovits, M., Ozin, G. A.: Can. J. Chem. *51*, 2710 (1973)
89. Green, D. W., Thomas, J., Gruen, D. M.: J. Chem. Phys. *58*, 5453 (1973)
90. Spiker, R. C. jr., Andrews, L., Trundle, C.: J. Amer. Chem. Soc. *94*, 2401 (1972)
91. Middleton, R., Hull, J. R., Simpson, S. R., Tomlinson, C. H., Timms, P. L.: J. C. S. Dalton *1973*, 120
92. Young, D.: unpublished results, 1977
93. Kruck, T.: Angew. Chem., Int. Ed. *6*, 53 (1967)
94. Swanson, B. I., Satija, S. K.: J. C. S. Chem. Comm. *1973*, 40
95. Ballivet, D., Billard, C., Tkatchenko, I. B., Young, D.: Proc. Nineteenth Int., Conf. on Coord. chem. Prague, C.S.S.R. 1978
96. Tevault, D. E., Andrews, L.: J. Phys. Chem. *77*, 1640 (1973)
97. Tevault, D. E., Andrews, L.: J. Phys. Chem. *77*, 1646 (1973); Andrews, L., Pimentel, G. C.: J. Chem. Phys. *44*, 2361 (1966)
98. Spiker, R. C., Andrews, L.: J. Chem. Phys. *58*, 702 (1973)
99. Spiker, R. C., Andrews, L.: J. Chem. Phys. *58*, 713 (1973)
100. Green, D. W., Reedy, G. T.: J. Chem. Phys. *65*, 2921 (1976)
101. Timms, P. L.: J. Chem. Soc. (A) *1970*, 2526
102. Middleton, R.: Ph. D. Thesis, Univ. Bristol 1974
103. Young, D.: D. Phil. Thesis, Univ. Oxford 1974
104. Timms, P. L.: Adv. Inorg. Chem. Radiochem. *14*, 121 (1972)
105. Timms, P. L., Turney, T. W.: unpublished, reported in Ref. 386
106. Trabelsi, M., Loutellier, A.: Proc. Eighteenth Int. Conf. on Coord. Chem., Sao Paulo, Brazil, 1977
107. Green, M. L. H., Young, D.: unpublished
108. Akhmedov, V. M., Anthony, M. T., Green, M. L. H., Young, D.: J. C. S. Dalton *1975*, 1412
109. Trabelsi, M., Loutellier, A., Bigorgne, M.: J. Organometal. Chem. *40*, C45 (1972)
110. Tolman, C. A., Yarbrough II, L. W., Verkade, J. G.: Inorg. Chem. *16*, 479 (1977)
111. Atkins, R. M., MacKenzie, R., Timms, P. L., Turney, T. W.: J. C. S. Chem. Comm. *1975*, 764
112. Tolman, C. A.: Chem. Rev. *77*, 313 (1977)
113. Scholer, F. R., Gladkowski, D.: Proc. Amer. Chem. Soc., Meeting, 171st, New York, INOR 133 (1976)
114. Vollhardt, K. P. C.: Acc. Chem. Res. *10*, 1 (1977); Maitlis, P. M.: Acc. Chem. Res. *9*, 93 (1976)
115. Skell, P. S., Williams-Smith, D. L., McGlinchey, M. J.: J. Amer. Chem. Soc. *95*, 3337 (1973)

116. Gladysz, J. A., Fulcher, J. G., Lee, S. J., Bocarsley, A. B.: Tetrahedron Lett. *1977*, 3421
117. Blackborow, J. R., Eady, C. R., Koerner v. Gustorf, E. A., Scrivanti, A., Wolfbeis, O.: J. Organometal. Chem. *108*, C32 (1976)
118. Skell, P. S., Havel, J. J., Williams-Smith, D. L., McGlinchey, M. J.: J. C. S. Chem. Comm. *1972*, 1098
119. Blackborow, J. R.: private communication; structures 4.6 and 4.7 private communication Krüger, C., Blackborow, J. R., Feldhoff, U., Grevels, F.-W., Grubbs, R. H., Miyashita, A.: to be published
120. Skell, P. S.: Int. Congr. Pure. Appl. Chem., 23rd, Boston (1971); Special Lectures, Vol. 4, p. 215, London: Butterworths 1971
121. Skell, P. S., Wolf, L. R.: J. Amer. Chem. Soc. *94*, 7919 (1972)
122. Wolf, L. R.: Ph. D., Pennsylvania State Univ. 1974
123. Kasai, P. H., McLeod, D. jr.: J. Amer. Chem. Soc. *97*, 5609 (1975)
124. Kasai, P. H., McLeod, D. jr.: Watanabe, T.: J. Amer. Chem. Soc. *99*, 3522 (1977)
125. Skell, P. S.: U. S. Government Report No AFOSR-75-0200 (1975)
126. Skell, P. S., Havel, J. J.: J. Amer. Chem. Soc. *93*, 6687 (1971)
127. Skell, P. S., McGlinchey, M. J.: Angew. Chem., Int. Ed. *14*, 195 (1975)
128. Timms, P. L.: Int. Conf. Coord. Chem., 17th (1976)
129. Atkins, R. M.: Ph. D. Thesis, Univ. Bristol 1976
130. Huber, H., McIntosh, D., Ozin, G. A.: J. Organometal. Chem. *112*, C50 (1976)
131. Huber, H., Ozin, G. A., Power, W. J.: J. Amer. Chem. Soc. *98*, 6508 (1976)
132. Ozin, G. A., Power, W. J.: Inorg. Chem. *16*, 212 (1977)
133. Ozin, G. A., Huber, H., McIntosh, D.: Inorg. Chem. *16*, 3070 (1977)
134. Kasai, P. H., McLeod, D. jr.: J. Amer. Chem. Soc. *97*, 6602 (1975)
135. McIntosh, D., Ozin, G. A.: J. Organometal. Chem. *121*, 127 (1976)
136. Muetterties, E. L.: Bull. Soc. Chim. Belg. *84,* 959 (1975); Ugo, R.: Catal. Rev. *11,* 225 (1975)
137. Prentice, J. D., Lesiunas, A., Sheppard, N.: J. C. S. Chem. Comm. *1976*, 76
138. Demuth, J. E., Eastman, D. E.: Phys. Rev. Lett. *32*, 1123 (1974)
139. Kitty, P. A., Sachtler, W. M. H.: Catal. Rev. *10*, 1 (1974)
140. Ozin, G. A., Power, W. J.: Inorg. Chem. *16*, 2864 (1977)
141. Klabunde, K. J.: Acc. Chem. Res. *8*, 393 (1975)
142. Blackborow, J. R., Grubbs, R., Miyashita, A., Scrivanti, A.: J. Organometal. Chem. *120*, C49 (1976)
143. Bönnemann, H.: Angew. Chem., Int. Ed. *9*, 736 (1970)
144. Byrne, J. W., Blaser, H. U., Osborn, J. A.: J. Amer. Chem. Soc. *97*, 3871 (1975)
145. Taylor, P. W.: B. Sc. Thesis, Univ. Bristol 1974
146. Cloke, F. G. N., Green, M. L. H., Young, D.: unpublished (1975)
147. Hoberg, H., Krause, S.: Angew. Chem., Int. Ed. *15*, 694 (1976)
148. Morrison, J. A., Chung, C., Lagow, R. J.: J. Amer. Chem. Soc. *97*, 5015 (1975)
149. Rautenstrauch, R.: Angew. Chem., Int. Ed. *14*, 259 (1975)
150. Mingos, D. M. P.: J. C. S. Dalton, *1977,* 26 and 31
151. Murdoch, H. D., Lucken, E. A.: Helv. Chim. Acta *47*, 1517 (1964)
152. Emerson, G. F., Mahler, J. E., Petit, R.: Chem. Ind. *1964*, 836
153. Akhmedov, V. M., Anthony, M. T., Green, M. L. H., Young, D.: J. C. S. Chem. Comm. *1974*, 777
154. Anthony, M. T., Green, M. L. H., Young, D.: unpublished (1975)
155. Martin, H., Vohwinkel, F.: Chem. Ber. *94*, 2416 (1961)
156. Vohwinkel, F.: Trans. N. Y. Acad. Sci. *26*, 446 (1964)
157. Akhmedov, V. M., Zakharin, L. J.: Kinet. Katal. *8*, 331 (1967)
158. Breil, H., Heimbach, P., Kröner, M., Müller, H., Wilke, G.: Makromol. Chem. *69*, 18 (1963)
159. Dzierzgowski, S., Giezyński, R., Pasynkiewicz, S., Niżyńska, M.: J. Mol. Catal. *2*, 243 (1977)
160. Young, D., Green, M. L. H.: J. Appl. Chem. Biotechnol. *25*, 641 (1975)
161. Wilke, G., Bogdanovič, B., Hardt, P., Heimbach, P., Keim, W., Kröner, M., Oberkirch, W., Tanaka, K., Steinrucke, E., Walter, D., Zimmermann, H.: Angew. Chem., Int. Ed. *5*, 151 (1966)
162. Bönnemann, H., Grard, G., Kopp, W., Pump, W., Tanaka, K., Wilke, G.: Angew. Chem., Int. Ed. *12*, 964 (1974)

163. Alderson, T., Jenner, E. L., Lindsay, R. V. jr.: J. Amer. Chem. Soc. *87*, 5638 (1965); James, B. R.: Coordination Chem. Rev. *1*, 505 (1966)
164. Hidai, M., Uchida, Y., Misono, A.: Bull. Chem. Soc. Japan *38*, 1243 (1965)
165. Carbonaro, A.: Chimica e Industria, *55*, 244 (1973)
166. Pruett, R. L., Myers, W. R.: US. P. 3,372,206 (Chem. Abs. *69*, 18714 (1968))
167. Carbonaro, A., Greco, A., Dall'Asta, G.: J. Organometal. Chem. *20*, 177 (1969); Tetrahedron Lett. *1967*, 2037
168. Baker, R.: Chem. Rev. *73*, 487 (1973)
169. Margrave, J. L.: Characterisation of High Temperature Vapours. New York: Wiley 1967
170. Ackerman, R. J., Thorn, R. J.: Progress. Ceram. Sci. *1*, 39 (1961)
171. Green, M. L. H., Knight, J.: personal communication (1974)
172. Tsutsui, M.: Trans. N. Y. Acad. Sci. *80*, 658 (1968); Fischer, E. O., Seeholzer, J.: Z. Anorg. Chem. *312*, 244 (1961)
173. For example, see Fischer, E. O.: Inorg. Synth. *1960*, 1033
174. Blackborow, J. R., Koerner, v. Gustorf, E. A., Scrivanti, A.: unpublished
175. Koerner v. Gustorf, E. A., Jaenicke, O., Polansky, O. E.: Angew. Chem., Int. Ed. *11*, 532 (1972)
176. Nasielski, J., Kirsch, P., Wilputte-Steinert, L.: J. Organometal. Chem. *27*, C13 (1971); Wilputte-Steinert, L., Kirsch, P.: Int. Conf. Organometal. Chem., Fifth., Moscow 1971, Abstract 165
177. Bennett, M. A., Wilkinson, G.: Chem. Ind. (London) *1959*, 1516; Fischer, E. O., Fröhlich, W.: Chem. Ber. *92*, 2995 (1959); Manuel, T. A., Stone, F. G. A.: Chem. Ind. (London) *1959*, 1349; Bennett, M. A., Pratt, L., Wilkinson, G.: J. Chem. Soc. *1961*, 2037; Fischer, E. O., Kreiter, C. G., Berngruber, W.: Angew. Chem., Int. Ed. *6*, 634 (1967); Huttner, G., Mills, O. S.: Chem. Comm. *1968*, 344
178. Churchill, M. R., Mason, R.: Adv. Organometal. Chem. *5*, 93 (1967); Cotton, F. A., Wilkinson, G.: Advanced Inorganic Chemistry, 3rd ed. New York: Wiley-Interscience 1972; Mingos, D. M. P.: J. C. S. Dalton *1977*, 20
179. Herberhold, M.: Chemie in unserer Zeit *10*, 120 (1976)
180. Rennington, B. T., Howell, J. E.: J. Organometal. Chem. *113*, 75 (1976)
181. Candlin, J. P., James, W. H.: J. Chem. Soc. (A) *1968*, 1856
182. Tkachenko, I. B.: J. Organometal. Chem. *124*, C39 (1977)
183. Ballivet, D., Billard, C., Tkatchenko, I. B.: to be published
184. Otsuka, S., Kikuchi, T., Taketomi, T.: J. Amer. Chem. Soc. *85*, 3709 (1963)
185. Havel, J. J.: Ph. D. Thesis, The Pennsylvania State Univ. 1972, see Ref. 127
186. Wilke, G.: Angew. Chem., Int. Ed. *2*, 105, 1963
187. Bogdanovič, B., Heimbach, P., Kröner, M., Wilke, G., Hoffman, E. G., Brandt, J.: Liebigs Ann. Chem. *727*, 143 (1969)
188. Heimbach, P., Jolly, P. W., Wilke, G.: Adv. Organometal. Chem. *8*, 29 (1970)
189. Fischer, K.: Ph. D. Dissertation, Univ. Bochum 1973; Fischer, K., Jonas, K., Misbach, P., Stabba, R., Wilke, G.: Angew. Chem., Int. Ed. *12*, 934 (1973)
190. Jolly, P. W., Wilke, G.: The Organic Chemistry of Nickel, Vol. I and II. New York: Academic Press 1974 and 1975
191. Young, W. G., et al.: J. Amer. Chem. Soc. *69*, 2046 (1947)
192. Nakano, Y., Natsukawa, K., Yasuda, H., Tani, H.: Tetrahedron Lett. *1972*, 2833
193. For example, see Akutagawa, S., Otsuka, S.: J. Amer. Chem. Soc. *98*, 7420 (1976)
194. Kundig, E. P., Timms, P. L.: unpublished, reported in Ref. 386
195. Hartman, J. S., Timms, P. L.: unpublished, reported in Ref. 386
196. Kuz'yants G. M.: Izv. Akad. Nauk. SSR, Ser. Khim. *1976*, 1895; Chem. Abs. *86*, 5510b
197. Wailes, P. C.: Organometallic Chemistry of Titanium, Zirconium and Hafnium. New York: Academic Press 1974; Green, J. C., Green, M. L. H., Prout, C. K.: Chem. Comm. *1972*, 421; Green, M. L. H.: Pure Appl. Chem. *30*, 373 (1972); Lauher, J. W., Hoffman, R.: J. Amer. Chem. Soc. *98*, 1729 (1976)
198. Jonas, K., Wilke, G.: Angew. Chem., Int. Ed. *8*, 519 (1969)
199. Englert, M., Jolly, P. W., Wilke, G.: Angew. Chem., Int. Ed. *10*, 77 (1971)
200. John, J. P., Timms, P. L., Turney, T. W.: unpublished, reported in Ref. 386
201. Tan, T. S., Fletcher, J. L., McGlinchey, M. J.: J. C. S. Chem. Comm. *1975*, 771

202. Shul'pin, G. B., Rybinskaya, M. I.: Russian Chem. Rev. *43*, 716 (1974)
203. Lentzner, H. L., Watts, W. E.: Tetrahedron *27*, 4343 (1971)
204. Blackborow, J. R., Grubbs, R. H., Miyashita, A., Scrivanti, A., Koerner v. Gustorf, E. A.: J. Organometal. Chem. *122*, C6 (1976)
205. Herberhold, M.: Diplomarbeit, Univ. München 1961
206. Fischler, I., Budzwait, M., Koerner v. Gustorf, E. A.: J. Organometal. Chem. *105*, 325 (1976)
207. Gmelin: Handbuch d. anorganischen Chemie, Vol. 3, Chrom-organische Verbindungen, p. 141. Weinheim: Verlag Chemie 1971
208. Fischer, E. O., Fröhlich, W.: Chem. Ber. *98*, 2995 (1959); Leigh, G. J., Fischer, E. O.: J. Organometal. Chem. *4*, 461 (1965)
209. Brauer, D. J., Krüger, C., Robert, P. J., Tsay, Yi.-H.: to be published, reported in Ref. 17
210. Bogdanovič, B., Kröner, M., Wilke, G.: Liebigs Ann Chem. *699*, 1 (1966)
211. Müller, J., Goser, P.: Angew. Chem., Int. Ed. *6*, 364 (1967)
212. Dierks, H., Dietrich, H.: Z. Krist. *122*, 1 (1965)
213. Arnet, J. E., Pettit, R.: J. Amer. Chem. Soc. *83*, 2954 (1961)
214. Hoffmann, R., Heilbronner, E., Gleiter, R.: J. Amer. Chem. Soc. *92*, 706 (1970)
215. Ballivet, D., Billard, C., Tkatchenko, I. B.: submitted for publication
216. Fraser, A. R., Bird, P. H., Bezman, S. A., Shapley, J. R., White, R., Osborn, J. A.: J. Amer. Chem. Soc. *95*, 597 (1973)
217. Haines, R. J., Leigh, G. J.: Chem. Soc. Rev. *4*, 155 (1975); Calderon, N., Olfstead, E. A., Judy, W. A.: Angew. Chem., Int. Ed. *15*, 401 (1976)
218. Casey, C. P., Tuinstra, M. E., Saeman, M. C.: J. Amer. Chem. Soc. *98*, 608 (1976); Ephritikhine, M., Green, M. L. H.: J. C. S. Chem. Comm. *1976*, 926; Gassman, P. G., Johnson, T. H.: J. Amer. Chem. Soc. *98*, 6057 (1976); McLain, S. J., Wood, C. D., Schrock, R. R.: J. Amer. Chem. Soc. *99*, 3519 (1977)
219. Cooper, N. J., Green, M. L. H.: J. C. S. Chem. Comm. *1974*, 761; Schrock, R. R.: J. Amer. Chem. Soc. *97*, 6577 (1975)
220. Laverty, D. T., McKervey, M. A., Rooney, J. J., Stewart, A.: J. C. S. Chem. Comm. *1976*, 193
221. Timms, P. L., Turney, T. W.: J. C. S. Dalton *1976*, 2021
222. Silverthorn, W.: Adv. Organometal. Chem. *13*, 48 (1975)
223. Cloke, F. G. N., Green, M. L. H., Price, D. H.: J. C. S. Chem. Comm. *1978*, 431
224. Klabunde, K. J., Efner, H. F.: J. Fluorine Chem. *4*, 114 (1976)
225. Graham, W. R. M., Duley, W. W.: J. Chem. Phys. *54*, 586 (1971); *55*, 2527 (1971)
226. Benfield, F. W. S., Green, M. L. H., Ogden, J. S., Young, D.: J. C. S. Chem. Comm. *1973*, 866
227. Kundig, E. P., Timms, P. L.: J. C. S. Chem. Comm. *1977*, 912
228. Radonovich, L. J., Zuerner, E. C., Efner, H. F., Klabunde, K. J.: Inorg. Chem. *15*, 2976 (1976)
229. Gruen, D. M.: unpublished reported in Ref. 386
230. Graves, V., Łagowski, J. J.: Inorg. Chem. *15*, 577 (1976)
231. Efner, H. F., Fox, W. B., Smardezewski, R. R., Tevault, O. E.: Inorg. Chim. Acta *24*, L93 (1977)
232. Nesmeyanov, A. N., Zaitseva, N. N., Domrachev, G. A., Zinov'ev, V. D., Yur'eva, L. P., Tverdokhlebova, I. I.: J. Organometal. Chem. *121*, C52 (1976)
233. Jackson, S. E., Young, D.: unpublished reported in Ref. 234
234. Evans, S., Green J. C., Jackson, S. E., Higginson, B.: J. C. S. Dalton *1974*, 304.
235. McGlinchey, M. J., Tan, T.-S.: Canad. J. Chem. *52*, 2439 (1974)
236. Domrachev, G. A., Zinov'ev, V. D., Yur'eva, L. P., Zaitseva, N. N.: Bull. Acad. Sci. USSR. Chem. *25*, 1553 (1976)
237. Razuvaev, G. A., Domrachev, G. A., Zinov'ev, V. D.: Dokl. Akad. Nauk. SSSR *223*, 617 (1975)
238. McGlinchey, M. J., Tan, T.-S.: J. Amer. Chem. Soc. *98*, 2271 (1976)
239. Simons, L. H., Riley, P. E., Davis, R. E., Lagowski, J. J.: J. Amer. Chem. Soc. *98*, 1044 (1976)
240. Middleton, R., Timms, P. L. Gibson, G. M. (Albright and Wilson Limited): Ger. Offen. 2,337,393 (Cl.C 07f, C23c), 28 Feb. 1974; Brit. Appl. 34, 551/72, 24 July 1972; Chem. Abs. *80*, 133631e (1974)
241. Tan, T.-S., McGlinchey, M. J.: J. C. S. Chem. Comm. *1976*, 155
242. Silvon, M. P., Van Dam E. M., Skell, P. S.: J. Amer. Chem. Soc. *96*, 1945 (1974)
243. Cloke, F. G. N., Green, M. L. H., Morris, G. E.: J. C. S. Chem. Comm. *1978*, 72

244. Evans, S., Green, J. C., Jackson, S. E.: J. C. S. Faraday II, *68*, 249 (1972)
245. Work by Green, J. C.: reported in Ref. 223, 243 and 374
246. Schustorovich, E. M., Dyatkina, M. E.: Doklady Akad. Nauk. SSSR *128*, 1234 (1959)
247. Barinsky, R. P.: Zhur. Strukt. Khim. *1*, 200 (1960)
248. Kenlen, K., Jellinek, F.: J. Organometal. Chem. *5*, 490 (1966); Albrecht, G., Forster, E., Sippel, D., Eichkorn, F., Kurras, E.: Z. Chem. *8*, 311 (1968)
249. Fischer, E. O., Joos, G., Meer, W.: Z. Naturforsch. *B 13*, 456 (1958)
250. Anthony, M. T., Green, M. L. H., Young, D.: unpublished (1974)
251. I. C. I. Ltd.: Belgian Pat. 848.768 (1977)
252. Blackborow, J. R., Cloke, F. G. N., Green, M. L. H., Osborn, J. A., Tkatchenko, I. B., Young, D.: unpublished (1976 and 1977)
253. Fischer, E. O., Hafner, W.: Z. Naturforsch. *B 10*, 665 (1955)
254. Calderazzo, F.: Inorg. Chem. *3*, 810 (1964); Fischer, E. O., Seeholzer, J.: Z. Anorg. Allg. Chem. *312*, 244 (1961)
255. Gmelin: Handbuch d. anorgan. Chemie, Ergänzungswerk, Chrom-organische Verbindungen (1971)
256. Sorokin, Y. A., Petukhov, G. G.: Zh. Obshch. Khim. *35*, 2135 (1966)
257. Fischer, E. O., Scherer, F., Stahl, H. O.: Chem. Ber. *93*, 2065 (1960)
258. Riley, P. E., Davis, R. E.: Inorg. Chem. *15*, 2735 (1976)
259. Yamada, S., Nakamura, N., Tsuchida, R.: Bull. Chem. Soc. Jap. *30*, 647 (1957)
260. Treichel, P., Essenmacher, G., Klabunde, K. J., Efner, H.: unpublished reported in Ref. 141
261. Maciel, G. E., Natterstad, J. J.: J. Chem. Phys. *42*, 2427 (1965)
262. Bodner, G. M., Todd, L. J.: Inorg. Chem. *13*, 1335 (1974)
263. Swain, C. G., Lupton, E. C. jr.: J. Amer. Chem. Soc. *90*, 4328 (1968)
264. Graves, V.: Ph. D. Thesis, Univ. Texas 1975; Diss. Abstr. Int. B. *36*, 2211 (1975)
265. Evans, J., Norton, J. R.: Inorg. Chem. *13*, 3042 (1974)
266. Banks, R. E.: Fluorocarbons and Their Derivatives, p. 140. London: Oldbourne Press 1964
267. Birchall, T.: unpublished reported in Ref. 238
268. Nesmeyanov, A. N., Yur'eva, L. P., Zaitseva, N. N., Mysov, E. I., Vainberg, A. M.: Izv. Akad. Nauk, SSSR Ser. Khim. *1973*, 165
269. Fischer, E. O., Schmidt, M. W.: Chem. Ber. *102*, 1954 (1969); Fischer, E. O., Stahl, H. O.: *Ibid. 89*, 1805 (1956)
270. Koerner v. Gustorf, E., Buchkremer, J., Pfaifer, Z., Grevels, F.-W.: Angew. Chem., Int. Ed. *10*, 260 (1971); Kelly, J. M., Bent, D. V., Hermann, H., Schultz-Frohlinde, D., Koerner v. Gustorf, E.: J. Organometal. Chem. *69*, 259 (1974)
271. Büchkremer, J.: Dissertation, Univ. Bochum 1973; see also Ref. 17
272. Yagubskii, E. B., Khidikel', M. L., Shchegolev, I. F., Buravov, L. I., Gribov, B. G., Makova, M. K.: Bull. Acad. Sci. USSR Chem. Sci. *1968*, 2013
273. Fischer, E. O., Elsenbroich, Ch., Kreiter, C. G.: J. Organometal. Chem. *7*, 481 (1967); Hein, F., Kleinwachter, H.: Monatsber. Deut. Akad. Wiss. Berlin *60*, 607 (1960)
274. Weber, S. R., Brintzinger, H. H.: J. Organometal. Chem. *127*, 45 (1977)
275. Churchill, M. R., Mason, R.: Proc. Roy. Soc. Ser. A. *292*, 61 (1966)
276. Browning, J., Cundy, C. S., Green, M., Stone, F. G. A.: J. Chem. Soc. A *1971*, 448
277. Browning, J., Green, M., Penfold, B. R., Spencer, J. L.: Stone, F. G. A.: J. C. S. Chem. Comm. *1973*, 31
278. Van Oven, H. O., de Liefde Meijer, H. J.: J. Organometal. Chem. *31*, 71 (1971)
279. Müller, J., Mertschenk, B.: J. Organometal. Chem. *34*, C41 (1972)
280. Groenenboom, C. J., Sawatsky, G., de Liefde Meijer, H. J., Jellinek, F.: J. Organometal. Chem. *76*, C4 (1974)
281. Zeinstra, J. D., de Boer, J. L.: J. Organometal. Chem. *54*, 207 (1973)
282. Engebretson, G., Rundle, R. E.: J. Amer. Chem. Soc. *85*, 481 (1963)
283. Green, M. L. H., Hart-Davies, A.: unpublished (1971)
284. Fischer, E. O., Reckziegel, A., Müller, J., Goser, P.: J. Organometal. Chem. *11*, P13 (1968)
285. Whitesides, S. H., Budnik, R. A.: Chem. Comm. *1971*, 1514
286. Blackborow, J. R., Hildebrand, K., Koerner v. Gustorf, E. A., Scrivanti, A., Eady, C. R., Ehnhart, D., Krüger, C.: J. C. S. Chem. Comm. *1976*, 16

287. Müller, J., Kreiter, C. G., Mertschenk, B., Schmitt, G.: Chem. Ber. *108*, 273 (1975); Whitesides, T. H., Budnik, R. A.: Chem. Comm. *1974*, 302
288. Mahler, J. E., Jones, D. A. K., Pettit, R.: J. Amer. Chem. Soc. *86*, 3589 (1964); Dauben, H. J. jr., Bertelli, D. J.: *Ibid.* *83*, 498 (1961)
289. Müller, J., Mertschenk, B.: Chem. Ber. *105*, 3346 (1972)
290. Burt, R., Green, M., Turney, T. W.: unpublished see Ref. 386
291. Brauer, D. S., Krüger, C.: Inorg. Chem. *15*, 2511 (1976)
292. Studiengesellschaft m. b. H.: British Pat. 1,128,128; Chem. Abs. *70*, 96963K (1972); Müller, S., Holzinger, W., Menig, M.: Proc. Eighth Internat. Conf. Organometal. Chem., Venice 1975
293. Pettit, R., Emerson, G. F.: Adv. Organometal. Chem. *1*, 1 (1964) and reference therein
294. Report on work by Graves, V., Lagowski, J. J., Simons, L. H.: Chem. Eng. News, Dec. 24, p. 26 (1973)
295. Ely, S. R., Hopkins, T. E., DeKock, C. W.: J. Amer. Chem. Soc. *98*, 1624 (1976)
296. Hodgson, K. O., Mares, F., Starks, P. F., Streitwieser, A.: J. Amer. Chem. Soc. *95*, 8650 (1973); Streitwieser, A., Müller-Westerhof, U., Sonnichsen, G., Mares, F., Morrell, D. G., Hodgson, K. O., Harmon, C. A.: *Ibid.* *95*, 8644 (1973)
297. Dierks, H., Dietrich, H.: Acta Cryst. *B 24*, 58 (1968)
298. Atwood, J. L., Smith, K. D.: J. Amer. Chem. Soc. *95*, 1488 (1973)
299. Kuzyants, G. M., Baronetsky, A. O., Chernyshova, T. M., Todres, Z. V.: J. Organometal. Chem. *142*, 139 (1977)
300. Fritz, H. P., Keller, H.: Z. Nat. *16B*, 848 (1961); Chem. Ber. *95*, 158 (1962); Fritz, H. P.: Adv. Organometal. Chem. *1*, 239 (1964)
301. Chivers, T., Timms, P. L.: J. Organometal. Chem. *118*, C37 (1976)
302. Hawthorne, M. F., Young, D. C., Andress, T. D., Howe, D. V., Pilling, R. L., Pitts, A. D., Reintjes, M., Warren, L. F., Wegner, P. A.: J. Amer. Chem. Soc. *90*, 879 (1968)
303. Stone, F. G. A.: J. Organometal. Chem. *100*, 257 (1975) and references therein
304. Miller, V. R., Sneddon, L. G., Beer, D. C., Grimes, R. N.: J. Amer. Chem. Soc. *96,* 3090 (1974)
305. Rudolph, R. W.: Acc. Chem. Res. *9,* 447 (1976) and references therein
306. Evans, W. J., Dunks, G. B., Hawthorne, M. F.: J. Amer. Chem. Soc. *95,* 4565 (1973)
307. Miller, V. R., Grimes, R. N.: J. Amer. Chem. Soc. *95,* 5078 (1973); J. Amer. Chem. Soc. *98,* 1600 (1976); Miller, V. R., Weiss, R., Grimes, R. N.: J. Amer. Chem. Soc. *99,* 5646 (1977)
308. Hall, L. W., Zimmerman, G. J., Sneddon, L. G.: J. C. S. Chem. Comm. *1977*, 45
309. Skell, P. S., Girard, J. E.: J. Amer. Soc. *94*, 5518 (1972)
310. Klabunde, K. J.: J. Fluor. Chem. *7*, 95 (1976)
311. Klabunde, K. J., Roberts, J. S.: J. Organometal. Chem. *137*, 113 (1977)
312. Kasai, P. H.: Acc. Chem. Res. *4*, 329 (1971)
313. Andrews, W. L. S., Pimentel, G. C.: J. Chem. Phys. *44,* 2527 (1966)
314. Tan, L. Y., Pimentel, G. C.: J. Chem. Phys. *48*, 5202 (1968)
315. Andrews, L.: J. Chem. Phys. *48*, 972 (1968); Tetrahedron Lett. *1968*, 1423; Andrews, L., Carver, T. G.: J. Phys. Chem. *72*, 1743 (1968)
316. Andrews, L., Carver, T. G.: J. Chem. Phys. *49*, 896 (1968)
317. Andrews, L., Carver, T. G.: J. Chem. Phys. *50*, 4223 and 4235 (1969)
318. Carver, T. G., Andrews, L.: J. Chem. Phys. *50*, 5100 (1969)
319. Andrews, L., Smith, D. W.: J. Chem. Phys. *53*, 2956 (1970)
320. Hutzenbühler, D. A., Andrews, L., Cavey, F. A.: J. Amer. Chem. Soc. *97*, 187 (1975)
321. Chung, C., Lagow, R. J.: Chem. Comm. *1972*, 1078
322. Sneddon, L. G., Lagow, R. J.: J. C. S. Chem. Comm. *1974*, 302
323. Skell, P. S., Goldstein, E. J., Petersen, R. J., Tingey, G. L.: Chem. Ber. *100*, 1442 (1967); Doerr, R. G., Skell, P. S.: J. Amer. Chem. Soc. *89*, 4684 (1967)
324. Girard, J. E.: Ph. D. Thesis, The Pennsylvania State Univ. (1971), see Ref. 127
325. Klabunde, K. J., Efner, H. F., Satek, L., Donley, W. J.: J. Organometal. Chem. *71*, 309 (1974)
326. Skell, P. S., Girard, J. E.: J. Amer. Chem. Soc. *94*, 5518 (1972)
327. Rieke, R. D., Bates, S. E.: J. C. S. Chem. Comm. *1973*, 789; J. Amer. Chem. Soc. *96*, 1775 (1974); Rieke, R. D.: Acc. Chem. Res. *10*, 301 (1977)
328. Skell, P. S., Girard, J. E.: Proc. Amer. Chem. Soc. 1971, ORGN. 006

329. Klabunde, K. J., Scott, M., Low, J. Y. F.: J. Amer. Chem. Soc. *94*, 999 (1972)
330. McGlinchey, M. J., Tan, T.-S.: Inorg. Chem. *14*, 1209 (1975)
331. Heck, R. F.: J. Amer. Chem. Soc. *90*, 5518, 5526, 5531, 5535 (1968); Dieck, H. A., Heck, R. F.: J. Amer. Chem Soc. *96*, 1133 (1974); Hosokawa, T., Calvo, C., Lee, H. B., Maitlis, P. M.: J. Amer. Chem. Soc. *95*, 4914, 4924 (1973); Fleury, P. M.: Tetrahedron Lett. *1968*, 2285; Heck, R. F., Nolley, J. P. jr.: J. Org. Chem. *96*, 1133 (1974); Tsuji, J., Monikawa, M., Kiki, J.: J. Amer. Chem. Soc. *86*, 4851 (1971); Tetrahedron Lett. *1963*, 1437
332. Roberts, J. S., Klabunde, K. J.: J. Organometal. Chem. *85*, C13 (1975)
333. Roberts, J. S., Klabunde, K. J.: J. Amer. Chem. Soc. *99*, 2509 (1977)
334. Klabunde, K. J., Murdock, T. O.: Chemtech. *5*, 624 (1975)
335. Houldin, R., Timms, P. L.: unpublished see Ref. 386
336. Felkin, H., Swierczewski, G.: Tetrahedron *31*, 2735 (1975)
337. Farády, L., Bencze, L., Markó, L.: J. Organometal. Chem. *10*, 505 (1967)
338. Finkbeiner, H. L., Cooper, G. D.: J. Org. Chem. *27*, 3395 (1962)
339. Anthony, M. T., Green, M. L. H., Young, D.: unpublished results 1976
340. Corriu, P. J. P., Massé, J. P.: Chem. Comm. *1972*, 144; Tamao, K., Sumitani, K., Kumada, M.: J. Amer. Chem. Soc. *94*, 4374 (1972)
341. Timms, P. L.: Chem. Comm. *1968*, 1525
342. Timms, P. L.: J. C. S. Dalton *1972*, 830
343. Ozin, G. A.: The Spex Speaker, Vol. XVI, No. 4, December 1971; Huber, H., Kündig, E. P., Ozin, G. A., Van der Voet, A.: Can. J. Chem. *52*, 95 (1974)
344. McIntosh, D., Ozin, G. A.: Inorg. Chem. *16*, 59 (1977)
345. Hanlan, A. J. L., Ozin, G. A.: Inorg. Chem. *16*, 2857 (1977)
346. Hanlan, A. J. L., Ozin, G. A.: Inorg. Chem. *16*, 2848 (1977)
347. Huber, H., Ozin, G. A.: Can. J. Chem. *60*, 3746 (1972)
348. Klotzbücher, W. E., Ozin, G. A.: J. Amer. Chem. Soc. *95*, 3790 (1973)
349. Huber, H., Klotzbücher, W., Ozin, G. A., Van der Voet, A.: Can. J. Chem. *51*, 2722 (1973)
350. Darling, J. H., Garton-Sprenger, M. B., Ogden, J. S.: J. C. S. Faraday *8*, 75 (1973)
351. Andrews, W. L. S.: J. Chem. Phys. *54*, 4935 (1971)
352. Hatzenbuhler, D. A., Andrews, L.: J. Chem. Phys. *56*, 3399 (1972)
353. Andrews, W. L. S.: J. Phys. Chem. *73*, 3922 (1971); J. Chem. Phys. *54*, 4935 (1971); Smardzewski, R. R., Andrews, L.: J. Chem. Phys. *57*, 1327 (1972); Smardzewski, R. R., Andrews, L.: J. Phys. Chem. *77*, 801 (1973); Andrews, L., Hwang, J. T., Trindle, C.: J. Phys. Chem. *77*, 1065 (1973); Jacox, M., Milligan, D.: Chem. Phys. Lett. *14*, 518 (1972); Andrews, L.: J. Mol. Spectrosc. *61*, 337 (1976)
354. Brewer, L., Wang, J. L.: J. Chem. Phys. *56*, 759 (1972)
355. Abramowitz, S., Acquista, N.: J. Res. Nat. Bur. Stand. A. *75*, 13 (1971)
356. Ogden, J. S.: unpublished, see Ref. 54; Kundig, E. P., Ozin, G. A.: unpublished, see Ref. 54
357. Anderson, J. S., Bos, A., Ogden, J. S.: J. Chem. Soc. D *1971*, 1381
358. Griffiths, M. J., Barrow, R. F.: J. C. S. Faraday II *73*, 943 (1977)
359. Howard, W. F., Andrews, W. L. S.: J. Amer. Chem. Soc. *95*, 2056 (1973)
360. Gladysz, J. A., Fulcher, J. G., Togashi, S.: J. Org. Chem. *41*, 3647 (1976)
361. Gladysz, J. A., Fulcher, J. G., Togashi, S.: Tetrahedron Lett. *1977*, 521
362. Gladysz, J. A., Bocarsly, A. B.: Proc. Amer. Chem. Soc., 171st, 132 (1976)
363. Wescott, L. D. jr., Williford, C., Parks, F., Dowling, M., Sublett, S., Klabunde, K. J.: J. Amer. Chem. Soc. *98*, 7853 (1976)
364. Friedman, L., Schechter, H.: J. Amer. Chem. Soc. *83*, 3159 (1961)
365. McMurry, J. E., Fleming, M. P.: J. Org. Chem. *41*, 896 (1976)
366. McMurry, J. E., Fleming, M. P.: J. Amer. Chem. Soc. *96*, 4708 (1974); McMurry, J. E., Krepski, L. R.: J. Org. Chem. *41*, 3929 (1976); McMurry, J. E., Kees, K. L.: J. Org. Chem. *42*, 2655 (1977)
367. McMurry, J. E., Fleming, M. P.: J. Org. Chem. *40*, 2555 (1975)
368. Ledon, H., Tkatchenko, I. B., Young, D.: Tetrahedron Lett. *1979*, 173
369. McMurry, J. E., Silvestri, M.: J. Org. Chem. *40*, 2687 (1975)
370. Chivers, T., Timms, P. L.: Can. J. Chem. *55*, 3509 (1977)

371. MacKenzie, R. A., Timms, P. L.: J. C. S. Chem. Comm. *1974*, 650
372. Francis, C. G., Timms, P. L.: J. C. S. Chem. Comm. *1977*, 446
373. Green, M. L. H., Young, D.: Phys. Bull. *1975*, 350
374. Anthony, M. T., Green, M. L. H., Young, D.: J. C. S. Dalton *1975*, 1419
375. Busby, R., Klotzbücher, W., Ozin, G. A.: Inorg. Chem. *16*, 822 (1977); Bassler, J. M., Timms, P. L., Margrave, J. L.: Inorg. Chem. *5*, 729 (1966); Timms, P. L., Stump, D. D., Kent, R. A., Margrave, J. L.: J. Amer. Chem. Soc. *88*, 940 (1966)
376. Boyd, J. W., Lavoie, J. M., Gruen, D. M.: J. Chem. Phys. *60*, 4088 (1974)
377. Ozin, G. A.: Appl. Spect. *30*, 573 (1976)
378. Kundig, E. P., Moskovits, M., Ozin, G. A.: Angew. Chem., Int. Ed. *14*, 292 (1975)
379. McIntosh, D., Ozin, G. A.: J. Amer. Chem. Soc. *98*, 3167 (1976)
380. Klabunde, K. J., Low, J. Y. F.: J. Amer. Chem. Soc. *96*, 7675 (1974)
381. Timms, P. L.: Chem. Comm. *1969*, 1033
382. Leach, G.: B. Sc. Thesis, Univ. Bristol 1974
383. Williams-Smith, D. L., Wolf, L. R., Skell, P. S.: J. Amer. Chem. Soc. *94*, 4042 (1972)
384. Francis, C.: B. Sc. Thesis, Univ. Bristol 1974
385. MacKenzie, R. E., Timms, P. L.: unpublished, reported in Ref. 386
386. Timms, P. L., Turney, T. W.: Adv. Organometal. Chem. *15*, 53 (1977)
387. Cable, R. A., Green, M., MacKenzie, R. E., Timms, P. L., Turney, T. W.: J. C. S. Chem. Comm. *1976*, 270
388. English, A. D., Jesson, J. P., Tolman, C. A.: Inorg. Chem. *15*, 1730 (1976)
389. Klabunde, K. J.: Angew. Chem., Int. Ed. *14*, 287 (1975)
390. Klabunde, K. J., Efner, H. F., Murdock, T. O., Ropple, R.: J. Amer. Chem. Soc. *98*, 1021 (1976)
391. Murdock, T. O., Klabunde, K. J.: J. Org. Chem. *41*, 1076 (1976)
392. Report on work by Klabunde, K. J. In: Metal Atom Research Spawns New Catalysts. Chem. Eng. New, January 24, p. 23 (1977)

# Recent Developments in Metal Vapour Synthesis

Since this book was written many new developments in MVS technology have taken place and also some new chemistry, mainly of refractory metals, has been published. Some major chemical companies have been investigating the preparation of catalysts by MVS techniques particularly with a view to depositing organometallic compounds on solid supports. The developments during the last eighteen months have been highlighted in the following sections.

## New Experimental Techniques

There has hitherto been a problem with the rate of evaporation of refractory metals (W, Nb, Ti, Zr, and Hf) which all have high heats of vaporisation: in particular tungsten has proved troublesome to evaporate at reasonable rates. A new electron gun with a water cooled positive hearth permits the evaporation of tungsten at the rate of several grams per hour[1–5]; even bigger guns are in development[6].

The positive water cooled hearth appears to present no problems as regards shorting of the very big voltages providing the insulated supply pipes for the cooling water are sufficiently long. Yields of bisarenemetal compounds from systems with positive hearths are invariably much higher than those obtained from systems with conventional earthed hearths (Table 1, see p. 180). A new commercial metal vapour synthesiser with twin electron beams and or resistance heated sources is now available[6].

The reactivity of the alkali metals with respect to reducing power is often limited by the cleanliness and divisibility of their surfaces and on the stability of the alkali metal complex produced as a coproduct in the reduction reaction: kinetics also seem to be important. These points have been recently illustrated quite damatically by the work of Jonas and also Wilke[7,8] and coworkers.

$$Cp_2Fe + 4Li + C_2H_4 \xrightarrow[\text{DME}]{\text{thf}} 2CpLi + Fe(C_2H_4)_4Li_2$$

$$MoCl_4 + 2Mg + 3Bd \xrightarrow[-\,20^\circ\,C]{\text{thf}} Mo(Bd)_3 + 2\,MgCl_2$$

**Table 1.** Examples of Experiments carried out with the Positive-hearth Furnace

| Metal | Reactant | Scale | | Product | Yield[c] | |
|---|---|---|---|---|---|---|
| | | Initial Metal Sample[a] (g) | Ligand[b] ($cm^3$) | | g | % |
| Tungsten | Benzene | 10 | 90 | $(\eta\text{-}C_6H_5)_2W$ | 1.2 | 40 |
| | Toluene | | 90 | $(\eta\text{-}C_6H_5Me)_2W$ | 1.2 | 40 |
| | Mesitylene | | 70 | $(\eta\text{-}C_6H_3Me_3)_2W$ | 1.5 | 40 |
| | Fluorobenzene | | 80 | $(\eta\text{-}C_6H_5F)_2W$ | 1.0 | 35 |
| Niobium | Benzene | 8 | 90 | $(\eta\text{-}C_6H_6)_2Nb$ | 1.0 | 40 |
| | Toluene | | 90 | $(\eta\text{-}C_6H_5Me)_2Nb$ | 1.2 | 40 |
| Titanium | Toluene | 6 | 90 | $(\eta\text{-}C_6H_5Me)_2Ti$ | 3.5 | 50 |
| Zirconium | Toluene + trimethylphosphine | 6 | 75+5 | $(\eta\text{-}C_6H_5Me)_2Zr(PMe_3)$ | 1.0 | 45 |
| Hafnium | Benzene + trimethylphosphine | 10 | 80+5 | $(\eta\text{-}C_6H_6)_2Hf(PMe_3)$ | 1.5 | 35 |
| | Toluene + trimethylphosphine | | 75+5 | $(\eta\text{-}C_6H_5Me)_2Hf(PMe_3)$ | 1.5 | 35 |
| Uranium | Cyclooctatetraene | | | $(\eta\text{-}Cot)_2U$ | 1.8 | 90 |

[a] In the form of a premelted ingot.
[b] Introduced during *ca.* 3 hours.
[c] Based on metal leaving the furnace (±10%). (Reproduced with permission of G. V. Planer Ltd. and F. G. N. Cloke and M. L. H. Green).

**Table 2**

| Starting materials | Initial solution colour | "End point" colour | Product isolated | Approximative yield[c] % |
|---|---|---|---|---|
| $CrCl_3 \cdot 3THF$ + 1-Methylnaphthalene | Purple | Red-brown | $Cr(\eta^6\text{–}C_{10}H_7Me)_2$ | 40 |
| $MoCl_5$ + Toluene | Red-brown | Red-brown[a] | $Mo(\eta^6\text{–}PhMe)_2$ | 40 |
| $MoCl_5$ + 1-Methylnaphthalene | Red-brown | Red-brown[a] | $Mo(\eta^6\text{–}C_{10}H_7Me)_2$ | 45 |
| $VCl_3 \cdot 3THF$ + Toluene | Pink | Black[b] | $V(\eta^6\text{–}PhMe)_2$ | 30 |
| $VCl_3 \cdot 3THF$ + 1-Methylnaphthalene | Pink | Black[b] | $V(\eta^6\text{–}C_{10}H_7Me)_2$ | 35 |
| $TiCl_3 \cdot 3THF$ + Toluene | Blue | Brown-black[b] | $Ti(\eta^6\text{–}PhMe)_2$ | 20 |
| $TiCl_3 \cdot 3THF$ + Mesitylene | Blue | Brown-black[b] | $Ti(\eta^6\text{–}C_6H_3Me_3)_2$ | 15 |

[a] Several Intermediate colours were observed.
[b] Excess potassium atoms used.
[c] Yields based on the transition metal halide. (Reproduced from J.C.S. Chem. Comm. 1978, 731)

It has been found that evaporation of alkali metals[9] in the presence of organic ligands and metal halides at low temperatures proceeds with the production of metal organometallics in relatively high yields; although no preparative details have been given for the preparation of tungsten and tantalum complexes. This method proves a very cheap and useful alternative to electron beam evaporation for Ti, V, Cr and Mo. Lithium, Sodium and Potassium have low heats of evaporation and any laboratory may, without undue cost, construct a simple evaporator of this type. Further energy savings can be made because the condensation temperature need not be that of liquid nitrogen (Table 2 and Fig. 1).

A Ruthenium paint has been made from ruthenium powder, epoxy-cement and a hydrocarbon, and has been used to coat tungsten filaments; after removal of the organic part by heating and oxidation, the filaments may be used to evaporate ruthenium without contamination from tungsten[10]. $Ru(\eta^4C_6H_6)(\eta^6C_6H_6)$ has been made in this way.

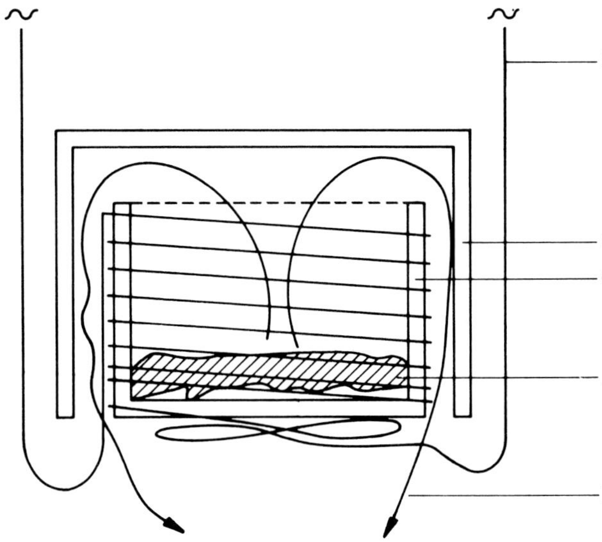


**Fig. 1.** Furnace for the downward evaporation of alkali metal vapour into a solution (Reproduced with the permission of Dr. P. L. Timms)

## Catalysts

There seems to be a need for high turnover supported polymerisation catalysts, such that the polymerisation takes place in a flow sequence and of necessity both the catalyst and support are incorporated in the product. For this to be an effective process turnovers of the order $10^7$ to $10^8$ are required, the supported catalysts must contribute only 1% to the total process cost and be incorporated in such small concentrations into the final product that further (costly) purifications are avoided. Throwaway supported catalysts of this type can be made by an MVS technique. Absorption of bisarenetitanium made by MVS[11] on silica and alumina supports have proved effective in this respect, though of limited applicability[11a]. In a very recent patent, cocondensation of Mg atoms and $TiCl_4$ leads to a highly active and flexible catalyst for the polymerisation of α-olefins (0.8 to 20 kg polymer/mmol Ti/atoms of olefin/hour)[11b].

Highly dispersed Ni, Pd, Pt and Ag on silica or alumina can be prepared by condensing the metal into weakly bonding substrates such as toluene or thf and impregnating the support with these "metal solutions" at low temperatures (−100 °C) followed by warming to room temperature and removal of organic material. The $Ni/Al_2O_3$ system is active toward olefin hydrogenation with a maximum activity at a loading of 0.4%. Heat treatment under vacuum increases the activity until metal sintering occurs. The shape, size and activity of the metal particles depends on the solvent used in the evaporation process[12,13].

Nickel aggregates react with pentane at low temperature and, depending on the ratio Ni:pentane, may produce black non-ferromagnetic powders (Ni:C:H = 2:1:2, crystalite size 30 Å, Ni:pentane = < 1:100) or black ferromagnetic powders (Ni:C:H = 5:1:2, crystalite size 80 Å, Ni:pentane = > 1:100). ESCA shows that the carbon is not carbide-like in either case[14].

Toluenenickelbispentaflorophenyl is an arene hydrogenation catalyst but only for about ten turnovers[15].

## Metal Vapour Chemistry of the Transition Metals

### *Ti, Zr, Hf*[1–5b]

Use has been made of positive hearth electron beam techniques to preparate arene complexes of Zr and Hf in yields of 40% and 30% based on metal (Table 1). The known compound $Cot_3Ti_2$ can be prepared by reaction of Ti atoms and Cot[5b].

Bisarenetitanium has been prepared by evaporating potassium into a cooled mixture of arene and $TiCl_3$3thf (Table 2).

Zirconium atoms oxidatively add to saturated hydrocarbons at cryoscopic temperatures: CH and CC bond fission is proposed[16].

### *V, Nb, Ta*

$(Arene)_2V$ complexes can be made by the potassium vapour technique (see Table 2). Bisareneniobium complexes made by the positive hearth electron beam technique have He 1 photoelectron spectra which are consistent with a $d^5$ ground state. Bisbenzeneniobium may also have been made by niobium sputtering onto an argon window with codeposition of benzene at 30 K. Tantalum atoms react with arenes to give a red crystaline complex which esr studies suggest is a paramagnetic $Ta^\circ$ complex: these complexes decompose at around 0 °C.

*Cr, Mo, W*

Yields of trisbutadiene-molybdenum and -tungsten are increased by an order of magnitude when prepared using a positive hearth electron gun. These compounds have since been made by a classical reduction route[8)].

The chemistry of bisarenetungsten can now be investigated without the arduous Fischer-Hafner synthesis used previously[17)].

$$W(a) + arene \xrightarrow{-196\,°C} (\eta^6 arene)_2W$$

$$(\eta^6 arene)_2W + HCl \xrightarrow{aqueous} (\eta^6 arene)_2WH^+Cl^-$$

$$(\eta^6 arene)_2W + C_3H_5Cl \xrightarrow{20\,°C} (\eta^6 arene)(\eta^3 allyl)W\begin{smallmatrix}Cl\\Cl\end{smallmatrix}W(\eta^6 arene)(\eta^3 allyl)$$

An Xray structure of $(\eta^6C_6H_5F)WH^+$ shows the two arene rings bent back off axis by 22.6° and a tungsten hydrogen bond length of 1.85 Å. There also appears to be a through space coupling of the hydride and fluorine nuclei which yields a triplet structure to the hydride resonance (J = 10 Hz). Bisarenetungsten and the bisarenetungstenhydride halide may be reacted with $FeCl_3$ and $NH_4Fe(SO_4)_2$ respectively to yield the bisarenetungsten-halide cation

$$(\eta^6 arene)W + FeCl_3 \longrightarrow (\eta^6 arene)_2WCl^+$$

Reaction of Cr atoms and the diphosphine $(PF_2)_2NMe$ forms the very stable trischelate $Cr[(PF_2)_2NMe]_3$ in good yield[21a)].

The novel 1:1 [2,2] paracyclophane complex has been formed in very low yields[21b)].

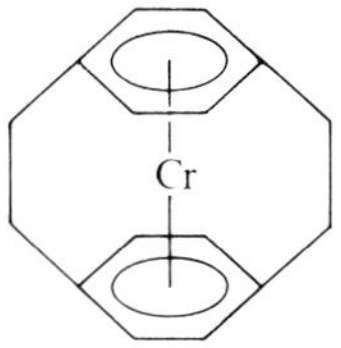


*Fe, Ru, Os*

Triphenylphosphite will displace cyclooctadiene from $Cod_2Fe$ to give a bis orthometalled $Fe^{II}$ species[18)].

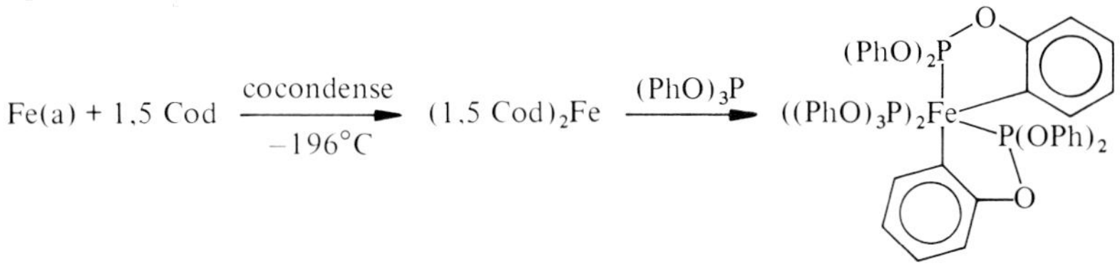

$Fe(PR_3)_5$ and $Fe[P(OR)_3]_5$ complexes synthetisable by MVS catalyse the hydrocyanation of ethylene[19].

Trimethylphosphite displaces one cyclooctadiene from $Cod_2Fe$ without metalation but a 1,5 to 1,3 isomerisation takes place in the remaining cyclooctadiene ligand; reaction of this 18 electron trisphosphitecyclooctadiene-iron complex with acid yields a 16 electron $\eta^3$-allyl compound:

$$\text{Fe(a)} + \text{1,5-Cod} \xrightarrow[\text{77 K}]{\text{cocondense}} \text{(1,5-Cod)}_2\text{Fe} \xrightarrow{(\text{MeO})_3\text{P}} \text{(1,3-Cod)Fe}[(\text{MeO})_3\text{P}]_3$$

$$\text{(1,3Cod)Fe}[(\text{MeO})_3\text{P}]_3 + \text{HX} \longrightarrow [(\text{MeO})_3\text{P}]_3\text{Fe}(\eta^3\text{C}_8\text{H}_{13})^+\text{X}^-$$

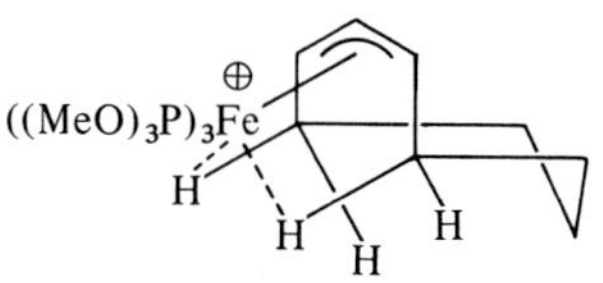


There is a strong interaction between the endo-hydrogens and the iron atom as evidenced by high field shifts of these hydrogen resonances in the pmr spectrum. Reaction of the allyl cation with other donor ligands (but not excess trimethyl-phosphite) yields 18 electron addition complexes. Reduction of the 16 electron cation with sodium amalgam gives the neutral 17 electron paramagnetic $\eta^3$ allyl complex[20].

Iron atoms react with the new ligand $Me_2NPF_2$ to yield $Fe(F_2PNMe_2)_5$ and with $(PF_2)_2NMe$ forming $Fe((F_2P)_2NMe)_4$, where one ligand bonds as a chelate[21].

*Co, Rh, Ir*

Cobalt atoms and $MeN(PF_2)_2$ react to give $Co(MeN(PF_2)_2)_5$ which may be hydrolised to yield $(MeN(PF_2)_2)_3Co(PF_2NHMe)_2$[22].

A mixture of $MeN(PF_2)_2$ and $Me_2NPF_2$ (1:4) reacts with cobalt atoms to yield

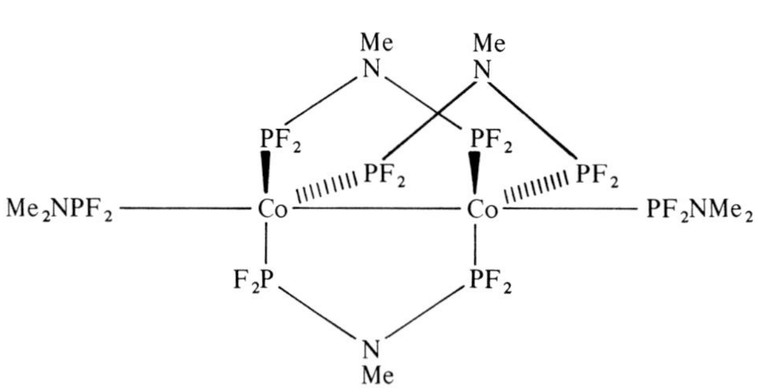


Cobalt atoms condensed in toluene react with carbon monoxide to yield $Co_2CO_8$, $Co_4CO_{12}$, $Co_6CO_{16}$ and higher clusters[23].

Cobaltacarboranes can be made from cobalt atoms, acetylenes and pentaboranes[24].

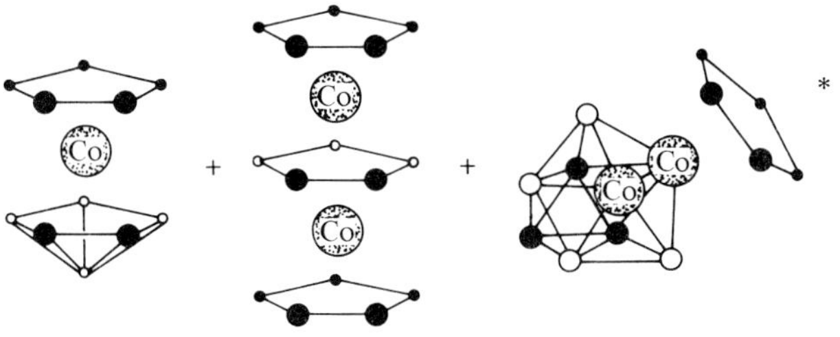


○ = BH
● = CMe
* one Cp ring omitted for clarity

*Ni, Pd, Pt*

Nickel atoms react with bromopentafluorobenzene as follows[25]:

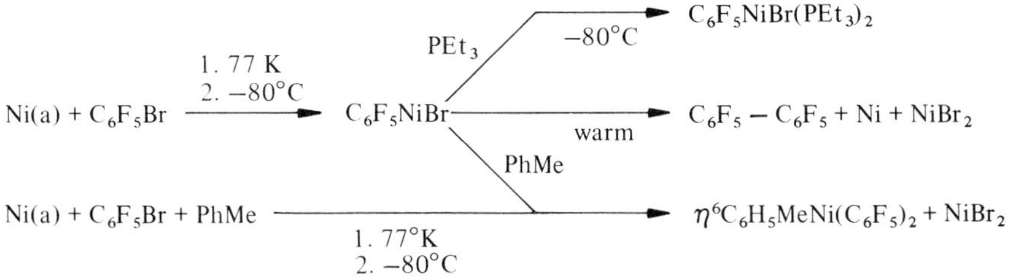


$Ni(C_6F_5)_2(\eta^6$toluene) is an arene hydrogenation catalyst.

Nickel and palladium atoms react with hexafluorobutyne-2(HFB):

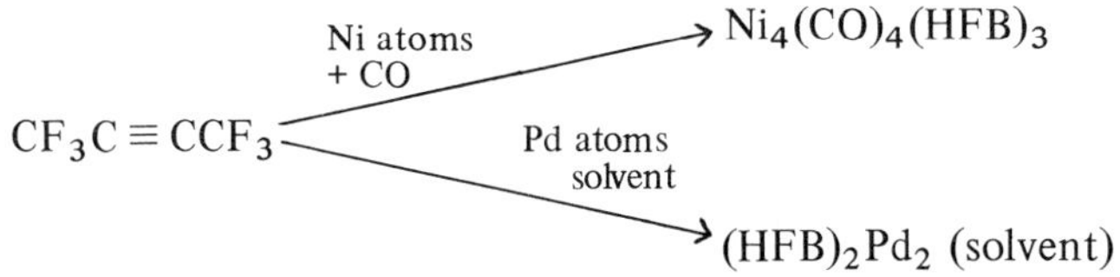


*Cu, Ag, Au and Main Group Elements*

Michael Faraday was one of the first men who recorded the making of gold clusters electrostatically and much work has been done since. Recently gold was evaporated in a rotating cryostat into an ethanol film at −100 °C containing $Au(PAr_3)X$ (X = SCN, I, CN) and $PAr_3$ at a pressure of below $10^{-2}$ torr. $Au_{11}(PAr_3)_7X_3$ was obtained in high yields (60–70%)[26].

Barium, strontium and calcium have been evaporated into alkylhalides in thf and alkyl metalhalides obtained; similarly dicyclopentadienyl and diindenyl metal compounds were obtained. The alkyl metal halides have some differing reactivity compared to Grignard reagents[27)]. Maintaining the M/RHal cocondensates at 10–20 °C below their mp for 1–8 h increases the yield of RMHal, as would be expected for a process with a relatively high activation energy (see Chap. 3).

**Fig. 2.** Commercial Metal Vapour Synthesiser (from KONTES)

*Matrix Isolation Spectroscopy*

Neutron bombardment of $^{40}Ar$ in a matrix produces $^{41}K$; esr measurements on the $^{41}K/^{40}Ar$ matrix so produced leads to the conclusion that there are different environments for the potassium atom. Between 4 and 20 K there is a predominance of eight coordination, annealing to between 20 and 30 K increases the sites of ten coordination and finally above 30 K both eight and ten coordination sites are displaced by sites of twelve coordination[28)].

A comprehensive review on cryophotoclustering has appeared; $Ni_n$, $Ag_n$, $Mo_n$, and $Cr_nMo_m$, $NiC_2H_4$, $Rh_2(O_2)_n$, $Ni_{1-3}(CO)$, $Cu_2(C_2H_4)_n$ are discussed with relevence to catalytic functions[29)]. Metal olefin complexes have been well studied and complexes of the type $M(CHR{=}CHR')_nM$=Ni, Pd have been prepared in low temperature matrices and show characteristic U.V. spectra depending on n and the substituent on the olefin[30)]: decomposition temperatures of the complexes are recorded. Interestingly internal oxidative addition of nickel in tris(allylchloride)Ni° does not occur below 120 K.

Cobalt atoms in ethylene containing matrices give polynuclear clusters but no $Co(C_2H_4)_3$: $Co(C_2H_4)_{1,2}$, $Co_2(C_2H_4)_{1,2}$ and $Co_4(C_2H_4)_4$ are formed[31)].

The range of spectroscopic techniques employed in MIS studies has been extended to include secondary ion mass spectroscopy (SIMS)[32)], pmr[33)] and natural abundance $^{13}C$ nmr[34)]. They will prove very useful for the better identification of hydrocarbons in matrices.

## References


1. G. V. Planer Ltd.: "Recent Developments in Metal Atom Chemistry", by Cloke, F. G. N., and Green, M. L. H.
2. Cloke, F. G. N.: D. Phil. Thesis Oxford University, 1978
3. Cloke, F. G. N., Green, M. L. H., Morris, G. E.: J.C.S. Chem. Comm. *1978,* 72
4. Cloke, F. G. N., Green, M. L. H., Price, D. H.: J.C.S. Chem. Comm. *1978,* 431
5. a) Cloke, F. G. N., Green, M. L. H.: J.C.S. Chem. Comm. *1979,* 127. b) Kolesnikov, S. P., Dobson, J. E., Skell, P. S.: J. Amer. Chem. Soc. *100,* 999 (1978)
6. G. V. Planer Ltd., Sunbury-on-Thames, Middlesex, England
7. Jonas, K.: presented at the annual congress, Chem. Soc., Bristol 3–5 April 1979: Jonas, K., Schieferstein, L., Krüger, C., Yi-Hung Tsay: Angew. Chem. *91,* 590 (1979)
8. Wilke, G.: presented at the annual congress, Chem. Soc., Bristol 3–5 April 1979
9. Hawker, P. N., Kundig, E. P., Timms, P. L.: J.C.S. Chem. Comm. *1978,* 730
10. King, R. B., Timms, P. L.: J.C.S. Chem. Comm. *1978,* 898
11. Candlin, J. P., presented at the annual congress, Chem. Soc., Bristol 3–5 April 1979: a) Candlin, J. P., Wilson, K. W., Pearce, R. (ICI Ltd.): US Patent 4,121,030 (1978). b) Snamprogetti SPA: Belgian Patent 850–878 (1979)
12. Klabunde, K. J., Ralston, D., Zoellner, R., Hattori, H., Tanaka, Y.: J. Catalysis *55,* 213 (1978)
13. Klabunde, K. J., Davis, S. C., Hattori, H., Tanaka, Y.: J. Catalysis *54,* 254 (1978)
14. Davis, S. C., Klabunde, K. J.: J. Amer. Chem. Soc. *100,* 5974 (1978)
15. Klabunde, K. J., Anderson, B. B., Bader, M., Radonovich, J. J.: J. Amer. Chem. Soc. *100,* 1313 (1978)
16. Remick, R. J., Asunta, T. A., Skell, P. S.: J. Amer. Chem. Soc. *101,* 1320 (1978)
17. Fischer, E. O.: Inorg. Synth. *6,* 132 (1960)
18. Tolman, C. A., English, A. D., Ittel, S. D., Jesson, J. P.: Inorg. Chem. *17,* 2374 (1978)

19. Jesson, J. P., Tolman, C. A.: US Patent 3,997,579 (1976)
20. Ittel, S. D., Van Catledge, F. A., Tolman, C. A., Jesson, J. P.: J. Amer. Chem. Soc. *100,* 1317 (1978). Ittel, S. D., Krugic, P. J., Meakin, P.: J. Amer. Chem. Soc. *100,* 3264 (1978)
21. Chang, M., Newton, M. G., King, R. B., J. Am. Chem. Soc. *100,* 998 (1978). a) Chang, M., Newton, M. G., King, R. B.: Inorg. Chim. Acta *30,* L 341 (1978). b) Elschenbroich, C., Mökkel, R., Zenneck, U.: Angew. Chem. Int. Ed. *17,* 531 (1978)
22. Chang, M., King, R. B., Newton, M. G.: Inorg. Chim. Acta. *28,* L153 (1978)
23. Klabunde, K. J.: private communication to Ozin, G. A., ref. 14 in Hanlan, A. J. L., Ozin, G. A., Power, W. J.: Inorg. Chem. *17,* 3648 (1978)
24. Zimmerman, G. J., Wilcznski, R., Sneddon, L. G.: J. Organometal. Chem. *154,* C 29 (1978)
25. Klabunde, K. J., Groshens, T., Brezinski, M., Kenelly, W.: J. Am. Chem. Soc. *100,* 4437 (1978)
26. Vollenbroek, F. A., Bouten, P. C. P., Trooster, J. M., Van der Berg, J. P., Bour, J. J.: Inorg. Chem. *17,* 1345 (1978)
27. Gowenlock, B. G., Lindsell, W. E., Singh, B.: J. Chem. Soc. Dalton *1978,* 657
28. Coufal, H., Nagel, U., Lüscher, E.: Ber. Bunsenges. Phys. Chem. *82,* 133 (1978)
29. Ozin, G. A.: Catal. Rev. Sci. Eng. *16,* 191 (1977)
30. Ozin, G. A., Power, W. J.: Inorg. Chem. *17,* 2836 (1978)
31. Hanlan, A. J. L., Ozin, G. A., Power, W. J.: Inorg. Chem. *17,* 3648 (1978)
32. Jonkman, H. J., Michl, J.: J.C.S. Chem. Comm. *1978,* 751
33. Kohl, J. E., Semack, M. G., White, D.: J. Chem. Phys. *69,* 5378 (1978)
34. Zilm, K. W., Contin, R. T., Grant, D. M., Michl, J.: J. Amer. Chem. Soc. *100,* 803 (1978)

## Technical Suppliers

*1. Vacuum Systems*

Edwards Vacuum, Manor Royal, Crawley, Sussex RH10 2LW England
Leybold-Hereaus, Graff-Adolf-Strasse 21, Düsseldorf 4000 B.R.D.

*2. Wires, filaments, metals*

Balzers A.G., F.L. 9496 – Fürstentum Liechtenstein
Goodfellow Metals Ltd., Science Park, Milton Road, Cambridge CB4 4DJ, England
R.D. Mathis Company 2840, Gundry Ave, Long Beach, California, U. S. A.

*3. Electron Guns*

G.V. Planer Ltd., Windmill Road, Sunbury-on-Thames, Middlesex, England
Balzers, Liechtenstein

*4. Ceramic seals, crucibles, oxide cements, etc.*

R.D. Mathis Company
Balzers

*5. Rotating Vacuum Seals*

Crane Seals, Ferrofluidic Seals, G.V. Planer, Ltd.

*6. Sputterers*

R.D. Mathis Company
Balzers A.G.

*7. Commercial Evaporators*

G.V. Planer Ltd.
Kontes Glass Company, Vineland N.J. 08360, U. S. A.

# Subject Index